Cancer through the Lens of Evolution and Ecology

New cancer cells exist in an ever-changing "ecology" and are subject to evolutionary pressures just like any species in nature. This edited book explores the following themes: 1) how the dynamics of mutation, epigenetics, and gene expression noise are sources of genetic diversity; 2) how scarce resources influence cancer therapy resistance; 3) how predator–prey dynamics are mirrored in immune-cancer cross-talk; 4) how cancer cells parallel niche construction theory; 5) how changing fitness landscapes enable cancer growth; and 6) how cancer cells interact within the body. This book is a resource for understanding cancer as a disease of multicellularity grounded in evolutionary principles. By using this knowledge, researchers are starting to exploit these behaviors for treatment paradigms.

KEY FEATURES

- Bridges disciplines exemplifying the ways disparate fields create new perspectives when integrated.
- Offers insights from leading scholars in cancer biology, ecology, and evolutionary biology.
- Provides a timely recognition by oncologists that evolutionary paradigms are crucial for breakthroughs in cancer treatment.
- Integrates basic and applied sciences of oncology and evolutionary biology.

Cancer through the Lens of Evolution and Ecology

Edited by
Jason A. Somarelli and Norman A. Johnson

CRC Press
Taylor & Francis Group
Boca Raton London New York

CRC Press is an imprint of the
Taylor & Francis Group, an **informa** business

Designed cover image: Cover illustration by Uko Gorter.

First edition published 2024
by CRC Press
2385 NW Executive Center Drive, Suite 320, Boca Raton FL 33431

and by CRC Press
4 Park Square, Milton Park, Abingdon, Oxon, OX14 4RN

CRC Press is an imprint of Taylor & Francis Group, LLC

© 2024 Taylor & Francis Group, LLC

Library of Congress Cataloging-in-Publication Data
Names: Somarelli, Jason, editor. | Johnson, Norman A., 1966– editor.
Title: Cancer and evolution / edited by Jason Somarelli and Norman Johnson.
Description: First edition. | Boca Raton, FL : CRC Press, 2024. | Includes
    bibliographical references and index.
Identifiers: LCCN 2023054049 (print) | LCCN 2023054050 (ebook) |
    ISBN 9781032310787 (hardback) | ISBN 9781032310770 (paperback) |
    ISBN 9781003307921 (ebook)
Subjects: LCSH: Cancer. | Mutation. | Epigenetics. | Evolution (Biology)
Classification: LCC RC261 .C26525 2024 (print) | LCC RC261 (ebook) |
    DDC 616.99/4—dc23/eng/20240405
LC record available at https://lccn.loc.gov/2023054049
LC ebook record available at https://lccn.loc.gov/2023054050

ISBN: 978-1-032-31078-7 (hbk)
ISBN: 978-1-032-31077-0 (pbk)
ISBN: 978-1-003-30792-1 (ebk)

DOI: 10.1201/9781003307921

Typeset in Times LT Std
by Apex CoVantage, LLC

Contents

Chapter 4   Multicellularity, Phenotypic Heterogeneity, and Cancer...................................42

Christopher Helenek, Jason A. Somarelli, and Gábor Balázsi

Chapter 5   Feedback Loops in Gene Regulatory Networks and Cell-Cell Communication
Networks: Drivers of Cancer Cell Plasticity 50

Yeshwanth Mahesh, Subbalakshmi Ayalur Raghu, and Mohit Kumar Jolly

Chapter 6   Polygenic Evolution of Germline Variants in Cancer 58

Ujani Hazra and Joseph Lachance

Foreword

This book reviews thrilling new research on how cancers evolve in bodily ecosystems. It shows how cancers are shaped by and shape diverse niches as they cope with predatory immune cells and shifting chemotherapeutic agents. This new theoretical foundation is already improving treatment, but until recently it was beyond our imagination.

When I was a medical student administering chemotherapy to patients with Hodgkin's disease in the 1970s, we were taught that cancer was caused by mutations that made a cell go rogue and replicate identical copies out of control. Treatment aimed at eliminating every malignant cell. Chemotherapy agents were used one at a time on the rationale that if one failed, others could then be tried. We had no idea that cancer cells differed and evolved in the body and that sequential administration of different drugs selected for the cells resistant to first one, then two, then all drugs. Evolution was not considered. Many unnecessary deaths resulted.

When George Williams and I wrote about evolution and cancer 30 years ago, we were astounded not by its prevalence, but by its rarity. We realized that natural selection must have shaped potent mechanisms to control cancer, and that cancers evolve within organisms, but we missed the profound point that this book makes so well—the principles of evolutionary ecology are essential for understanding how cancer evolves as an invasive species within the body. Several specialized areas of knowledge are useful: comparative studies, species invasions and extinctions, predator–prey dynamics, niche construction, coping with shifting environments, cooperation and competition, population dynamics, island biogeography, and migrations that become metastases. The chapters in this book provide eye-opening applications of how these evolutionary ecological principles can explain cancer and suggest new treatment options.

These principles are best considered in the context of "the paradox of the organism" (a forthcoming book with that title will make a fine companion to this one). The paradox addresses the central evolutionary question about cancer: given that genes in slower replicating cells become less prevalent in every generation, how could natural selection preserve genes that inhibit cell replication and control cancer?

The general answer is kin selection. Natural selection shapes traits that reduce individual reproductive success when that results in sufficient benefits to identical genes in other individuals. The same process operates at the cellular level. Genes that rigidly control the replication of individual cells are selected because they increase the inclusive fitness of all other cells in the body because they all share the same genes. This made metazoans possible.

Metazoans came on the scene only after three billion years of unicellular life on Earth. Natural selection has since gone to enormous lengths to ensure that all cells in an individual start off as identical twins whose interests are therefore completely aligned. This requires stripping the codex down to a single strand of DNA in meiosis and rigidly separating somatic cells from germ cells. However, cells do not all stay genetically identical; somatic mutations cannot all be prevented. So natural selection also shaped mechanisms that eliminate mutated and malignant cells. Telomeres shorten with every cell division; when they are used up, the cell dies, eliminating most clones growing out of control . . . unless they have mutations that extend telomeres. The p-53 system also stands ready to eliminate dysfunctional cells via apoptosis; killing themselves off is no sacrifice for a cell, it increases its own inclusive fitness. Multiple immune mechanisms surveil the cellular landscape and attack invasive cell lines as well as pathogens, setting in motion predator–prey dynamics that have been well studied by ecologists and infectious disease experts. These and other mechanisms provide protection against cancer that is astoundingly effective. Alas, they are all too often insufficient, not just for our species, but for all species.

Why didn't natural selection provide better protection? The inevitability of mutations and genetic drift are important, but in technological societies more than half of cancer vulnerability results from mismatch with modern environments. Arms races with pathogens that induce cell replication and inflammation cause additional vulnerability. Benefits to gene transmission at the expense of robustness contribute to some cancers, especially in reproductive systems. Finally, and ripe for full investigation, systems that protect against cancer are constrained by trade-offs that limit their effectiveness. Short telomeres decrease cancer risk but speed aging. Aggressive immune responses provide protection at the risk of autoimmune disease and inflammation. More stem cells speed healing and slow aging at the cost of more cancer. And stronger selection early in the life span maintains genes and traits like fast wound healing, even if they cause big problems later in life.

The advances described in this book already have clinical applications. Instead of trying to kill all malignant cells immediately, adaptive therapy adjusts the dose of chemotherapy to keep tumor volume within bounds, a strategy that is proving to extend lifespans as well as reduce side-effects. Predator–prey dynamics suggest strategies for luring malignant cells into states where they can be recognized and attacked. Checkpoint inhibitors bypass mechanisms that protect against autoimmune disease to stimulate exhausted immune cells. New strategies that make chemotherapy resistance expensive for cells can foster faster replication by non-resistant cells, setting them up for the kill. Strategies for controlling niche environments can inhibit cells from metastasizing. And the wonderful description of how heath hens went extinct offers an inspiring model for eliminating a population of cancer cells; first cut down the population size and then add one more stress.

Readers of this book will get a preview of the future of cancer research and justification for hopes of rapid progress. Somarelli and Johnson have made superb choices of authors and topics that provide a nearly live view of a major scientific transition. I hope those who appreciate it will also turn to *Darwin's Reach*, Johnson's wonderful 2022 book about the practical advances evolutionary thinking is spurring in diverse fields. And we can all hope that these ideas spread far and fast and that they inspire new strategies for cancer prevention and treatment.

Randolph M. Nesse

Preface

Ecologists and evolutionary biologists are usually housed in different departments than cancer researchers. Sometimes they are even in different colleges within a university. They rely on different sources of funding. They have different training. They go to different meetings. They read different journals. They have different frames of reference, with each having their own jargon.

And yet despite the partitioning of researchers in these fields of study, cancer progression is fundamentally an evolutionary process that takes place within ecological theaters. With each division, tumor cells pass on their genetic information to descendants—this inheritance is not perfect, allowing genetic differences to accumulate. The cancer cells differ in their ability to replicate, to acquire and use resources, and to survive the various insults in their environments. Inheritance, variation, and differential survival and reproduction, as Darwin taught us, are the essential postulates for evolution by natural selection. Ecology is also of essence: the ability of different genotypes of cells to survive and replicate will vary across different microenvironments. Moreover, much like beavers (or oak trees) shape their environment, so do tumor cells through communication—locally and distally—with both cancer cells and non-cancer cells. And the ecology of the body also matters with respect to how well tumor cells can survive therapeutic treatments and disperse (metastasize).

The increasing specialization of academia is a natural consequence of the growth of disciplines: no one has the capacity to be an expert in everything or even everything within biology or even subfields within biology. And yet this increasing specialization impedes communication across scientists studying interdisciplinary topics, such as the ecology and evolution of cancer.

The first two decades of this century have seen more communication between evolutionary biologists and cancer biologists. The interplay between these two groups has not yet reached its full potential, but there are clear signs of progress across several realms, such as the establishment of a cancer evolution working group by the American Association for Cancer Research. Communication between ecologists and cancer biologists has generally lagged further behind the burgeoning connectivity between evolutionary biologists and cancer researchers, but some green shoots are sprouting.

This book is the fruit of seeds planted at a 2019 satellite meeting on cancer and evolution sponsored by the Society for Molecular Biology and Evolution and organized by Jeff Townsend and Jason A. Somarelli. It was at this meeting where we met: Somarelli, a molecular biologist and cancer researcher, and Johnson, an evolutionary geneticist and science writer. Herein, we aim to sow more seeds of conversation and collaboration and to provide exciting and thought-provoking examples where the fields of ecology, evolutionary biology, and cancer research can come together.

We begin with a bit of a roadmap, first with an overview of the intersection of the three fields, followed by an understanding of the influence of therapy on cancer evolution and ecology and the genetic underpinnings of tumor evolution. Next, we provide a series of chapters on molecular evolution of cancer, including the connections between cancer and multicellularity, phenotypic plasticity in cancer, and cancer genetics. Then, we transition to ecology, covering cancer-immune system cross-talk, and the potential for landscape genetics, island biogeography, and invasive species ecology to inform cancer research and vice versa. The last section covers the relationship of cancer to multicellular life through comparative oncology and a discussion of ways in which comparative oncology can be leveraged to speed translation of new therapeutic strategies for cancer. Finally, we attempt to synthesize how viewing cancer through an ecological and evolutionary lens may benefit cancer researchers, ecologists, and evolutionary biologists alike. We hope this book encourages others to engage across disciplines for new solutions to benefit the lives of patients and their families.

Acknowledgments

We thank all authors for their contributions and helpful discussions, Randy Nesse for writing the foreward and sowing many of the seeds of evolutionary medicine, Jeffrey Townsend for his leadership during the SMBE satellite meeting in 2019 and for continuing to support and champion the cancer eco/evo field, Lisa Abegglen for thoughtful reviews, Charles Nunn and the Triangle Center for Evolutionary Medicine for support of the Uko Gorter for the cover illustration, and Chuck Crumly, Kara Roberts, and the production team at CRC Press.

Jason A. Somarelli and Norman A. Johnson
October 2023

About the Editors

Jason A. Somarelli is a molecular biologist who received his doctoral degree from Florida International University in Miami, FL, and was an American Cancer Society postdoctoral fellow at Duke University. He is currently an assistant professor in the department of medicine at Duke University. Somarelli serves as the director of research for the Duke Cancer Institute Comparative Oncology Group and is the director of the Duke University Marine Laboratory Scholars in Marine Medicine Program. Dr. Somarelli's research uses evolutionary and ecology paradigms to understand molecular adaptations to extreme environments. His work spans diverse topics, such as the evolutionary pressures faced by cancer cells during drug treatment and metastatic spread, the molecular adaptations of whales and other marine mammals to low oxygen conditions, and the use of evolutionary strategies to adapt microbes to use plastic waste as a nutrient source.

In addition to his scholarly activities, Dr. Somarelli is passionately involved in training and education, leading multiple outreach and research training programs for high school students and undergraduates in an effort to enhance diversity in STEM careers.

Norman A. Johnson is an evolutionary geneticist who received his doctoral degree from the University of Rochester and was a postdoctoral fellow at the University of Chicago. He is now an adjunct professor in the biology department at the University of Massachusetts at Amherst. Johnson has also taught at the University of Chicago, the University of Texas at Arlington, and Hampshire College. Most of his research has been on the genetics and evolution of why hybrids between species are often sterile or inviable. Other research interests include the evolution of sex chromosomes, the evolution of extremely large dietary niches in insects, conflict between different parts of the genome, and the interplay between the relaxation of selection and the loss of traits. Johnson's first book, *Darwinian Detectives: Revealing the Natural History of Genes and Genomes* (Oxford University Press, 2007), examines how biologists infer the action of natural selection and other evolutionary processes from DNA sequence information. More recently, he wrote *Darwin's Reach: 21st Century Applications of Evolutionary Biology* (CRC Press, 2022). This book explores how ideas and information from evolutionary biology are being used in diverse ways in the realms of health, food, the environment, and society at large. Dr. Johnson has been active in science communication and outreach; notably, he co-led the National Evolutionary Synthesis Center working group on communicating the relevance of human evolution.

Contributors

Sarah R. Amend
Johns Hopkins University
Baltimore, Maryland USA

Dana Ataya
Moffitt Cancer Center
Tampa, Florida USA

Gábor Balázsi
Stony Brook University
Stony Brook, New York USA

Joel S. Brown
Moffitt Cancer Center
Tampa, Florida USA

Antonia Chroni
NYU School of Medicine
New York, New York USA

Veronica Colmenares
Duke University Medical Center
Durham, North Carolina USA

Zachary T. Compton
University of Arizona Cancer Center
and
University of Arizona College of Medicine
Tucson, Arizona USA

William C. Eward
Duke University Medical Center
Durham, North Carolina USA

Robert A. Gatenby
Moffitt Cancer Center
Tampa, Florida USA

Laurie A. Graves
Duke University Medical Center
Durham, North Carolina USA

Ujani Hazra
Georgia Institute of Technology
Atlanta, Georgia USA

Christopher Helenek
Stony Brook University
Stony Brook, New York USA

Henry H. Heng
Wayne State University School of
    Medicine
Detroit, Michigan USA

Rohini Janivara
Georgia Institute of
    Technology
Atlanta, Georgia USA

Norman A. Johnson
University of Massachusetts
Amherst, Massachusetts USA

Mohit Kumar Jolly
Indian Institute of Science
Bangalore, India

Irina Kareva
Northeastern University
Boston, Massachusetts USA

Andrzej Kasperski
University of Zielona Góra
Zielona Góra, Poland

Joseph Lachance
Georgia Institute of Technology
Atlanta, Georgia USA

Erin L. Landguth
University of Montana
Missoula, Montana USA

Yeshwanth Mahesh
Indian Institute of Science
Bangalore, India

Kenneth J. Pienta
Johns Hopkins University
Baltimore, Maryland USA

Subbalakshmi Ayalur Raghu
Indian Institute of Science
Bangalore, India

Jason A. Somarelli
Duke Cancer Institute
Durham, North Carolina USA

1 A Species within a Species

Jason A. Somarelli and Norman A. Johnson

Life on Earth is estimated to be approximately 3.5 to 3.8 billion years old. For the vast majority of this time, life on Earth consisted of single-celled prokaryotes. Then, something changed. Some of these single-celled organisms began to form multicellular complexes, eventually giving rise to more complex multicellular life. Although the great diversification of metazoans occurred only about 600–650 million years ago, the timing of the first multicellular eukaryote is uncertain; it could be as far back as a billion years ago (or further). Put another way, if we could squish the entire timeline of life on Earth into a single hour, the presence of multicellular life would arise within the last 15 minutes of the hour. The exact details and mechanisms through which multicellularity arose remain an issue of debate. Some posit that single cells began to divide while maintaining contact between sister cells while others propose a mechanism in which separate cells began to aggregate together (Brunet and King 2017). This transition from single-celled to multicellular arose multiple independent times throughout the history of life (Parfrey and Lahr 2013). Indeed, this transition—or something reminiscent of it—occurs today in multiple organisms, such as protists (unicellular ancestors of animals) (Sebé-Pedrós et al. 2013), yeast (Fisher and Regenberg n.d.), and bacteria (Ebrahimi, Schwartzman, and Cordero 2019), most often in response to stress or resource depletion.

The transition from single-celled to multicellular brings with it both benefits and costs. Multicellular organisms leverage division of labor in times of stress to ensure survival (Kuzdzal-Fick, Chen, and Balázsi 2019) or to protect from predation (Ebrahimi, Schwartzman, and Cordero 2019; Kapsetaki and West 2019). Yet costs often accompany these benefits. For instance, aggregate formations often grow at slower rates as compared to single-celled formations (Kuzdzal-Fick, Chen, and Balázsi 2019), and multicellular groups need to share their resources as opposed to hoarding resources in the single cell context. These trade-offs—along with the environmental contexts in which they arose—reveal the evolutionary scenarios in which stable multicellularity emerged. But these observations also suggest that stable multicellular organisms harbor within their cells the potential for single-celled behavior. This behavior arises in the context of mutation that enables a single cell to undergo a speciation event, breaking free from the confines of its role in the multicellular body and beginning a new life as a single-celled organism. This single-celled organism is cancer. Its freedom can wreak havoc; as an invasive species within a species, it cripples the otherwise healthy ecosystem of the body, often leaving the body ecosystem uninhabitable for the multicellular organism from which it arose. If we remove these cancer species from the body and allow them to grow in the laboratory, we find them to be capable of sustaining life indefinitely. They reproduce by cell division and evolve, just as any other species does; their evolution fueled by mutational diversity and subsequent selection. In this way, cancer represents a speciation event that is inexorably linked to multicellularity and the return to unicellularity.

These cancer species arise independently across nearly all multicellular organisms (Aktipis et al. 2015) through unique combinations of genetic alterations in single somatic cells to form sub-clonal populations of cancer cells. Subsequent mutations and genomic alterations enhance genotypic diversity within these populations, which are subjected to evolutionary selection within their diverse environmental contexts. The ecological fitness landscapes of the body induce convergent evolution onto cancer hallmarks (Somarelli 2021; Hanahan and Weinberg 2000, 2011), key phenotypes critical for cancer cell survival and reproduction (Somarelli 2021; Gatenby, Gillies, and Brown 2011). These phenotypic convergences are akin to convergent evolution observed in the evolution of flight, streamlined body plans, and vision at independent times across different taxonomic groups (reviewed in (Gatenby, Gillies, and Brown 2011) and (Somarelli 2021)).

DOI: 10.1201/9781003307921-1

Phenotypic convergence in cancer cells occurs in the context of often-substantial genetic and non-genetic heterogeneity at the cellular level (Gupta et al. 2020; Jamal-Hanjani et al. 2017; Raynaud et al. 2018), underscoring the strong selective pressures that exist within the body. Similar to natural systems, the ecological fitness landscapes of the body are ever-changing. Factors such as age (Aktipis et al. 2015; Laconi, Marongiu, and DeGregori 2020), chemical exposures (International Agency for Research on Cancer 2016), and changes in immunity/inflammation (Gonzalez, Hagerling, and Werb 2018) are constantly in flux and can contribute to a cancer-permissive environment. Then, after cancer cells establish themselves within the body, the cancer cells themselves exist within a dynamic ecosystem. Cancer cells encounter non-uniform distributions of resources (Y.-Q. Wang et al. 2021; Y. Wang et al. 2021; Yoshimura et al. 2021; Yuan 2016), including variations in physical-mechanical strains and space (Northcott et al. 2018), growth factor and nutrient availability (Hensley et al. 2016; Heaster, Landman, and Skala 2019), oxygen availability (Zaidi et al. 2019), and immune attack (Y.-Q. Wang et al. 2021; Y. Wang et al. 2021; Yoshimura et al. 2021). The complex and dynamic environment of the body induces evolutionary selection pressure on the cancer cell population, restricting the growth of cancer cells that are unable to survive and selecting for the phenotypes necessary for continued survival and reproduction.

The origin story of the new cancer species is often a quiet one from the perspective of the host environment—the environment that is to become the future cancer patient. While cancer cells are dividing, dying and being killed, and mutating, the unknowing host continues life unaware. While new and emerging technologies are providing important new methods to improve cancer detection (Chen et al. 2020), most often it is only after many cancer cell divisions that a cancer becomes detectable on modern imaging systems (Chen et al. 2020; Erdi 2012; Countercurrents Series and Narod 2012), and it is likely that many more cell doublings are required before the cancer causes any symptoms. By this time, a tumor comprised of many billions of cells, some of which have likely already spread to other sites (Hu et al. 2020), would be exposed to some form of treatment. During conventional treatment, the cancer cell population finds itself in an environment in which everything has been upended. Many cancer cells are removed surgically or die when exposed to radiation or chemotherapy and oftentimes resources that were once scarce become abundant. Targeted therapies have the opposite effect: abundant pro-survival resources become scarce; and dramatic shifts take place in the makeup of the non-cancer cells and the extracellular components of the tumor. These changing conditions change the outcome, with a population that is now suddenly facing a harsh, resource-poor environment. While many cancer cells die as a result of this dramatic environmental change, the few survivors often end up possessing phenotypic traits that allow them to survive in the face of the new, therapy-induced scarcity. The cells that emerge with features of treatment resistance can also carry with them phenotypes of other aggressive traits, such as increased self-renewal and invasive properties (Shibue and Weinberg 2017) and/or immune-evasive characteristics (Haas et al. 2021; Ware et al. 2020).

The cancer cell population is, no doubt, shaped by the environment, but it is not simply a passive bystander in its interaction with the environment. Quite to the contrary, mounting evidence portrays a scenario in which cancer cell populations actively shape both local and distant environments within the body. This niche construction occurs in several ways, including through continuous release of soluble factors (Vinay et al. 2015; Chin and Wang 2016), remodeling of extracellular matrix components (Winkler et al. 2020), secretion of extracellular vesicles (Zeng et al. 2018; Zhou et al. 2014; Wen et al. 2016), and cell-cell cross-talk with various non-cancer cell types (Pardoll 2012; Mitra et al. 2012). Like beavers arriving at a stream for the first time, the cancer cell population reconstructs the landscape to suit its needs (reviewed in (Somarelli 2021)).

As we can see, the invasive cancer species within a species is highly-adaptable, fueled by genetic and epigenetic dysfunction that can create substantial population diversity. In the chapters that follow, we will explore in detail the ways in which cancer species evolve, hearing from experts across multiple fields of science, from mathematics and modeling to human medicine. These authors will examine the following topics: 1) our current understanding of the genetic and non-genetic mechanisms by which cancer cell population diversity occurs, 2) the various selection pressures that exist in the body, 3) ways in which ecological paradigms can inform our understanding of cancer, and 4) emerging approaches to leverage ecological and evolutionary principles for new cancer treatments. We hope this collection serves as a resource for anyone interested in better understanding cancer as a consequence of multicellular life and the evolutionary paradigms that dictate its behavior. Most of all, we hope the present work illustrates the importance of transdisciplinary work and encourages scientists to engage with others across disciplines in an effort to stop cancer's deadly return to unicellularity.

REFERENCES

Aktipis, C. Athena, Amy M. Boddy, Gunther Jansen, Urszula Hibner, Michael E. Hochberg, Carlo C. Maley, and Gerald S. Wilkinson. 2015. "Cancer Across the Tree of Life: Cooperation and Cheating in Multicellularity." *Philosophical Transactions of the Royal Society of London. Series B, Biological Sciences* 370 (1673). https://doi.org/10.1098/rstb.2014.0219.

Brunet, Thibaut, and Nicole King. 2017. "The Origin of Animal Multicellularity and Cell Differentiation." *Developmental Cell* 43 (2): 124–40.

Chen, Xingdong, Jeffrey Gole, Athurva Gore, Qiye He, Ming Lu, Jun Min, Ziyu Yuan, et al. 2020. "Non-Invasive Early Detection of Cancer Four Years Before Conventional Diagnosis Using a Blood Test." *Nature Communications* 11 (1): 3475.

Chin, Andrew R., and Shizhen Emily Wang. 2016. "Cancer Tills the Premetastatic Field: Mechanistic Basis and Clinical Implications." *Clinical Cancer Research: An Official Journal of the American Association for Cancer Research* 22 (15): 3725–33.

Countercurrents Series, and S. A. Narod. 2012. "Disappearing Breast Cancers." *Current Oncology* 19 (2): 59–60.

Ebrahimi, Ali, Julia Schwartzman, and Otto X. Cordero. 2019. "Multicellular Behaviour Enables Cooperation in Microbial Cell Aggregates." *Philosophical Transactions of the Royal Society of London: Series B, Biological Sciences* 374 (1786): 20190077.

Erdi, Yusuf Emre. 2012. "Limits of Tumor Detectability in Nuclear Medicine and PET." *Molecular Imaging and Radionuclide Therapy* 21 (1): 23–8.

Fisher, Roberta, and Birgitte Regenberg. n.d. "Multicellular Group Formation in Saccharomyces Cerevisiae." *Proceedings: Biological Sciences* 286 (1910). https://doi.org/10.20944/preprints201812.0220.v1.

Gatenby, Robert A., Robert J. Gillies, and Joel S. Brown. 2011. "Of Cancer and Cave Fish." *Nature Reviews, Cancer* 11 (4): 237–8.

Gonzalez, Hugo, Catharina Hagerling, and Zena Werb. 2018. "Roles of the Immune System in Cancer: From Tumor Initiation to Metastatic Progression." *Genes & Development* 32. https://doi.org/10.1101/gad.314617.118.

Gupta, Santosh, Daniel H. Hovelson, Gabor Kemeny, Susan Halabi, Wen-Chi Foo, Monika Anand, Jason A. Somarelli, et al. 2020. "Discordant and Heterogeneous Clinically Relevant Genomic Alterations in Circulating Tumor Cells vs Plasma DNA from Men with Metastatic Castration Resistant Prostate Cancer." *Genes, Chromosomes & Cancer* 59 (4): 225–39.

Haas, Lisa, Anais Elewaut, Camille L. Gerard, Christian Umkehrer, Lukas Leiendecker, Malin Pedersen, Izabela Krecioch, et al. 2021. "Acquired Resistance to Anti-MAPK Targeted Therapy Confers an Immune-Evasive Tumor Microenvironment and Cross-Resistance to Immunotherapy in Melanoma." *Nature Cancer* 2. https://doi.org/10.1038/s43018-021-00221-9.

Hanahan, Douglas, and Robert A. Weinberg. 2000. "The Hallmarks of Cancer." *Cell* 100. https://doi.org/10.1016/s0092-8674(00)81683-9.

———. 2011. "Hallmarks of Cancer: The Next Generation." *Cell* 144. https://doi.org/10.1016/j.cell.2011.02.013.

Heaster, Tiffany M., Bennett A. Landman, and Melissa C. Skala. 2019. "Quantitative Spatial Analysis of Metabolic Heterogeneity Across and Tumor Models." *Frontiers in Oncology* 9 (November): 1144.

Hensley, Christopher T., Brandon Faubert, Qing Yuan, Naama Lev-Cohain, Eunsook Jin, Jiyeon Kim, Lei Jiang, et al. 2016. "Metabolic Heterogeneity in Human Lung Tumors." *Cell* 164. https://doi.org/10.1016/j.cell.2015.12.034.

Hu, Zheng, Zan Li, Zhicheng Ma, and Christina Curtis. 2020. "Multi-Cancer Analysis of Clonality and the Timing of Systemic Spread in Paired Primary Tumors and Metastases." *Nature Genetics* 52 (7): 701–8.

International Agency for Research on Cancer. 2016. *Outdoor Air Pollution: IARC Monographs on the Evaluation of Carcinogenic Risks to Humans*. International Agency for Research on Cancer.

Jamal-Hanjani, Mariam, Gareth A. Wilson, Nicholas McGranahan, Nicolai J. Birkbak, Thomas B. K. Watkins, Selvaraju Veeriah, Seema Shafi, et al. 2017. "Tracking the Evolution of Non-Small-Cell Lung Cancer." *The New England Journal of Medicine* 376 (22): 2109–21.

Kapsetaki, Stefania E., and Stuart A. West. 2019. "The Costs and Benefits of Multicellular Group Formation in Algae." *Evolution; International Journal of Organic Evolution* 73 (6): 1296–308.

Kuzdzal-Fick, Jennie J., Lin Chen, and Gábor Balázsi. 2019. "Disadvantages and Benefits of Evolved Unicellularity Versus Multicellularity in Budding Yeast." *Ecology and Evolution* 9 (15): 8509–23.

Laconi, Ezio, Fabio Marongiu, and James DeGregori. 2020. "Cancer as a Disease of Old Age: Changing Mutational and Microenvironmental Landscapes." *British Journal of Cancer* 122 (7): 943–52.

Mitra, Anirban K., Marion Zillhardt, Youjia Hua, Payal Tiwari, Andrea E. Murmann, Marcus E. Peter, and Ernst Lengyel. 2012. "MicroRNAs Reprogram Normal Fibroblasts into Cancer-Associated Fibroblasts in Ovarian Cancer." *Cancer Discovery* 2 (12): 1100–8.

Northcott, Josette M., Ivory S. Dean, Janna K. Mouw, and Valerie M. Weaver. 2018. "Feeling Stress: The Mechanics of Cancer Progression and Aggression." *Frontiers in Cell and Developmental Biology* 6 (February): 17.

Pardoll, Drew M. 2012. "The Blockade of Immune Checkpoints in Cancer Immunotherapy." *Nature Reviews Cancer* 12. https://doi.org/10.1038/nrc3239.

Parfrey, Laura Wegener, and Daniel J. G. Lahr. 2013. "Multicellularity Arose Several Times in the Evolution of Eukaryotes (response to DOI 10.1002/bies.201100187)." *BioEssays: News and Reviews in Molecular, Cellular and Developmental Biology* 35 (4).

Raynaud, Franck, Marco Mina, Daniele Tavernari, and Giovanni Ciriello. 2018. "Pan-Cancer Inference of Intra-Tumor Heterogeneity Reveals Associations with Different Forms of Genomic Instability." *PLoS Genetics* 14 (9): e1007669.

Sebé-Pedrós, Arnau, Manuel Irimia, Javier del Campo, Helena Parra-Acero, Carsten Russ, Chad Nusbaum, Benjamin J. Blencowe, and Iñaki Ruiz-Trillo. 2013. "Regulated Aggregative Multicellularity in a Close Unicellular Relative of Metazoa." *eLife* 2. https://doi.org/10.7554/elife.01287.

Shibue, Tsukasa, and Robert A. Weinberg. 2017. "EMT, CSCs, and Drug Resistance: The Mechanistic Link and Clinical Implications." *Nature Reviews. Clinical Oncology* 14 (10): 611–29.

Somarelli, Jason A. 2021. "The Hallmarks of Cancer as Ecologically Driven Phenotypes." *Frontiers in Ecology and Evolution* 9 (April). https://doi.org/10.3389/fevo.2021.661583.

Vinay, Dass S., Elizabeth P. Ryan, Graham Pawelec, Wamidh H. Talib, John Stagg, Eyad Elkord, Terry Lichtor, et al. 2015. "Immune Evasion in Cancer: Mechanistic Basis and Therapeutic Strategies." *Seminars in Cancer Biology* 35 Suppl (December): S185–98.

Wang, Ya-Qin, Xu Liu, Cheng Xu, Wei Jiang, Shuo-Yu Xu, Yu Zhang, Ye Lin Liang, et al. 2021. "Spatial Heterogeneity of Immune Infiltration Predicts the Prognosis of Nasopharyngeal Carcinoma Patients." *Oncoimmunology* 10 (1): 1976439.

Wang, Youyu, Xiaohua Li, Shengkun Peng, Honglin Hu, Yuntao Wang, Mengqi Shao, Gang Feng, Yu Liu, and Yifeng Bai. 2021. "Single-Cell Analysis Reveals Spatial Heterogeneity of Immune Cells in Lung Adenocarcinoma." *Frontiers in Cell and Developmental Biology* 9 (August): 638374.

Ware, Kathryn E., Santosh Gupta, Jared Eng, Gabor Kemeny, Bhairavy J. Puviindran, Wen-Chi Foo, Lorin A. Crawford, et al. 2020. "Convergent Evolution of p38/MAPK Activation in Hormone Resistant Prostate Cancer Mediates Pro-Survival, Immune Evasive, and Metastatic Phenotypes." *bioRxiv*. 2020–4. https://doi.org/10.1101/2020.04.22.050385.

Wen, Shu Wen, Jaclyn Sceneay, Luize Goncalves Lima, Christina S. F. Wong, Melanie Becker, Sophie Krumeich, Richard J. Lobb, et al. 2016. "The Biodistribution and Immune Suppressive Effects of Breast Cancer-Derived Exosomes." *Cancer Research* 76 (23): 6816–27.

Winkler, Juliane, Abisola Abisoye-Ogunniyan, Kevin J. Metcalf, and Zena Werb. 2020. "Concepts of Extracellular Matrix Remodelling in Tumour Progression and Metastasis." *Nature Communications* 11 (1): 5120.

Yoshimura, Kanako, Takahiro Tsujikawa, Junichi Mitsuda, Hiroshi Ogi, Sumiyo Saburi, Gaku Ohmura, Akihito Arai, et al. 2021. "Spatial Profiles of Intratumoral PD-1 Helper T Cells Predict Prognosis in Head and Neck Squamous Cell Carcinoma." *Frontiers in Immunology* 12 (October): 769534.

Yuan, Yinyin. 2016. "Spatial Heterogeneity in the Tumor Microenvironment." *Cold Spring Harbor Perspectives in Medicine* 6 (8). https://doi.org/10.1101/cshperspect.a026583.

Zaidi, Mark, Fred Fu, Dan Cojocari, Trevor D. McKee, and Bradly G. Wouters. 2019. "Quantitative Visualization of Hypoxia and Proliferation Gradients Within Histological Tissue Sections." *Frontiers in Bioengineering and Biotechnology* 7 (December): 397.

Zeng, Zhicheng, Yuling Li, Yangjian Pan, Xiaoliang Lan, Fuyao Song, Jingbo Sun, Kun Zhou, et al. 2018. "Cancer-Derived Exosomal miR-25–3p Promotes Pre-Metastatic Niche Formation by Inducing Vascular Permeability and Angiogenesis." *Nature Communications* 9 (1): 5395.

Zhou, Weiying, Miranda Y. Fong, Yongfen Min, George Somlo, Liang Liu, Melanie R. Palomares, Yang Yu, et al. 2014. "Cancer-Secreted miR-105 Destroys Vascular Endothelial Barriers to Promote Metastasis." *Cancer Cell* 25 (4): 501–15.

2 Therapy as a Driver of Evolutionary Selection

Dana Ataya, Joel S. Brown, and Robert A. Gatenby

2.1 CANCER: THE ORIGIN EVENT

In normal human tissue, individual cells are "building blocks" that contribute to the fitness of the whole organism. Because survival and proliferation of normal somatic cells are entirely determined by internal controls that maintain tissue homeostasis, they do not undergo evolution by natural selection. In other words, in multicellular organisms, the organism is the evolutionary unit of selection, and its individual cellular components have the same fitness as the organism [1].

Cancer cells, in contrast, no longer follow the homeostatic directives of the whole organism. This transition from tissue control to proliferative independence (i.e., carcinogenesis) represents a transition of the evolutionary unit of selection starting from a cell that has the same fitness as the multicellular organism to one that has a "self-defined fitness function" independent of the organism [1]. In fact, in far too many cases the cancer, by evolving and proliferating, will result in a patient's death.

Three conditions permit cells to evolve by natural selection [1, 2]:

1) A population of cells with variable heritable phenotypic characteristics
2) Capacity for population growth that surpasses the environmental limits, so that not all cells can survive and propagate exponentially (a "struggle for existence")
3) Heritable phenotypic characteristics impact survival and proliferation.

Therefore, a cancer cell can evolve because it gains independence from normal tissue growth constraints. Survival and proliferation are determined by its inherited phenotypic characteristics and their interactions with local populations of other cancer cells, normal cells of diverse types, resources, metabolites, and pH. Collectively, these properties of a cancer cell's environment represent the selective forces within the tumor.

In the widely accepted model of carcinogenesis, the transition of normal cells to cancer is presumed to occur via a series of genetic or epigenetic alterations that generate premalignant phenotypes. Conceptually, these alterations generate transient premalignant states that eventually reach a fully malignant phenotype [3–5]. This widely accepted theory is gene-centric in that "somatic evolution" is seen principally as a genetic process governing the random accumulation of mutations [4–7].

Although cancer is clearly associated with accumulating genetic mutations, there are several observations that challenge this widely accepted gene-centric model of cancer [1]. Notably, normal non-cancerous cells can have multiple genetic mutations, some of which would generally be viewed as oncogenic (meaning cancer promoting) [8, 9]. In addition, these "oncogenic mutations" have been identified within normal cells bordering malignant tumors [10]. Other important observations include the following: the number of mutations in various cancer types can vary by two orders of magnitude [11] and known genetic mutations that result in defects in DNA repair genes can increase the risk of cancer in some—but not all—organs [12, 13]. It has also been recognized that the incidence of cancer increases later in life, even though most cell divisions and somatic mutations occur early in life. Ultimately, cells with accumulating genetic mutations will mostly not jump levels of selection and become a self-defined fitness function. If this were the case, hundreds if not thousands of cancers would emerge clinically during a person's lifetime. Specifically, cells with mutations

DOI: 10.1201/9781003307921-2

mostly remain phenotypically normal as observed in normal epithelial cells [1, 11, 14, 15]. They do not cross the threshold into being subject to natural selection.

So how does a cell develop a self-defined fitness function? Exposure to a mutagenic environment may increase the quantity of genetic mutations, subsequently increasing the probability of reaching a threshold of mutational burden required to achieve separation from tissue controls and into a self-defined fitness function. Alternatively—and perhaps more compelling—is the idea that a cell and its progeny gain a self-defined fitness function from a breakdown of tissue control signals resulting from tissue injury, inflammation, or infection [1]. In this scenario, the cells have acquired a self-defined fitness function that allows the cell to evolve—at least until normal tissue function is reestablished. In cases where normal tissue function is restored, healing allows for reestablished tissue control over normal cells. However, some cells with pre-existing mutations or those that acquire new heritable phenotypic properties from long runs of uncontrolled cell proliferation may evolve sufficiently to successfully resist reestablishment of normal tissue control [1]. From these cells, a clinically relevant cancer clade can develop.

The cell continues to evolve via "evolutionary triage" where cells with advantageous mutations (resulting in higher survival and proliferation) outcompete those with disadvantageous or less successful mutations. Such mutation will ultimately be lost and consequently not observed in the population [1, 16]. As an optimization process, natural selection operates through creative destruction by removing from the population mutations that adversely impact fitness. Thus, many of the genetic changes discovered in cancer cells may have transpired before the beginning of somatic evolution and actually may have no part in cancer formation.

2.2   EARLY TUMOR GROWTH

Initial cancer cell growth is often depicted simply as a growing ball of cells. However, the first phase of cancer growth is within a well-defined cylindrical geometry, such ductal carcinoma *in situ* (DCIS) of the breast or prostatic intraepithelial neoplasia (PIN). Here, the cancer population is initially constrained to the lumen of a duct and confined by the basement membrane, which physically separates them from the blood supply that does not enter the duct. Thus, the early evolution and development of breast cancer occurs in an avascular environment with relatively minimal exposure to the immune system—space, resources, and hazards are all minimal. Oxygen, glucose, and other serum growth factors such as EGF or estrogen must diffuse through the basement membrane before they can reach the *in situ* cancer cells. Furthermore, as substrate and growth factors diffuse into the intraductal tumors, it is consumed by cells resulting in a progressive decrease in concentration. Mathematical models of these reaction-diffusion dynamics as well as empirical studies show, for example, oxygenated cells are limited to a distance of about 150 μm from a blood vessel [17–19]. As the *in situ* cells continue to proliferate into an enlarging epithelial layer, the inner-most cells are pushed even further away from the supply of nutrients. This creates spatial gradients that result in changing evolutionary selection pressures along the radius of the duct—tumor cells in the center of the duct lumen will have very different levels of growth factors and substrate compared to cells at the edge of the lumen. The *in situ* lesion—composed of heterogeneous cellular phenotypes—is also subject to microenvironmental selection pressures and competition for sparse resources such as oxygen and glucose in which phenotypes adaptable to harsh environments and resistance to hypoxia are favored by natural selection [20, 21]. Upregulation of hypoxia-inducible factor (HIF) and other hypoxia-related proteins (HRPs) such as carbonic anhydrase IX and GLUT1 have been observed in DCIS lesions [19, 22–25] (Figure 2.1). Indeed, microenvironmental hypoxia, acidity, and resource scarcity result in a strong selection pressure on *in situ* cancer cell's survival and proliferation.

It is important to note these dynamics differ significantly from the conventional models of cancer evolution, often described as "branching clonal evolution" caused by accumulating random mutations. As shown in Figure 2.1, cancer cells evolve in response to specific environmental circumstances that characteristically vary across time and space. Of course, these phenotypic adaptations

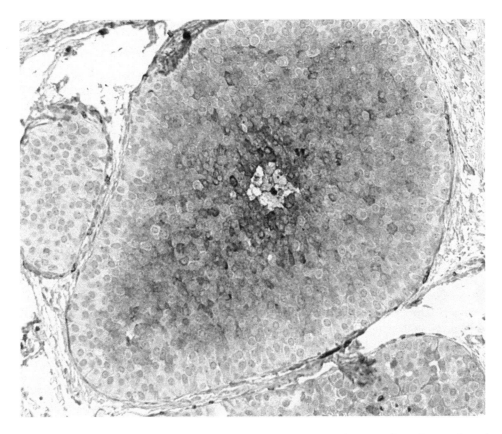

FIGURE 2.1 Ductal carcinoma *in situ* (DCIS) stained for GLUT1 showing a gradient of expression related to increasing hypoxia along the radius of the duct due to cellular consumption.

are governed by intracellular molecules encoded in the genome as mutations and epigenetic changes; but, here, the molecular events represent the "history" of evolution rather than the cause.

Development of a Warburg Effect (WE) phenotype—defined as aerobic glycolysis, despite the presence of oxygen—contributes to the acidity of the microenvironment and is commonly observed in aggressive cancers [26, 27]. DCIS cells near the basement membrane reside closer to nutrient-carrying capillaries and therefore are under little selection to develop a WE phenotype. Indeed, mathematical modeling demonstrates that a WE phenotype predominately evolves in the greatest metabolically depleted regions—near the necrotic core and distant from capillaries [28]. Development of a WE phenotype resistant to acid-induced toxicity results in a powerful evolutionary proliferative advantage. Persistent aerobic glycolysis alters the local microenvironment so that it is toxic to other cellular populations (which remain vulnerable to acid-mediated toxicity) but is harmless to itself. In addition, a persistently acidic microenvironment may enable tumor invasion by degrading the extracellular matrix (ECM), destroying neighboring normal cellular populations, and promoting angiogenesis. Once evolutionary pressures select for a WE phenotype, cells tend to migrate to the edge of a tumor [26, 28, 29]. Even in the absence of therapy and during the very earliest growth of the emerging tumor, the cancer cells are evolving adaptations to their circumstances.

The second phase of early cancer growth begins with cancer cells breaking through the basement membrane. The dynamics of this tumor spread match the range expansion of a species into novel habitats. Once outside the duct, the cancer cells experience a different composition of normal cells and nutrient supply. There are several examples of species first colonizing and then speciating to utilize a novel habitat. These include the apple maggot fly that colonized apple trees brought to North

America. The progenitor species was the hawthorn maggot fly (already present on native hawthorns [30, 31]). Pea aphids also showed a breakout into a new habitat (shifting to alfalfa from red clover) followed by speciation to better adapt to this new habitat [32].

Like these examples from nature, spatial variation in environmental selection forces drive continued evolution of cancer cells as they spread via tumor expansion. This is, perhaps, most obvious in the variations of local environmental selection forces due to temporal and spatial variations in blood flow. Furthermore, studies of invasive species, such as ants, plants, and birds, consistently demonstrate a competition-colonization trade-off within the population [33–35]. That is, individuals at the leading edge of an invasive species must adapt to different selection forces compared to those species in the geographic center or "core" of the population [36]. Thus, cancer cells at the tumor edge are characteristically more motile, invasive, and glycolytic while cells in the core are more angiogenic, less motile, and use more aerobic glucose metabolism [37, 38]. Interestingly, there appears to be some level of competition between these populations and manipulation of the tumor microenvironment by, for example, increasing the buffering capacity of the blood, can decrease the population at the tumor edge resulting in a less invasive and more slowly growing state (Figure 2.2).

Once the tumor cells break through the basement membrane and come into direct contact with new elements of the host mesenchyme and immune system, adaptations include angiogenesis and immune avoidance. As previously noted, a glycolytic acid-producing phenotype is selected for at the edge of tumors. There are several evolutionary benefits for cancer cells inhabiting the edge to have a pseudohypoxic strategy relative to those inhabiting the tumor's core. For cancer cells on the edge natural selection favors traits promoting migration and invasion into the stroma and surrounding normal tissue. Secondly, a pseudohypoxic environment also allows for the suppression of immune attack, analogous to the cane toads native to Central and South America. Cane toads were brought to Australia, and the population flourished due to lack of natural predators and vulnerability to

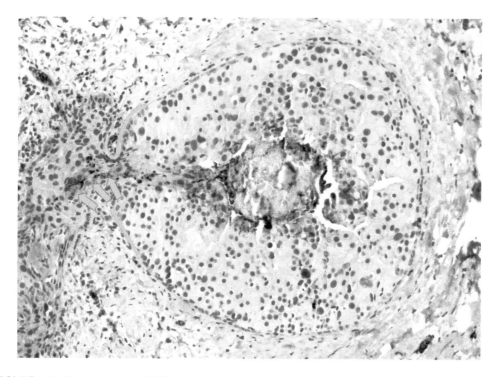

FIGURE 2.2 Transition from DCIS to invasive cancer. Note cells in the central portion of the duct evolved an invasive phenotype allowing them to invade the more peripheral cells and breech the basement membrane.

diseases [39]. Similar to the acidic "toxin" secreted by a WE tumor phenotype facilitating immune evasion, cane toads are toxic—facilitating evasion from predators. Food-safety trade-offs where one phenotype is favored in safer habitats (e.g., tumor core) and another in riskier habitats (e.g., tumor edge) is a near universal property across the tree of life [40, 41].

Analogous to antipredator adaptations in nature, cancer cells evolve a number of ways to evade and ultimately thwart the immune system. The immune surveillance theory defines the immune system's role as the patrolling entity tasked with recognizing and destroying cancer cells while sparing normal cells that may have very similar properties. This patrol is not perfect, as evidenced by the occurrence of malignancies. The cancer immunoediting theory sees a process where the immune system may be able to eliminate or at least contain the small emerging cancer cell population. If the cancer cells can begin to evade the immune system and grow in number, then a period of relative balance may emerge between the immune cells and cancer cells. Eventually, as the tumor grows in size and the cancer cells evolve stronger and more diverse immune-evasion phenotypes the immune response ceases to be effective at containing tumor growth [42, 43]. There are three main strategies used by a tumor in immune evasion [44, 45]:

1) Evading immune recognition by fooling cytotoxic immune cells into treating the cancer cells as off-limit normal cells
2) Presenting noxious inhibitory factors or acidity that either permanently or temporarily disable cytotoxic T-cells
3) Releasing signals and chemicals that activate other component cells that then suppress the survival, activation, and/or proliferation of cytotoxic immune cells.

With regard to evading immune recognition, a major mechanism utilized by cancer cells involves MHC class I presentation. MHC class I molecules are important in presentation and recognition of tumor antigens by cytotoxic CD8 T cells, and therefore tumors often downregulate MHC class Ia expression to escape CD8 T-cell-mediated immune detection and eradication [46]. This is similar to Batesian mimicry in nature where a species desirable to the predator mimics the color pattern of an undesirable prey [45, 47]. Monarch butterflies (the model) contain poisons that birds must avoid while the viceroy butterfly (the mimic) is valuable prey for birds. Fear of consuming monarchs causes birds to forgo viceroys. Similarly, "fear" of killing normal cells (the model) allows cancer cells (the mimic) to avoid being killed by T-cells that encounter them.

With respect to suppressing cytotoxic immune cells, cancer cells upregulate inhibitory surface receptors and/or produce inhibitory factors. A notable example of an inhibitory receptor expressed by many cancer cells is PD-L1, the ligand for programmed cell death 1 (PD-1), which is a transmembrane protein on the surface of T cells, which, upon ligation, prompts the death (apoptosis) of the targeted T cell. Tumors may also secrete immunosuppressive factors, such as TGF-β, IL-6, interleukin (IL)-10, nitric oxide, and reactive oxygen species [44, 48]. Such responses are widespread in nature where prey possessing noxious chemicals or the capacity to fight back as seen in tourist videos of African buffaloes taking on lions or hyenas [49].

With respect to manipulating other components of the community of immune cells (the immune-web [50]), cancer cells can promote a suppressive microenvironment by recruiting various suppressor leukocytes—macrophages, tolerogenic dendritic cells, myeloid-derived suppressor cells (MDSCs), and regulatory T cells [44, 51]. These cells then suppress the cytotoxic T-cells, thus providing safety for the cancer cells. This is similar to systems where a plant, for instance, will release chemicals that attract wasps that will then prey upon the aphids or other insects that may be consuming the plant [52].

In addition to immune evasion, successful adaptations of cancer cells during the early stages of tumor growth includes angiogenesis whereby cancer cells emit factors that attract blood vasculature, thus bringing increased nutrient supply to the neighborhood of the cancer cells. While producing VEGF (vascular endothelial growth factor) is directly beneficial to the individual cancer cells it also represents a public good, aiding nearby cancer cells [53]. So even as the cancer cells compete with others for space and nutrients [54], angiogenesis becomes an emergent property of this competition. The dynamics of angiogenesis in early tumor growth resembles the ecological process of "niche construction" [55, 56]. Through niche construction organisms alter their environment to increase their own fitness while having potentially beneficial (or negative) effects on other individuals of the same or different species. Niche construction impacts the complexity of eco-evolutionary dynamics by introducing "ecological inheritance"—the concept that the altered habitat improves fitness of the organism's progeny [57, 58].

An example in nature of niche construction and ecological inheritance would be the construction of dams by beavers, which would subsequently increase the odds of survival of the beaver's offspring [56]. In angiogenesis, malignant cells or groups of cells individually stimulate vascular proliferation in their environment via secretion of paracrine angiogenic factors like VEGF and angiopoietins [59, 60]. Since the growth factor signaling gradients are confined to the spatial habitat of the malignant cell(s), neovascularization will arise from the closest vascular structure (Figure 2.3). As there is no evolutionary advantage for competing malignant cells to form an organized functional vascular network (which may improve survival of competing neighboring cells) neovascularization will be disordered and without the highly organized configuration of the vascular network in normal organs. This will consequently result in an unstable vascular network and often produce complex and hectic hemodynamics with fluctuating oxygen, nutrient, and metabolic gradients [56, 61, 62]. These variations in oxygen and metabolite concentrations generate noticeable spatial distinctions in tumor cell densities and tumor habitats [63–66]. Temporal changes in regional tumoral blood flow also occur and result in cyclic hypoxia [67]. These spatial and temporal dynamics of angiogenesis have been fittingly described as "waves and tides" [56, 68, 69], emphasizing the important concept that while one

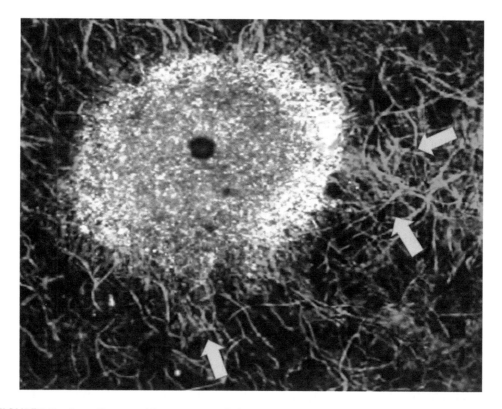

FIGURE 2.3   A small tumor with a cancer population evolving the capacity to attract blood vessels.

tumor region experiences a surge of blood flow and nutrients, another region is experiencing diminished concentrations of oxygen, metabolites, growth factors, and hormones. Indeed, intermittent hypoxia in early cancers can select for aggressive phenotypes that promote tumor progression and have been proposed to play an important role in the metastatic dissemination of cancer [56, 67, 70].

Angiogenic and immune adaptations impact the early tumor microenvironment creating eco-evolutionary feedbacks where the current environment selects for traits in the cancer cells that result in changes to the environment and so on. Several studies demonstrate evidence that the stromal microenvironment of *in situ* lesions contributes to the progression of invasive carcinoma [71, 72]. Reactive stroma has often been observed adjacent to ductal carcinoma *in situ* (DCIS) lesions prior to the emergence of microinvasion, and additional studies have shown that cancer-associated fibroblasts may further progression to invasion [73–75]. Macrophage infiltration in the stroma bordering DCIS lesions correlates with early metastasis and recurrence [76, 77]. Clearly, selection pressures impacting the evolution of the spatial, vascular, metabolic, and immune microenvironment play a role in early tumor growth and progression. This begins to set the stage for effects of therapy on an already sophisticated community of cancer cells proliferating and evolving within their emerging tumor environment.

2.3 SMALL INVASIVE POPULATIONS

Whether in nature or cancers, small invasive populations are governed by different ecological forces than large populations of the same species. Small populations are vulnerable to extinctions due to stochastic perturbations (small changes in birth and death rates) and Allee effects. The unique eco-evolutionary forces observed in small populations have been observed and examined through the lens of Anthropocene extinctions [78–80], which are discussed later in this chapter. Stochastic effects and Allee effects invite different approaches to cancer therapy.

Stochastic effects—defined as unpredictable probabilistic events affecting individuals or whole populations. Such events in natural populations, including cancer populations, have their greatest effects on small populations dynamics [79, 81, 82]. Demographic stochasticity represents the variability in birth and death rates associated with the probability of births and deaths associated with each individual. Even if the probability of proliferation is higher than the probability of death for a cancer cell (or any individual within natural populations), a small population may decline if there are an unexpectedly high number of deaths. As populations become smaller, variations in birth and death rates become more significant on variability in population growth. The impact of demographic stochasticity is inversely proportional to population size, and it may result in the stochastic extinction of small populations. Environmental stochasticity represents random fluctuations in environmental conditions that influence the expected birth and death rates of an entire population or subpopulation. The impact of environmental stochasticity on population dynamics is inversely proportional to the number of subpopulations experiencing fluctuations somewhat independently of each other. With more subpopulations, each can be recolonized and rescued following a stochastic extinction. With just one subpopulation such an ecological rescue is not possible. For small populations, environmental stochasticity can compound the impact of demographic stochastic effects [83]. In the context of small cancer populations, environmental stochasticity may pertain to the fluctuations in angiogenic blood flow—the random fluctuations of oxygen, nutrients, and growth factor delivery—and the stochasticity of the immune response [79].

The importance of population size in defining fitness of individuals within a population was recognized by Darwin and later expanded by Warder Clyde Allee [84–86]. In Darwin's theory of natural selection, the proliferation rate of individuals within a population must at some point decline with population size due to competition for space and resources. Nevertheless, Darwin also observed that for many species with small population sizes, the threat of extinction (via predation) also declined as the population became larger. Similarly, Allee observed that survival and proliferation of a population declined if the population was smaller, increasing the risk for extinction

[85, 87, 88]. Allee observed that goldfish had a greater survival rate when more fish were in the tank. Beneficial aggregation effects can include group foraging, safety in numbers, public goods, and cooperative behaviors [89–91]. Furthermore, solitary individuals may be more unfavorably impacted by a perturbation than a group of individuals [92, 93].

The dynamics of stochastic perturbation and the Allee effect may be used to eradicate small invasive populations of cancers. As previously described, cancer populations usually arise from a single cell or collection of cells (a "clade"), with the heterogeneous environment producing several ecological niches inhabited by phenotypically diverse cancer cell "species" [94]. As these cancer cell "species" compete amongst one another in a spatially, temporally, and metabolically diverse and fluctuating environment, they are "hunted" by the predatory host immune system and subsequently utilize niche construction strategies like angiogenesis to increase fitness [94]. In fact, angiogenesis may be considered an example of the Allee effect—the collaboration of neighboring cancer cells produces an adequate signal to initiate new blood vessel development [94]. Therefore, although cancer populations likely begin with one cell, proliferation of that cell will predictably generate group dynamics such as the Allee effect (i.e., cancer cell proliferation is an increasing function of cell density at low densities) [92]. This cooperation of individuals within a population resulting in an increased individual fitness has been observed in cultured cancer cells and growth dynamic thresholds of small metastases [95]. Additionally, the probability of tumor formation in preclinical testing of immunocompetent and immunosuppressed mice correlates directly with the number of cells injected—this emphasizes the importance of small population tumor dynamics [94].

Allee effects and the occurrence of stochastic perturbations ensures that small populations of cancer cells respond differently than larger ones to the same treatment(s) [96]. Exploiting these small population eco-evolutionary dynamics can be leveraged in the prevention of primary and metastatic tumors.

2.3.1 PREVENTION OF PRIMARY TUMORS

Cancers are thought to arise from a series of premalignant states before emerging as primary invasive cancers [97–99]. For example, atypical ductal hyperplasia (ADH) is a non-obligate precursor to ductal carcinoma *in situ* (DCIS) and invasive ductal carcinoma. Eco-evolutionary strategies may be applied in treating premalignant states to manage or eliminate small cell populations of premalignant cells that contain lower degrees of cellular heterogeneity before they progress to form a malignant invasive tumor. Chemoprevention is the utilization of pharmacologic (or dietary) agents in an effort to prevent or reduce the progression of cancer. For example, chemoprevention with tamoxifen—a selective estrogen receptor modulator (SERM)—is utilized for women at higher risk for developing breast cancer. In the BCPT trial—which recruited over 13,000 high-risk women between 1992 and 1997—women were randomized to receive either 20 mg per day of tamoxifen or placebo with a median follow up of 54.6 months [100]. The study found a 49% risk reduction in invasive breast cancer incidence among the participants who were randomized to receive tamoxifen [100]. Notably, although tamoxifen reduced the occurrence of estrogen receptor positive invasive cancers, there was no effect on estrogen receptor negative malignancies. Importantly, there was no evidence that preventative treatment with tamoxifen results in an increase in drug resistant cancers in malignancies that do eventually emerge years after chemoprevention [101, 102]. This is likely due to the fact that preventative therapeutics are typically low-dose and aimed at a single epigenetically altered clone that has a higher probability of initiating tumorigenesis [102].

Chemoprevention has also been studied and applied in prostate cancer. The PCPT was the first randomized large-scale study to demonstrate that finasteride use could reduce the risk of prostate cancer [103, 104]. The study evaluated over 18,000 men ages 55 and older for 7 years with prostate-specific antigen (PSA) equal to or less than 3.0 ng/mL and a normal digital rectal exam (DRE). Men were randomly assigned to take 5 mg of finasteride daily or a placebo [103]. The study found that finasteride use reduced the risk of prostate cancer by 25% [103].

Other potential methods of eco-evolutionary preventative strategies include altering glucose and lipid metabolism, which are characteristics of all malignancies [105, 106], and targeting the tumor microenvironment and local acidosis [97, 107–109]. A study evaluating premalignant conditions showed that local control of acidosis increased overall survival in mice [110]. Other preventative evolutionary methods include vaccination, and there has been increasing efforts to develop pro-phylactic and therapeutic vaccines designed to target cancer causing pathogens and cancer cells [111–113]. In fact, cervical cancer rates have declined by up to 22.5% per year following the intro-duction of the HPV vaccine [114].

These preventative strategies have parallels in nature. When faced with a novel invasive pest species, early eradication efforts have a much higher likelihood of success than those implemented later. Just as in cancer, early detection and eradication can exploit the vulnerabilities of populations that are small and less likely to be subdivided into numerous dispersed subpopulations. The newly invasive species has not yet evolved adaptations to their novel surroundings, making them less fit than they shall become. And the likelihood of pre-existing resistance mechanisms is much lower as the early population will possess much less phenotypic and genetic variability.

2.3.2 PREVENTION OF METASTATIC TUMORS

The metastatic cascade is composed of a series of events that includes malignant cell invasion into the vascular and/or lymphatic systems as circulating tumor cells at the primary tumor site and sub-sequent circulation and seeding into a distant organ site where colony formation results in popula-tion expansion [115]. Fortunately, far less than 1% of circulating tumor cells create clinically evident metastases [116–120]. The disseminated tumor cells and/or micrometastases that are successful in seeding distant organs either have pre-adaptations or evolve properties that allow them to survive and proliferate [121]. In fact, disseminated tumor cells (DTCs) have been noted in the bone marrow of patients with early breast cancer and represent an independent prognostic factor in early breast cancer [122, 123]. Dormant DTCs may go undetected for long periods of time before proliferating and evolving into clinically detectable metastatic disease [121, 124]. Exploiting small population eco-evolutionary dynamics seen in pest management strategies can be applied in the prevention of metastatic tumors. For example, since small populations of tumors (such as DTCs and micrometas-tases) may be relatively homogeneous—and consequently more likely to be uniformly sensitive to therapy—there is a window of opportunity to eradicate a small population while it is susceptible to extinction [120]. This is consistent with the established clinical benefits of adjuvant therapy. Once a cancer population proliferates in size and phenotypic and environmental heterogeneity, treatment becomes more challenging with the rise of resistant subpopulations and phenotypes. Evolution-based treatments of these disseminated and heterogeneous cancer populations will be discussed next.

2.4 EVOLUTION-BASED TREATMENT OF DISSEMINATED CANCERS

Evolution-based treatment of metastatic cancer represents a departure from the traditional onco-logic construct of cancer therapy in which maximum tolerated dose is applied until cancer progres-sion. For decades, the clinical treatment of disseminated cancer has focused on development of "the magic bullet"—new drugs as the primary strategy to improve outcomes for metastatic cancer. Despite new drug developments, however, metastatic cancer remains fatal. For example, in the treatment of metastatic prostate cancer, many classes of drugs have been developed and utilized—hormonal therapy, chemotherapy, immunotherapy, angiogenesis inhibitors, radiopharmaceuticals—yet metastatic prostate cancer that requires systemic therapy continues to be fatal. Although improved and novel drugs may be necessary for curative treatment, suboptimal drug treatment strategies remain a substantial barrier to cure for many metastatic cancers [125]. Innovative new constructs for drug treatment strategies can be found by observing the eco-evolutionary dynamics of species extinctions.

2.4.1 Response to Treatment and the Cost of Resistance

As described earlier in this chapter, the current clinical applications of systemic therapy to treat cancer have been based upon the assumption that maximum patient benefit is achieved by killing the maximum number of cancer cells and inducing the greatest possible cell death [125–127]. This theory is so deep-rooted that the first phase in cancer drug development involves identifying the maximum tolerated dose, which is limited only by fatal toxicities. When this approach is viewed with an evolutionary lens, several flaws in this conventional assumption arise. By obliterating an entire population of treatment-sensitive cancer cells, maximum tolerated dose applies an intense selection for resistant cells and eliminates all potential competitors—thereby maximizing the proliferation of resistant clones. This is a recognized evolutionary phenomenon called "competitive release" [128, 129].

Additional insights and parallels into these various treatment response dynamics can be found in pest and weed management [128, 129], where treatments for cancer can each be viewed as a form of pest management, each representing a selection force (Table 2.1). Historically, the use of high-dose pesticides was the routine method for pest management, and this approach promoted the development of overpowering and resistant pest strains. Recognition of this has led to an emphasis on limited use of pesticides to reduce crop damage and prevent the emergence of resistant strains [128, 129]. In fact, the United States Department of Agriculture mandates incorporation of temporal data sampling and evolutionary Darwinian dynamics when handling invasive species. Indeed, the use of computational models to guide pest management has been employed to assist in formulating optimized pest treatment strategies—similar to models used in current evolutionary-based cancer therapy trials [130–132].

In evolutionary cancer therapy, a critical component of evolutionary dynamics is the *cost of resistance*. The cost of resistance refers to the concept that although adaptation of a resistant cancer cell phenotype lends a fitness advantage while exposed to treatment, it may come at a fitness cost in a drug-free environment [133–135]. For example, cancer cells frequently utilize multidrug resistance pumps, which require a substantial investment of energy resources to synthesize, maintain, and operate [136, 137]. These pumps are required for survival in the harsh tumor microenvironment but necessitate diversion of resources that would typically be allocated to tumor invasion or propagation. In a drug-free environment, the diversion of resources to maintain the multidrug resistance pumps would likely be too great of a cost. Therefore, cancer cells may have the mechanism to develop therapy resistance, but the cost of resistance—along with other complex interactions with other tumor subpopulations—may govern the rate of resistant clone proliferation [135, 137].

Strategies to increase the cost of resistance can be applied in innovative ways to cancer treatment. Verapamil, a calcium-channel blocker, is used to treat hypertension, angina, and arrhythmias. In a preclinical model of breast cancer, cancer cells demonstrated reduced proliferation and invasion when exposed to Verapamil [136]. In this model, cancer cells detect the presence of Verapamil

TABLE 2.1

Pest Management Strategies for Cancer

Treatment	Pest Management Method
Surgery	Physical Pest Control
Radiation Therapy (XRT)	Field Burning
Chemotherapy	Chemical Pest Control
Immunotherapy	Biological Pest Control
Hormonal Therapy	Endocrine Disruptor
Cryoablation Therapy	Physical Pest Control

and use its resistant mechanisms to pump the drug from the cytosol, increasing the phenotypic cost of the cell's resistance with limited drug toxicity from a non-chemotherapeutic drug [125, 136].

The concept of cost of resistance can be used to leverage evolutionary strategies to treat disseminated cancer by creating an "evolutionary double bind" [120]. An "evolutionary double bind" creates a situation where evasion or resistance to the cytotoxic effects of a drug compromises the cancer cell's other fitness traits, thereby preventing tumor proliferation of the resistant clone [120, 138]. An ecological example would be the impact on a mouse population after the introduction of a new predator. The mice may limit scavenging patterns for food, resulting in less sustenance and subsequently fewer offspring. Although the dwindling mouse population may successfully avoid the predator, the unavoidable cost of that success means a dwindling population. As in this ecological example, the cost must be significant enough to create an evolutionary double bind for a population. In cancer therapy, if the cancer cell's adaptation to a drug can be accomplished by minimal energy expenditure (for example, via a redundant signaling pathway), the original malignant phenotype will be maintained, and proliferation of the resistant clones will continue. Cancer control in this setting is possible if resistance to therapy requires a substantial energy investment that significantly reduces the cancer cell's fitness in the tumor microenvironment [120, 138].

Not all resistance mechanisms come at a cost, however, and the extent of the cost depends on environmental and genetic factors [139–141]. Interestingly, the cost of resistance has been extensively studied in agricultural pest management and antibiotic resistance [142] with variable results. Out of 88 studies examining the cost of resistance in plants, 44 demonstrate a cost, 4 studies showed a benefit to resistance, and 40 found no difference [143], emphasizing that resistance adaptation is complex—relying not only on the resistance mechanism but on environmental and genetic factors [134, 142, 143]. One such additional factor includes a tumor's spatial architecture. If resistant cells are spatially encircled by sensitive cells, modeling studies have shown that the tumor may be controlled for an extended period of time, negating the fitness penalty of resistant cells [137, 144].

2.4.2 DYNAMIC OF ANTHROPOCENE EXTINCTIONS

Much can be gleaned from observing historical ecological and evolutionary changes in species extinctions. As described previously in this chapter, most drugs used in cancer therapy are applied at maximum tolerated dose until progression [126]. The framework of this conventional and widely accepted method of oncologic treatment is analogous to the extinction of dinosaurs following the enormous ecological and evolutionary dynamics incited by the meteor impact at the Cretaceous-Tertiary period (K-T impact). Although the application of devastating force to eliminate a population can be very effective, it is also quite indiscriminate. In addition to the mass extinction of dinosaurs, the K-T impact resulted in the extinction of 1 in 3 other species [145]. Because of this, the K-T extinction event is not an ideal paradigm for application in oncologic therapy.

Less dramatic extinctions caused by humans in the Anthropocene era provide a more successful blueprint for evolutionary-based cancer treatments. These Anthropocene extinctions include the unintended extinction events of the heath hen and passenger pigeon [146, 147] and the intentional extinction of the Galápagos Islands goats [145]. Domesticated goats were introduced to the Galápagos Islands in the 1800s as a food source but became an ecological threat to the native Galápagos tortoise. In order to reduce the ecological devastation caused by the rapidly expanding goat population, an environmental restoration project called Project Isabela was initiated in 1997 and began with eradication of the goat population via hunting efforts on trucks and helicopters [145, 148]. Hunting the goats eliminated approximately 90% of the goat population; the small population of uncatchable survivor goats remained. These survivor goats were likely more adept at identifying noises made by helicopters and trucks and escaping into the forest before being detected by hunters. As a result, these "resistant" goats began to breed and evade hunters successfully. The initial strategy of hunting these "resistant" goats became ineffective, so an alternative approach was used to target this population. "Judas goats"—sterilized female goats covered with hormones and fitted with radio collars—were released on the island and used to attract and mingle with the small surviving groups, ultimately allowing the hunters to locate and eradicate the population by 2006 [148].

The Galápagos Islands goat Anthropocene extinction emphasizes the two-strike dynamic observed with other extinction events [145–147] and serves as an important framework for successful evolutionary-based cancer treatment strategies. As seen with Project Isabela, a successful extinction event of a heterogeneous and large population begins with an initial assault (or series of assaults) that dramatically reduces the size, diversity, and spatial distribution of a population [145]. In the case of the Galápagos goats, the initial assault or "strike" caused by the hunting expeditions dramatically reduced the population but was not sufficient in causing extinction because of "evolutionary rescue"—the concept that populations can survive and recover from a massive assault because of diverse genetic, phenotypic, or behavioral traits [145]. These diverse traits rescue a large population from extinction when hit by a devastating force [145, 149, 150]. Importantly, the initial "strike"—which significantly reduced the goat population—was no longer effective after evolutionary rescue created a resistant population. As seen in Project Isabela, extinction of the remaining resistant small populations was still possible but required a different tactic (Judas goats). Even more critical to recognize is the success and effectiveness of a particular tactic or "strike" relies on the state of the population. Engaging the Judas goats as an initial "strike" would have been highly ineffective when the goat population was large and widely distributed, but as a second "strike" drove the small population of goats to extinction.

Estimation of the minimal viable population (MVP) and performing a population viability analysis (PVA) are ecological tools that can be applied in the treatment of disseminated metastatic cancer. MVP is an estimate of the minimum number of individuals required to save a population susceptible to extinction. An extinction event becomes exceedingly probable below the MVP due to small population vulnerability to stochastic and environmental changes [145, 151]. Population viability analysis (PVA) has been examined in conservation studies and estimates the probability of extinction given an existing number of individuals and a particular time frame [145, 152–154]. Although utilized in ecology to prevent extinction of endangered species, these principles and tools can be used to leverage extinction of small populations of disseminated cancer in the oncologic realm.

2.4.3 APPLYING GAME THEORY TO ONCOLOGIC TREATMENT

Treatment failure following conventional MTD therapy of metastatic cancer is attributable to the evolution and proliferation of resistant clones. Although a blueprint for evolutionary-based therapies may be derived from observation of Anthropocene extinctions, construction of a mathematically framed contest—game theory—allows for application of these principles to oncologic therapy where the oncologist and cancer cells engage in a predator–prey like game [2, 155, 156].

Game theory was initially developed to inform conflict and cooperation in economics by describing the choices, consequences, and interactions involving populations and individuals [157–161]. Maynard Smith and Price subsequently applied this economics concept to evolutionary dynamics [161, 162]. When game theory is applied in the evolutionary setting, the "players" inherit (instead of choose), and the "consequence" of inheritance is survival versus extinction [2]. Game theory has been successfully utilized to control agricultural pests [163] and has been applied to the management of antibiotic resistance [164].

Interestingly, game theory can be applied to cancer therapy, as therapy is administered episodically in cycles. Current conventional oncologic treatment of metastatic cancer involves the continuous or repeated use of drug or drug combinations at maximum tolerated dose [126]. Even metronomic drug strategies are structured so that maximum tumor cell killing is achieved [1, 155]. Although extermination of the maximum number of tumor cells would seem favorable, it is evolutionarily imprudent. As described earlier in this chapter, maximum cell killing allows for the development of resistant clones if the same drug regimen is constantly applied. By repeatedly dispensing the same treatment until disease progression, the oncologist is relinquishing control of the Stackelberg (leader-follower) dynamics to the cancer cells, and treatment failure becomes inescapable.

In order to apply game theory to successfully improve disseminated cancer treatment outcomes, the conventional maximum tolerated dose framework must be relinquished. Instead, the treating physician must exploit the following two critical points in game theory [155, 156, 165]:

1) Cancer cells cannot anticipate or evolve for treatments that have not been applied
2) Stackelberg (leader-follower) dynamics will always emerge.

The oncologist is the only rational player that can anticipate future events. On the other hand, cancer cells—like all evolving organisms—can only respond to current conditions. Cancer cells are not sentient beings that can anticipate and adapt to future environmental assaults. The oncologist always makes the first move by applying therapy. Only then can cancer cells engage in the "game" by reacting to the administered therapy and evolving resistance strategies. Because of these conditions, cancer therapy will always be a leader-follower game—also known as Stackelberg games [166, 167]—and a significant advantage will always be in the hands of the leader who initiates both the initial *and subsequent* "moves". The treating oncologist has the advantage of applying the initial "move" (i.e., treatment) and also has the ability to anticipate the consequent cancer cell response. This competitive advantage and information obtained from initial treatment cycles can be exploited by the oncologist in steering or restricting cancer cell resistance strategies [168, 169].

When the physician does not exploit the advantage of being a rational game theory player, the opportunity to anticipate and steer treatment outcomes is lost [155]. This occurs commonly in the conventional oncologic treatment of metastatic cancer—awaiting cancer cells to evolve and demonstrate progression before adjusting treatment to a second line therapy. In this scenario—when therapy is altered only when the tumor evolves resistance and demonstrates progression—the physician has relinquished control and becomes the "follower", reacting only to the evolution of the "leader" tumor cells. This is an imprudent strategic method in treating cancer.

A paradigm shift is required, where the oncologist regains control of the Stackelberg dynamics and applies therapies prior to the proliferation of resistant clones. Foreknowledge of the cancer cell's best response curves, treatment strategies available, estimates of disease burden, and observed past and estimated future changes in cancer populations [155, 156] allows for this informed treatment decision-making within a construct of adaptive and extinction ("first strike, second strike") therapies.

2.4.4 ADAPTIVE AND EXTINCTION (FIRST STRIKE, SECOND STRIKE) THERAPIES

As previously described, standard-of-care oncologic therapy of metastatic cancer involves the application of maximum tolerated dose (MTD) of a drug per a static dose and schedule until either unequivocal progression or unacceptable toxicity arises. This method places the oncologist as a "follower" and the tumor cells as the "leader" when viewed with the Stackelberg dynamic game theory lens. Importantly, the current schema of oncologic therapy utilizing a MTD approach represents a static method of treating metastatic cancer. In contrast, disseminated cancers are exceedingly dynamic and contain a great amount of spatial, temporal, and cellular heterogeneity that can rapidly change following the application of therapy. Indeed, the tumor treated in the second or third cycle of chemotherapy is altered compared to the tumor at diagnosis treated in the first cycle [125]. Acknowledgment of tumor evolution or alteration with application of therapy emphasizes the importance of a dynamic approach to cancer therapy, where treatment drug selection, drug dose, and treatment sequence are strategically selected in order to "stay ahead" of intra-tumoral evolution.

When applying eco-evolutionary principles to cancer therapy, it is important to define the goals of therapy. There are two main evolutionary approaches in the treatment of metastatic cancer: adaptive therapy (AT) and extinction therapy (ET). The goal of AT is to maximize progression free survival—not necessarily reduce cancer burden—with the application of the minimum drug dose necessary to maintain tumor stability and patient quality of life. With AT, drugs and dose schedules are constantly adjusted to take advantage of evolutionary dynamics and maintain a stable tumor burden. With ET, the goal is to apply a sequence of strikes to produce extinction. We will discuss both strategies.

2.4.4.1 Adaptive Therapy (AT)

This treatment strategy was initially framed mathematically with feasibility demonstrated via computational simulations [170]. The first *in vivo* testing occurred using an ovarian cancer xenograft preclinical model (OVCAR-3 cells) treated with carboplatin, where the algorithm was based solely on tumor volumes measured by calipers. If the tumor increased in size during two consecutive measurements, the dose of carboplatin was increased. Conversely, carboplatin dose was reduced if the tumor volume decreased in the same period of time. Compared to the standard high-dose treatment of carboplatin, AT was able to keep the tumor controlled for an extended period of time compared to the standard therapeutic approach, with an increased overall survival in the mice undergoing AT [170].

AT has also been applied to two orthotopic preclinical models in triple-negative and estrogen receptor positive breast cancer [171]. In this study, mice were treated with paclitaxel with treatment algorithms based on tumor volumes measured by magnetic resonance imaging (MRI). Two different treatment algorithms were tested. One model only applied a fixed drug dose if the tumor grew more than 25% from the previous measurements (Model 1). The second model began with applying maximum tolerated dose, followed by an increase or decrease of the dose based upon tumor volume increase/decrease (Model 2). In Model 2, a low tumor threshold was used in which therapy was withheld if tumor volume dropped below the threshold. These two models were compared to the standard continuous maximum tolerated dose therapy with paclitaxel. The study found that Model 2 was more successful with maintaining tumor control. Even more importantly, results from this study and the OVCAR study revealed that AT could be divided into two phases: 1) an initially applied aggressive therapy, with decreasing dose over time, to reduce exponential tumor growth followed by 2) a maintenance phase after tumor growth was controlled, using progressively lower and/ or less frequent drug doses to maintain control [171].

AT has also demonstrated superior outcomes in recent clinical trials. In a study treating metastatic castrate-resistant prostate cancer (mCRPC) with abiraterone, a game theory model to guide on and off abiraterone therapy has showed significantly prolonged response and improved outcomes compared to standard therapy [131, 132]. Median radiographic progression free survival (rPFS) was 30.4 months and median overall survival (OS) was 58.5 months in the AT group compared to the 14.3 months median rPFS and 31.3 months median OS in a historical comparison cohort receiving standard-of-care treatment [132]. The mathematical model used to design this trial enabled the development of analytic techniques to assess the longitudinal trial results and revise the role of key parameters in AT [92, 172–175]. The model was subsequently updated and used to simulate the intra-tumoral dynamics leading to the clinical outcome in each patient within the cohorts [132]. This updated model is currently being applied in an ongoing AT trial in metastatic castration sensitive prostate cancer (NCT02415621), utilizing either on-and-off androgen deprivation therapy (ADT) with luteinizing hormone releasing hormone (LHRH) analog, an NHA (abiraterone, enzalutamide, or apalutamide), or in combination, based upon a patient's testosterone and PSA levels [130].

2.4.4.2 Extinction Therapy (ET)

Although results from clinical trials utilizing ET in the treatment of adult metastatic cancer have yet to be published, a phase IIA study for high-risk metastatic castration sensitive prostate cancer is currently accruing patients (NCT05189457). In this section, we will review important eco-evolutionary concepts that must be applied in any ET trial. There are two critical ecological evolutionary concepts required to see extinction of a large, diverse, heterogeneous and spatially dispersed population [94]:

1) Multiple sequential perturbations or "strikes"
2) The concept of minimum viable population (MVP).

The Anthropocene extinction of the heath hen illustrates these critical ecologic evolutionary concepts. Extensive over-hunting by European settlers in the 1800s was the "first strike" (initial perturbation) that initiated the extinction of this North American bird, dwindling the heath hen population to 50 hens [94, 146]. The "first strike" is an event—or sequence of several events—that considerably reduce(s) the size, diversity, and spatial distribution of a large and heterogeneous population. The "first strike" for the heath hen significantly reduced a large, spatially distributed, and diverse population of hens to a small, geographically restricted group with limited phenotypic and genetic diversity. Due to the intentional safeguarding of the heath hen population by the local community, the heath hen population recovered to approximately 2,000 hens in the subsequent four decades. Then, a series of various stochastic perturbations—or "second strikes"—pushed the heath hen into extinction. These various "strikes" included several harsh winters, destruction of their habitat by a fire, and the spread of an infectious poultry disease [94, 146]. The final extinction was a consequence of several small "second strikes" that would not have significantly impacted the original large, diverse population. The concept of minimum viable population (MVP)—the number of individuals necessary for a species to survive—is an important concept in these small population dynamics. Populations at or near MVP are sensitive to demographic changes (variations in birth and death rates) and environmental/ecological habitat disruptions [94, 145, 146].

Application of the "first strike-second strike" tactics to cancer therapy entails a departure from conventional oncologic practices in two key ways [94]:

1) Therapy must be switched despite it being effective
2) Therapy must be applied even in the absence of measurable disease.

There is an oncologic precedent in the application of "first strike-second strike" extinction therapy—the treatment of pediatric ALL. The exceedingly effective curative therapy for pediatric ALL was empirically-developed and involves an initial "induction" therapy, followed by a "treatment intensification" and subsequent "intermediate dose intensification" and "maintenance", all utilizing different drugs. Framed in evolutionary terms, the initial "induction" represents the first strike, quickly followed by a second strike ("treatment intensification") and subsequent third and fourth strikes. The over-arching concept is simple: deliver the strongest damage to the cancer cell population with the first strike and continue assaulting the remaining small populations with various diverse strikes to drive cancers to extinction. Just as with AT, mathematical modeling may be utilized to inform and guide the sequence, length, and dose of different therapeutic strikes utilized in ET.

2.5 CONCLUSION

Cancers are dynamic, complex systems that evolve resistance strategies when therapies are applied. The current treatment paradigm of static maximum tolerated dose until the protocol ends or tumor progression is observed inevitably results in the proliferation of resistant clones. To improve cancer treatment outcomes, application of novel therapies must be performed within an evolutionary-based strategic framework. Identification of additional dynamic datapoints and biomarkers that measure the internal evolutionary dynamics of the cancer during treatment will be critical. Future directions should include incorporation of a multitude of dynamic datapoints into sophisticated patient-specific mathematical models to provide oncologists with decision support tools to optimize cancer therapy.

REFERENCES

1. Gatenby, R.A. and J. Brown, *Mutations, evolution and the central role of a self-defined fitness function in the initiation and progression of cancer.* Biochim Biophys Acta Rev Cancer, 2017. **1867**(2): p. 162–6.
2. Brown, J.S., *Why Darwin would have loved evolutionary game theory.* Proc Biol Sci, 2016. **283**(1838).
3. Nowell, P.C., *The clonal evolution of tumor cell populations.* Science, 1976. **194**(4260): p. 23–8.
4. Greaves, M. and C.C. Maley, *Clonal evolution in cancer.* Nature, 2012. **481**(7381): p. 306–13.
5. Fearon, E.R. and B. Vogelstein, *A genetic model for colorectal tumorigenesis.* Cell, 1990. **61**(5): p. 759–67.
6. Vaux, D.L., *In defense of the somatic mutation theory of cancer.* Bioessays, 2011. **33**(5): p. 341–3.
7. Gerlinger, M., et al., *Cancer: evolution within a lifetime.* Annu Rev Genet, 2014. **48**: p. 215–36.
8. Martincorena, I., et al., *Tumor evolution: high burden and pervasive positive selection of somatic mutations in normal human skin.* Science, 2015. **348**(6237): p. 880–6.
9. Yadav, V.K., J. DeGregori and S. De, *The landscape of somatic mutations in protein coding genes in apparently benign human tissues carries signatures of relaxed purifying selection.* Nucleic Acids Res, 2016. **44**(5): p. 2075–84.
10. Deng, G., et al., *Loss of heterozygosity in normal tissue adjacent to breast carcinomas.* Science, 1996. **274**(5295): p. 2057–9.
11. Alexandrov, L.B., et al., *Signatures of mutational processes in human cancer.* Nature, 2013. **500**(7463): p. 415–21.
12. Levy-Lahad, E. and E. Friedman, *Cancer risks among BRCA1 and BRCA2 mutation carriers.* Br J Cancer, 2007. **96**(1): p. 11–15.
13. Brose, M.S., et al., *Cancer risk estimates for BRCA1 mutation carriers identified in a risk evaluation program.* J Natl Cancer Inst, 2002. **94**(18): p. 1365–72.
14. Alexandrov, L.B., et al., *Deciphering signatures of mutational processes operative in human cancer.* Cell Rep, 2013. **3**(1): p. 246–59.
15. Alexandrov, L.B. and M.R. Stratton, *Mutational signatures: the patterns of somatic mutations hidden in cancer genomes.* Curr Opin Genet Dev, 2014. **24**: p. 52–60.
16. Gatenby, R.A., J.J. Cunningham and J.S. Brown, *Evolutionary triage governs fitness in driver and passenger mutations and suggests targeting never mutations.* Nat Commun, 2014. **5**: p. 5499.
17. Krogh, A., *The number and distribution of capillaries in muscles with calculations of the oxygen pressure head necessary for supplying the tissue.* J Physiol, 1919. **52**(6): p. 409–15.
18. Thomlinson, R.H. and L.H. Gray, *The histological structure of some human lung cancers and the possible implications for radiotherapy.* Br J Cancer, 1955. **9**(4): p. 539–49.
19. Gatenby, R.A. and R.J. Gillies, *Why do cancers have high aerobic glycolysis?* Nat Rev Cancer, 2004. **4**(11): p. 891–9.
20. Graeber, T.G., et al., *Hypoxia-mediated selection of cells with diminished apoptotic potential in solid tumours.* Nature, 1996. **379**(6560): p. 88–91.
21. Gatenby, R.A. and E.T. Gawlinski, *A reaction-diffusion model of cancer invasion.* Cancer Res, 1996. **56**(24): p. 5745–53.
22. Gatenby, R.A., et al., *Cellular adaptations to hypoxia and acidosis during somatic evolution of breast cancer.* Br J Cancer, 2007. **97**(5): p. 646–53.
23. Wykoff, C.C., et al., *Expression of the hypoxia-inducible and tumor-associated carbonic anhydrases in ductal carcinoma in situ of the breast.* Am J Pathol, 2001. **158**(3): p. 1011–19.
24. Wykoff, C.C., et al., *The HIF pathway: implications for patterns of gene expression in cancer.* Novartis Found Symp, 2001. **240**: p. 212–25; discussion 225–31.
25. Brown, R.S., et al., *Expression of hexokinase II and Glut-1 in untreated human breast cancer.* Nucl Med Biol, 2002. **29**(4): p. 443–53.
26. Gillies, R.J. and R.A. Gatenby, *Adaptive landscapes and emergent phenotypes: why do cancers have high glycolysis?* J Bioenerg Biomembr, 2007. **39**(3): p. 251–7.
27. Gillies, R.J. and R.A. Gatenby, *Hypoxia and adaptive landscapes in the evolution of carcinogenesis.* Cancer Metastasis Rev, 2007. **26**(2): p. 311–17.
28. Damaghi, M., et al., *The harsh microenvironment in early breast cancer selects for a Warburg phenotype.* Proc Natl Acad Sci USA, 2021. **118**(3).
29. Lloyd, M.C., et al., *Darwinian dynamics of intratumoral heterogeneity: not solely random mutations but also variable environmental selection forces.* Cancer Res, 2016. **76**(11): p. 3136–44.
30. Feder, J.L., et al., *Host fidelity is an effective premating barrier between sympatric races of the apple maggot fly.* Proc Natl Acad Sci, 1994. **91**(17): p. 7990–4.

31. Jiggins, C.D. and J.R. Bridle, *Speciation in the apple maggot fly: a blend of vintages?* Trends Ecol Evol, 2004. **19**(3): pp. 111–14.
32. Via, S., A.C. Bouck and S. Skillman, *Reproductive isolation between divergent races of pea aphids on two hosts. II: selection against migrants and hybrids in the parental environments.* Evolution, 2000. **54**(5): p. 1626–37.
33. Stanton, M.L., T.M. Palmer and T.P. Young, *Competition-colonization trade-offs in a guild of African acacia-ants.* Ecol Monogr, 2002. **72**: p. 347–63.
34. Rodriguez, A., G. Jansson and H. Andren, *Composition of an avian guild in spatially structured habitats supports a competition-colonization trade-off.* Proc Biol Sci, 2007. **274**(1616): p. 1403–11.
35. Turnbull, L.A., D. Coomes, A. Hector and M. Rees, *Seed mass and the competition/colonization trade-off: competitive interactions and spatial patterns in a guild of annual plants.* J Ecol, 2004. **92**: p. 97–109.
36. Phillips, B.L., G.P. Brown, J.K. Webb and R. Shine, *Invasion and the evolution of speed in toads.* Nature, 2006. **439**: p. 803.
37. Orlando, P.A., R.A. Gatenby and J.S. Brown, *Tumor evolution in space: the effects of competition colonization tradeoffs on tumor invasion dynamics.* Front Oncol, 2013. **3**: p. 45.
38. Aktipis, C.A., C.C. Maley and J.W. Pepper, *Dispersal evolution in neoplasms: the role of disregulated metabolism in the evolution of cell motility.* Cancer Prev Res (Phila), 2012. **5**(2): p. 266–75.
39. Shine, R., *Invasive species as drivers of evolutionary change: cane toads in tropical Australia.* Evol Appl, 2012. **5**(2): p. 107–16.
40. Kotler, B.P., J.S. Brown and W.A. Mitchell, *The role of predation in shaping the behavior, morphology and community organization of desert rodents.* Aust J Zool, 1994. **42**(4): p. 449–66.
41. Kotler, B.P. and J.S. Brown, *Cancer community ecology.* Cancer Control, 2020. **27**(1): p. 1073274820951776.
42. Dunn, G.P., L.J. Old and R.D. Schreiber, *The three Es of cancer immunoediting.* Annu Rev Immunol, 2004. **22**: p. 329–60.
43. Mittal, D., et al., *New insights into cancer immunoediting and its three component phases—elimination, equilibrium and escape.* Curr Opin Immunol, 2014. **27**: p. 16–25.
44. Gabrilovich, D.I., S. Ostrand-Rosenberg and V. Bronte, *Coordinated regulation of myeloid cells by tumours.* Nat Rev Immunol, 2012. **12**(4): p. 253–68.
45. Peplinski, J. et al., *Ecology of fear: spines, armor and noxious chemicals deter predators in cancer and in nature.* Front Ecol Evol, 2021: p. 495.
46. Zitvogel, L., A. Tesniere and G. Kroemer, *Cancer despite immunosurveillance: immunoselection and immunosubversion.* Nat Rev Immunol, 2006. **6**(10): p. 715–27.
47. Pfennig, D.W., W.R. Harcombe and K.S. Pfennig, *Frequency-dependent Batesian mimicry.* Nature, 2001. **410**(6826): p. 323.
48. Lippitz, B.E., *Cytokine patterns in patients with cancer: a systematic review.* Lancet Oncol, 2013. **14**(6): p. e218–28.
49. Scheel, D., *Profitability, encounter rates, and prey choice of African lions.* Behav Ecol, 1993. **4**(1): p. 90–7.
50. Kareva, I., et al., *Predator-prey in tumor-immune interactions: a wrong model or just an incomplete one?* Front Immunol, 2021. **12**: p. 668221.
51. Quail, D.F. and J.A. Joyce, *Microenvironmental regulation of tumor progression and metastasis.* Nat Med, 2013. **19**(11): p. 1423–37.
52. Halitschke, R., et al., *Shared signals 'alarm calls' from plants increase apparency to herbivores and their enemies in nature.* Ecol Lett, 2008. **11**(1): p. 24–34.
53. Merlo, L.M., et al., *Cancer as an evolutionary and ecological process.* Nat Rev Cancer, 2006. **6**(12): p. 924–35.
54. Maley, C.C., et al., *Classifying the evolutionary and ecological features of neoplasms.* Nat Rev Cancer, 2017. **17**(10): p. 605–19.
55. Barker, G. and F.J. Odling-Smee, *Entangled life: organisms and environment in the biological and social sciences: history, philosophy and theory of the life sciences* (eds. G.E. Desjardins, G. Barker and T. Pearce). Springer, 2014. p. 187–211.
56. Gillies, R.J., et al., *Eco-evolutionary causes and consequences of temporal changes in intratumoural blood flow.* Nat Rev Cancer, 2018. **18**(9): p. 576–85.
57. You, L., et al., *Spatial vs. non-spatial eco-evolutionary dynamics in a tumor growth model.* J Theor Biol, 2017. **435**: p. 78–97.
58. Laland, K.N., F.J. Odling-Smee and M.W. Feldman, *Evolutionary consequences of niche construction and their implications for ecology.* Proc Natl Acad Sci USA, 1999. **96**(18): p. 10242–7.

59. Fukumura, D. and R.K. Jain, *Tumor microvasculature and microenvironment: targets for anti-angiogenesis and normalization.* Microvasc Res, 2007. **74**(2–3): p. 72–84.

60. Gillies, R.J., et al., *Causes and effects of heterogeneous perfusion in tumors.* Neoplasia, 1999. **1**(3): p. 197–207.

61. Khramtsov, V.V. and R.J. Gillies, *Janus-faced tumor microenvironment and redox.* Antioxid Redox Signal, 2014. **21**(5): p. 723–9.

62. Skala, M.C., et al., *Longitudinal optical imaging of tumor metabolism and hemodynamics.* J Biomed Opt, 2010. **15**(1): p. 011112.

63. Wang, J.W., et al., *Quantitative assessment of tumor blood flow changes in a murine breast cancer model after adriamycin chemotherapy using contrast-enhanced destruction-replenishment sonography.* J Ultrasound Med, 2013. **32**(4): p. 683–90.

64. Milosevic, M.F., A.W. Fyles and R.P. Hill, *The relationship between elevated interstitial fluid pressure and blood flow in tumors: a bioengineering analysis.* Int J Radiat Oncol Biol Phys, 1999. **43**(5): p. 1111–23.

65. Jain, R.K., *Determinants of tumor blood flow: a review.* Cancer Res, 1988. **48**(10): p. 2641–58.

66. Matsumoto, S., et al., *Imaging cycling tumor hypoxia.* Cancer Res, 2010. **70**(24): p. 10019–23.

67. Cairns, R.A., T. Kalliomaki and R.P. Hill, *Acute (cyclic) hypoxia enhances spontaneous metastasis of KHT murine tumors.* Cancer Res, 2001. **61**(24): p. 8903–8.

68. Dewhirst, M.W., *Relationships between cycling hypoxia, HIF-1, angiogenesis and oxidative stress.* Radiat Res, 2009. **172**(6): p. 653–65.

69. Zhang, G., et al., *A dual-emissive-materials design concept enables tumour hypoxia imaging.* Nat Mater, 2009. **8**(9): p. 747–51.

70. Verduzco, D., et al., *Intermittent hypoxia selects for genotypes and phenotypes that increase survival, invasion, and therapy resistance.* PLOS One, 2015. **10**(3): p. e0120958.

71. Sgroi, D.C., *Preinvasive breast cancer.* Annu Rev Pathol, 2010. **5**: p. 193–221.

72. Hu, M. and K. Polyak, *Microenvironmental regulation of cancer development.* Curr Opin Genet Dev, 2008. **18**(1): p. 27–34.

73. Osuala, K.O., et al., *Il-6 signaling between ductal carcinoma in situ cells and carcinoma-associated fibroblasts mediates tumor cell growth and migration.* BMC Cancer, 2015. **15**: p. 584.

74. Strell, C., et al., *Impact of epithelial-stromal interactions on peritumoral fibroblasts in ductal carcinoma in situ.* J Natl Cancer Inst, 2019. **111**(9): p. 983–95.

75. Hu, M., et al., *Regulation of in situ to invasive breast carcinoma transition.* Cancer Cell, 2008. **13**(5): p. 394–406.

76. Chen, X.Y., et al., *Higher density of stromal M2 macrophages in breast ductal carcinoma in situ predicts recurrence.* Virchows Arch, 2020. **476**(6): p. 825–33.

77. Linde, N., et al., *Macrophages orchestrate breast cancer early dissemination and metastasis.* Nat Commun, 2018. **9**(1): p. 21.

78. Otto, S.P., *Adaptation, speciation and extinction in the Anthropocene.* Proc Biol Sci, 2018. **285**(1891).

79. Artzy-Randrup, Y., et al., *Novel evolutionary dynamics of small populations in breast cancer adjuvant and neoadjuvant therapy.* NPJ Breast Cancer, 2021. **7**(1): p. 26.

80. Raup, D., *Extinction: bad genes or bad luck?* New Sci, 1991. **131**(1786): p. 46–9.

81. Sibly, R.M., et al., *On the regulation of populations of mammals, birds, fish, and insects.* Science, 2005. **309**(5734): p. 607–10.

82. Brook, B.W. and C.J. Bradshaw, *Strength of evidence for density dependence in abundance time series of 1198 species.* Ecology, 2006. **87**(6): p. 1445–51.

83. Melbourne, B.A. and A. Hastings, *Extinction risk depends strongly on factors contributing to stochasticity.* Nature, 2008. **454**(7200): p. 100–3.

84. Allee, W.C. *Animal aggregations, a study in general sociology.* The University of Chicago Press, 1931.

85. Odum, H.T.A. and W. C. Allee. *A note on the stable point of populations showing both intraspecific cooperation and disoperation.* Ecology, 1954. **35**: p. 95–7.

86. Darwin, C. and T.G. Bonney, *The structure and distribution of coral reefs.* 3rd ed. D. Appleton and Company, 1897.

87. Stephens, P.A. and W.J. Sutherland, *Consequences of the Allee effect for behaviour, ecology and conservation.* Trends Ecol Evol, 1999. **14**(10): p. 401–5.

88. Allee, W.C. and P. Frank, *The utilization of minute food particles by goldfish.* Physiol Zool, 1949. **22**(4): p. 346–58.

89. Boukal, D.S. and L. Berec, *Single-species models of the Allee effect: extinction boundaries, sex ratios and mate encounters.* J Theor Biol, 2002. **218**(3): p. 375–94.

90. Bottger, K., et al., *An emerging Allee effect is critical for tumor initiation and persistence.* PLoS Comput Biol, 2015. **11**(9): p. e1004366.

91. Allee, M.H. *Jane's island.* Houghton Mifflin Company, 1931.

92. Brown, J.S., J.J. Cunningham and R.A. Gatenby, *Aggregation effects and population-based dynamics as a source of therapy resistance in cancer.* IEEE Trans Biomed Eng, 2017. **64**(3): p. 512–18.

93. Allee, W.C. and E. Bowen, *Studies in animal aggregations: mass protection against colloidal silver among goldfishes.* J Exp Zool, 1932. **61**: p. 185–207.

94. Gatenby, R.A., J. Zhang and J.S. Brown, *First strike-second strike strategies in metastatic cancer: lessons from the evolutionary dynamics of extinction.* Cancer Res, 2019. **79**(13): p. 3174–7.

95. Axelrod, R., D.E. Axelrod and K.J. Pienta, *Evolution of cooperation among tumor cells.* Proc Natl Acad Sci USA, 2006. **103**(36): p. 13474–9.

96. Korolev, K.S., J.B. Xavier and J. Gore, *Turning ecology and evolution against cancer.* Nat Rev Cancer, 2014. **14**(5): p. 371–80.

97. Gatenby, R.A. and R.J. Gillies, *A microenvironmental model of carcinogenesis.* Nat Rev Cancer, 2008. **8**(1): p. 56–61.

98. Vogelstein, B. and K.W. Kinzler, *The multistep nature of cancer.* Trends Genet, 1993. **9**(4): p. 138–41.

99. Fang, J.S., R.D. Gillies and R.A. Gatenby, *Adaptation to hypoxia and acidosis in carcinogenesis and tumor progression.* Semin Cancer Biol, 2008. **18**(5): p. 330–7.

100. Fisher, B., et al., *Tamoxifen for prevention of breast cancer: report of the national surgical adjuvant breast and bowel project P-1 study.* J Natl Cancer Inst, 1998. **90**(18): p. 1371–88.

101. Metcalfe, K.A. and S.A. Narod, *Breast cancer prevention in women with a BRCA1 or BRCA2 mutation.* Open Med, 2007. **1**(3): p. e184–90.

102. Hochberg, M.E., et al., *Preventive evolutionary medicine of cancers.* Evol Appl, 2013. **6**(1): p. 134–43.

103. Thompson, I.M., et al., *The influence of finasteride on the development of prostate cancer.* N Engl J Med, 2003. **349**(3): p. 215–24.

104. Thompson, I.M., et al., *Prevention of prostate cancer with finasteride: US/European perspective.* Eur Urol, 2003. **44**(6): p. 650–5.

105. Porporato, P.E., et al., *Anticancer targets in the glycolytic metabolism of tumors: a comprehensive review.* Front Pharmacol, 2011. **2**: p. 49.

106. Dang, C.V., *Links between metabolism and cancer.* Genes Dev, 2012. **26**(9): p. 877–90.

107. Gatenby, R.A., R.J. Gillies and J.S. Brown, *Of cancer and cave fish.* Nat Rev Cancer, 2011. **11**(4): p. 237–8.

108. Gillies, R.J., D. Verduzco and R.A. Gatenby, *Evolutionary dynamics of carcinogenesis and why targeted therapy does not work.* Nat Rev Cancer, 2012. **12**(7): p. 487–93.

109. Rothwell, P.M., et al., *Short-term effects of daily aspirin on cancer incidence, mortality, and non-vascular death: analysis of the time course of risks and benefits in 51 randomised controlled trials.* Lancet, 2012. **379**(9826): p. 1602–12.

110. Ibrahim-Hashim, A., et al., *Systemic buffers inhibit carcinogenesis in TRAMP mice.* J Urol, 2012. **188**(2): p. 624–31.

111. Finn, O.J., *Cancer vaccines: between the idea and the reality.* Nat Rev Immunol, 2003. **3**(8): p. 630–41.

112. Finn, O.J., *Premalignant lesions as targets for cancer vaccines.* J Exp Med, 2003. **198**(11): p. 1623–6.

113. Palucka, K., H. Ueno and J. Banchereau, *Recent developments in cancer vaccines.* J Immunol, 2011. **186**(3): p. 1325–31.

114. Mix, J.M., et al., *Assessing impact of HPV vaccination on cervical cancer incidence among women aged 15–29 years in the United States, 1999–2017: an ecologic study.* Cancer Epidemiol Biomarkers Prev, 2021. **30**(1): p. 30–7.

115. Pantel, K. and R.H. Brakenhoff, *Dissecting the metastatic cascade.* Nat Rev Cancer, 2004. **4**(6): p. 448–56.

116. Weiss, L., *Metastatic inefficiency.* Adv Cancer Res, 1990. **54**: p. 159–211.

117. Luzzi, K.J., et al., *Multistep nature of metastatic inefficiency: dormancy of solitary cells after successful extravasation and limited survival of early micrometastases.* Am J Pathol, 1998. **153**(3): p. 865–73.

118. Fidler, I.J., *Metastasis: quantitative analysis of distribution and fate of tumor emboli labeled with 125 I-5-iodo-2'-deoxyuridine.* J Natl Cancer Inst, 1970. **45**(4): p. 773–82.

119. Cameron, M.D., et al., *Temporal progression of metastasis in lung: cell survival, dormancy, and location dependence of metastatic inefficiency.* Cancer Res, 2000. **60**(9): p. 2541–6.

120. Gatenby, R.A., J. Brown and T. Vincent, *Lessons from applied ecology: cancer control using an evolutionary double bind.* Cancer Res, 2009. **69**(19): p. 7499–502.

121. Blasco, M.T., I. Espuny and R.R. Gomis, *Ecology and evolution of dormant metastasis.* Trends Cancer, 2022. **8**(7): p. 570–82.

122. Borgen, E., et al., *NR2F1 stratifies dormant disseminated tumor cells in breast cancer patients.* Breast Cancer Res, 2018. **20**(1): p. 120.

123. Ring, A., et al., *Clinical and biological aspects of disseminated tumor cells and dormancy in breast cancer.* Front Cell Dev Biol, 2022. **10**: p. 929893.

124. Gui, P. and T.G. Bivona, *Evolution of metastasis: new tools and insights.* Trends Cancer, 2022. **8**(2): p. 98–109.

125. Enriquez-Navas, P.M., J.W. Wojtkowiak and R.A. Gatenby, *Application of evolutionary principles to cancer therapy.* Cancer Res, 2015. **75**(22): p. 4675–80.

126. Norton, L. and R. Simon, *The Norton-Simon hypothesis revisited.* Cancer Treat Rep, 1986. **70**(1): p. 163–9.

127. Rodrigues, D.S. and P.F. de Arruda Mancera, *Mathematical analysis and simulations involving chemotherapy and surgery on large human tumours under a suitable cell-kill functional response.* Math Biosci Eng, 2013. **10**(1): p. 221–34.

128. Neve, P., M. Vila-Aiub and F. Roux, *Evolutionary-thinking in agricultural weed management.* New Phytol, 2009. **184**(4): p. 783–93.

129. Oliveira, E.E., et al., *Competition between insecticide-susceptible and resistant populations of the maize weevil, Sitophilus zeamais.* Chemosphere, 2007. **69**(1): p. 17–24.

130. Zhang, J., et al., *A phase 1b adaptive androgen deprivation therapy trial in metastatic castration sensitive prostate cancer.* Cancers (Basel), 2022. **14**(21).

131. Zhang, J., et al., *Integrating evolutionary dynamics into treatment of metastatic castrate-resistant prostate cancer.* Nat Commun, 2017. **8**(1): p. 1816.

132. Zhang, J., et al., *Evolution-based mathematical models significantly prolong response to abiraterone in metastatic castrate-resistant prostate cancer and identify strategies to further improve outcomes.* Elife, 2022. **11**.

133. Gatenby, R.A., *A change of strategy in the war on cancer.* Nature, 2009. **459**(7246): p. 508–9.

134. Lenormand T, H.N. and R. Gallet, *Cost of resistance: an unreasonably expensive concept.* Rethink Ecol, 2018. **3**: p. 51–70.

135. Strobl, M.A.R., et al., *Turnover modulates the need for a cost of resistance in adaptive therapy.* Cancer Res, 2021. **81**(4): p. 1135–47.

136. Kam, Y., et al., *Sweat but no gain: inhibiting proliferation of multidrug resistant cancer cells with "ersatzdroges".* Int J Cancer, 2015. **136**(4): p. E188–96.

137. Gallaher, J.A., et al., *Spatial heterogeneity and evolutionary dynamics modulate time to recurrence in continuous and adaptive cancer therapies.* Cancer Res, 2018. **78**(8): p. 2127–39.

138. Basanta, D., R.A. Gatenby and A.R. Anderson, *Exploiting evolution to treat drug resistance: combination therapy and the double bind.* Mol Pharm, 2012. **9**(4): p. 914–21.

139. Smalley, I., et al., *Leveraging transcriptional dynamics to improve BRAF inhibitor responses in melanoma.* EBioMedicine, 2019. **48**: p. 178–90.

140. Jensen, N.F., et al., *Establishment and characterization of models of chemotherapy resistance in colorectal cancer: towards a predictive signature of chemoresistance.* Mol Oncol, 2015. **9**(6): p. 1169–85.

141. Kaznatcheev, A., et al., *Fibroblasts and alectinib switch the evolutionary games played by non-small cell lung cancer.* Nat Ecol Evol, 2019. **3**(3): p. 450–6.

142. Andersson, D.I. and D. Hughes, *Antibiotic resistance and its cost: is it possible to reverse resistance?* Nat Rev Microbiol, 2010. **8**(4): p. 260–71.

143. Bergelson, J. and C.B. Purrington, *Surveying patterns in the cost of resistance in plants.* Am Nat, 1996. **148**: p. 536–58.

144. Bacevic, K., et al., *Spatial competition constrains resistance to targeted cancer therapy.* Nat Commun, 2017. **8**(1): p. 1995.

145. Gatenby, R.A., et al., *Eradicating metastatic cancer and the eco-evolutionary dynamics of anthropocene extinctions.* Cancer Res, 2020. **80**(3): p. 613–23.

146. Pannell, J.H., *The heath hen.* Science, 1943. **98**(2538): p. 174.

147. Murray, G.G.R., et al., *Natural selection shaped the rise and fall of passenger pigeon genomic diversity.* Science, 2017. **358**(6365): p. 951–4.

148. Carrion, V., et al., *Archipelago-wide island restoration in the Galapagos Islands: reducing costs of invasive mammal eradication programs and reinvasion risk.* PLOS One, 2011. **6**(5): p. e18835.

149. Gonzalez, A., et al., *Evolutionary rescue: an emerging focus at the intersection between ecology and evolution.* Philos Trans R Soc Lond B Biol Sci, 2013. **368**(1610): p. 20120404.

150. Gomulkiewicz, R. and R.D. Holt, *When does evolution by natural selection prevent extinction?* Evolution, 1995. **49**(1): p. 201–7.

151. Foley, P., *Predicting extinction times from environmental stochasticity and carrying capacity.* Conserv Biol, 1994. **8**: p. 124–37.

152. Chirakkal, H. and L.R. Gerber, *Short- and long-term population response to changes in vital rates: implications for population viability analysis.* Ecol Appl, 2010. **20**(3): p. 783–8.

153. Beissinger, S.R., *Modeling approaches in avian conservation and the role of field biologists.* American Ornithologists' Union, 2006. p. 56.

154. Beissinger, S.R. and D.R. McCullough, *Population viability analysis.* University of Chicago Press, 2002. p. 577.

155. Stankova, K., et al., *Optimizing cancer treatment using game theory: a review.* JAMA Oncol, 2019. **5**(1): p. 96–103.

156. Gatenby, R.A. and T.L. Vincent, *Application of quantitative models from population biology and evolutionary game theory to tumor therapeutic strategies.* Mol Cancer Ther, 2003. **2**(9): p. 919–27.

157. Von Neumann, J. and O. Morgenstern, *Theory of games and economic behavior.* 3rd ed. Princeton University Press, 1953.

158. Nash, J.F., *Equilibrium points in N-person games.* Proc Natl Acad Sci USA, 1950. **36**(1): p. 48–9.

159. Holt, C.A. and A.E. Roth, *The Nash equilibrium: a perspective.* Proc Natl Acad Sci USA, 2004. **101**(12): p. 3999–4002.

160. Myerson, R., *Game theory: analysis of conflict.* Harvard University Press, 1991.

161. Aumann, R.J. et al., *Handbook of game theory with economic applications.* Elsevier. 1992.

162. Maynard Smith, J. and G.R. Price, *The logic of animal conflict.* Nature, 1973. **246**: p. 15–18.

163. Brown, J.S. and K. Stankova, *Game theory as a conceptual framework for managing insect pests.* Curr Opin Insect Sci, 2017. **21**: p. 26–32.

164. Conlin, P.L., J.R. Chandler and B. Kerr, *Games of life and death: antibiotic resistance and production through the lens of evolutionary game theory.* Curr Opin Microbiol, 2014. **21**: p. 35–44.

165. Orlando, P.A., R.A. Gatenby and J.S. Brown, *Cancer treatment as a game: integrating evolutionary game theory into the optimal control of chemotherapy.* Phys Biol, 2012. **9**(6): p. 065007.

166. Leyffer, S. and T. Munson, *Solving multi-leader-common-follower games.* Optim Methods Softw, 2010. **25**(4): p. 601–23.

167. vonStackelberg, H., *The theory of the market economy.* Oxford University Press, 1952.

168. Simaan, M. and J.B. Cruz, *On the Stackelberg strategy in nonzero-sum games.* J Optim Theory Appl, 1973. **11**(5): p. 533–55.

169. Baysar, T. and G.J. Olsder, *Dynamic noncooperative game theory.* 2nd ed. Society for Industrial and Applied Mathematics, 1998.

170. Gatenby, R.A., et al., *Adaptive therapy.* Cancer Res, 2009. **69**(11): p. 4894–903.

171. Enriquez-Navas, P.M., et al., *Exploiting evolutionary principles to prolong tumor control in preclinical models of breast cancer.* Sci Transl Med, 2016. **8**(327): p. 327ra24.

172. West, J.B., et al., *Multidrug cancer therapy in metastatic castrate-resistant prostate cancer: an evolution-based strategy.* Clin Cancer Res, 2019. **25**(14): p. 4413–21.

173. West, J., et al., *Towards multidrug adaptive therapy.* Cancer Res, 2020. **80**(7): p. 1578–89.

174. Brady-Nicholls, R., et al., *Predicting patient-specific response to adaptive therapy in metastatic castration-resistant prostate cancer using prostate-specific antigen dynamics.* Neoplasia, 2021. **23**(9): p. 851–8.

175. Brady-Nicholls, R., et al., *Prostate-specific antigen dynamics predict individual responses to intermittent androgen deprivation.* Nat Commun, 2020. **11**(1): p. 1750.

3 The Genetic Hitchhiker's Guide to Tumor Evolution

Rohini Janivara and Joseph Lachance

3.1 INTRODUCTION

Cancer is a heterogeneous class of diseases characterized by uncontrolled growth and spread of abnormal cells in the body. It can affect almost any organ or tissue and is caused by genetic alterations (mutation, karyotypic dysregulation, extrachromosomal DNA, etc.) that disrupt the normal regulation of cell division and growth. In multicellular organisms, cells are organized in modularized functional units like tissues and organs that maintain the organism's physiology. Individual cells experience checks that prevent them from disrupting this established coordination. However, cancer cells accumulate genetic changes and epigenetic marks that help them overcome these checks and exhibit abnormal behaviors, such as uncontrolled cell division and tissue invasion. If we consider tissues as populations of evolving cells, a healthy body is like an ecosystem in equilibrium. Tumor cells can then be compared to invasive species that disturb this equilibrium when present in a high density. Viewing tumor progression through the lens of evolutionary biology can bring common evolutionary dynamics to the forefront and aid in better understanding disease progression. Predictions of evolutionary events made with such an understanding can also inform the development and choice of therapeutic agents for treating tumor progression.

3.1.1 A TUMOR'S EYE VIEW OF EVOLUTION

Not all neoplasms become cancerous (Patel 2020). Though neoplasms exhibit abnormal cellular growth, they require additional growth benefits to grow larger and become malignant. These growth benefits arise from the accumulation of mutations that occur during cell division. When a normal cell replicates, DNA repair pathways identify and rectify errors caused during the replication process. Mutations within these genes can increase the rate of mutation accumulation. To develop complete autonomy in the body, cancer needs to overcome several other intrinsically programmed limitations that keep cells in check. For example, cells experience contact inhibition from surrounding cells that limit their replicative potential. Furthermore, individual cells have a limited lifespan in terms of the number of times they can replicate and still be viable. To overcome this, some tumor cells evolve mechanisms to reverse telomere shortening, thus gaining replicative immortality. Following a sudden expansion in growth, tissues can experience increased resource requirements, thus eliciting mechanisms to deregulate metabolism.

Each tumor also experiences fitness hurdles specific to its environment. A large proportion of selection forces acting on young neoplasms is immune-driven (Martin et al. 2021). When the immune system recognizes abnormal antigens produced by tumor cells, it strengthens the negative selection experienced by the cells. Hence, the component cells of a tumor often accumulate mutations that also evade immune recognition. Another mechanism to counter selection forces is phenotypic plasticity, which helps tumor cells to become robust to tumultuous changes within their local environment.

As a tumor evolves by accumulating advantageous mutations and purging deleterious mutations (from the tumor's perspective), cells interact with and influence their microenvironment. For example, once the tumor is big enough in size it induces the growth of blood vessels to access the

DOI: 10.1201/9781003307921-3

circulatory system for its increased energy requirements. Other non-genetic factors also play a role in increasing its chances of malignancy. If tumor growth is initiated near pre-existing blood vessels, the requirement for mutations that promote angiogenesis is not as pressing in the earlier stages of tumor progression. Therefore, the site of tumor initiation and the tissue of origin largely impact clinical outcomes. Similarly, tumor cells have a higher chance of successful metastasis if their primary site of origin is proximal to lymph nodes, since that would facilitate their entrance into the circulatory system (Nathanson 2003). Tumor cells also sculpt their microenvironment by manipulating the gene regulation of stromal cells by secreting growth signals. Examples of this include accumulating immune cells that promote inflammation and interacting with microbial communities that potentially have a symbiotic relationship with the tumor.

In this chapter, we unpack several factors that influence the progression of a neoplasm into a malignant tumor and draw parallels to relevant evolutionary phenomena. We elucidate key concepts from ecology and evolution that govern the growth and dynamics of any population in its given environment and describe their relevance in tumor progression. This involves taking a closer look at the onset of tumor-promoting mutations and how these interact with selection pressures within the host system, how these selection pressures cause auxiliary effects within the tumor through genetic hitchhiking, and how a growing cancer cell population can alter its environment to make it more conducive for its progression. Finally, we examine how tumor cells metastasize to different organs in the body, a process that is similar to individuals migrating to distant ecosystems.

3.2 MOSTLY HARMLESS: THE DISTRIBUTION OF FITNESS EFFECTS OF SOMATIC MUTATIONS

3.2.1 BACKGROUND: MUTATIONS VARY IN THEIR FITNESS EFFECTS

Mutations are a universal source of new variation in the genomes of all living organisms, and the distribution of fitness effects holds a central place in evolutionary theory. Depending on their effects on fitness, mutations can be classified as deleterious, neutral, or advantageous. In evolutionary biology, selection coefficients are used to quantify the relative effect of each mutation on fitness. The distribution of fitness effects describes the probability that a new mutation will have a given selection coefficient (i.e., it quantifies how likely new mutations are to decrease or increase fitness). On one end of the spectrum are lethal mutations, and on the other end are mutations that lead to large increases in Darwinian fitness. While highly deleterious mutations tend to be purged quickly and highly advantageous mutations often rapidly reach fixation, neutral mutations can segregate at intermediate frequencies for many generations. This is because neutral mutations are governed by genetic drift—changes in allele frequencies that are due to random chance. Effective population size also has a large impact in determining whether evolution is driven by natural selection or by genetic drift (Eyre-Walker and Keightley 2007). When effective population sizes are large, natural selection is a highly effective evolutionary force. However, when effective population sizes are small, the fates of nearly neutral mutations (i.e., mutations that are slightly disadvantageous or slightly beneficial) are due to genetic drift. In small cellular populations, mildly deleterious mutations are unable to be eliminated by natural selection, resulting in their gradual accumulation. Mutation frequencies and population sizes of clonal lineages can be measured in tumor biopsies, facilitating an evolutionary genetic understanding of tumor progression (Somarelli et al. 2020).

Cancers develop due to the accumulation of multiple somatic mutations. Rates of somatic mutations in healthy tissues can range from 3.5×10^{-9}/bp/division in the small intestine to 1.6×10^{-7}/bp/division in the skin (Werner and Sottoriva 2018). This wide spectrum could be due to the exposure to external agents like UV radiation or sampling bias of adult stem cells in different tissues. Mutation rates also vary across different layers of each tissue. Tissue-specific rates of somatic mutations may even be under weak selection (Lynch 2010). We also note that many somatic mutations in cancers are structural variants. Not all mutations lead to the formation of a neoplasm, and not all neoplasms progress to become cancerous. Although most somatic mutations in cancer cells are neutral (Ling et al. 2015), natural selection is still relevant to cancer evolution (Lean and Plutynski 2016).

Many core cellular functions have alternative pathways (Tomasetti and Vogelstein 2015), which means that cells are generally robust to random perturbations, including somatic mutations. However, whole-genome doubling events in somatic cell lines have been observed to be highly correlated with poor relapse-free survival (Dewhurst et al. 2014). This phenomenon arises due to cancer cells' ability to leverage the surplus copies of each gene, yielding modular genetic systems that are more resilient (Huminiecki and Conant 2012). Genetic systems that confer increased robustness, such as having multiple copies of essential genes, can still fail in the context of cancer.

Here, we focus on the distribution of fitness effects from the point of view of individual cancer cells. Importantly, mutations that are deleterious or beneficial to individual cells can have the opposite effect on organismal fitness (Michod 2005). Many mildly deleterious mutations (from a cell's perspective) can continue to persist in otherwise healthy tissues. Highly deleterious mutations often trigger both intrinsic and extrinsic cytotoxic mechanisms that remove cells harboring such mutations. While many instances of abnormal cell division and mutation accumulation can occur over an organism's lifespan, the body's intrinsic mechanisms are often able to eliminate these rogue colonies of cells. Only a select few of these cells evolve characteristics that result in malignancy. Here, we explore the selection pressures experienced by cancer cells and identify multiple types of beneficial mutations (from a cell's perspective).

3.2.2 MECHANISMS OF SELECTION: THE HALLMARKS OF CANCER

To classify mutations in terms of fitness we must first understand the selective forces that are experienced by individual cells. Adaptation can be viewed using the metaphor of a fitness landscape (Wright 1932; Stadler 2002). As a tumor evolves, it changes the environment around it, leading to changes in fitness landscapes over time. Most tumors progress with an accumulation of several beneficial functionalities documented as the *hallmarks of cancer* (Figure 3.1) (Hanahan 2022). In the following, we classify 14 extended hallmarks of cancer in terms of their evolutionary role in tumor progression. Though the hallmarks listed are functionally separable, we note that pleiotropic mutations can impart multiple fitness advantages.

3.2.3 INTRINSIC SELECTION PRESSURES

When tumor cells divide rapidly, they can overcome intrinsic checks and balances. Healthy tissues have a constant housekeeping mechanism that can remove and replace damaged cells. Hence these healthy cells observe constrained birth and death rates that typically balance each other to maintain homeostasis. This means that growth-promoting signals are only expressed for a limited period and that growth suppressors are released to prevent needless proliferation. One intrinsic safety check is that differentiated cells have a limit on the number of times they can replicate, known as the Hayflick limit. Once this limit is reached, the cells undergo apoptosis. Another mechanism that can trigger apoptosis is contact inhibition. While this complex system of checks and balances maintains the healthy functioning of tissues, cancer cells overcome many of these restrictions by accumulating a range of mutations. Damaging mutations in DNA repair pathway genes can result in faster mutation accumulation, especially if coupled with evasion of intrinsically triggered apoptosis. This increase in the genetic diversity of the cancer cell population that can initiate faster changes is also an important hallmark of cancer. Intrinsic selection pressures yield multiple ways to increase fitness (population growth, increased viability, and greater diversity), each of which map to multiple hallmarks of cancer (Figure 3.1).

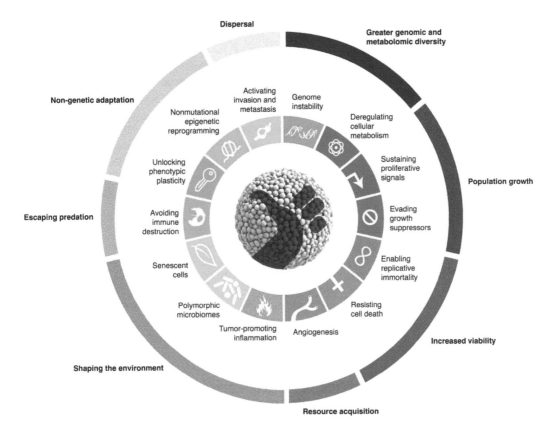

FIGURE 3.1 The extended hallmarks of cancer and their relevance to evolution. This figure was inspired by Hanahan's new dimensions in cancer biology (Hanahan 2022).

Population growth

- Mutations in growth signal receptors can lead to *sustaining proliferative signals in cancer cells. RUNX1* is a cell cycle regulator, and therefore its constitutive expression has been a major driver of malignancy in lymphoblastomas (Friedman 2009).
- Uninhibited growth often triggers several mechanisms informed by local regulators like space and oxygen constraints to suppress growth. Mutations that deactivate genes in the Notch signaling pathway can facilitate *evasion of growth suppressors.* An example of such a common mutation in hematological, lung, and gastric cancers is in the *NOTCH2* gene (Amin et al. 2015).

Increased viability

- Certain mutations that lead to the elongation of telomeres can help cells overcome the Hayflick limit, thereby *enabling replicative immortality.* Excessive mutation accumulation triggers early senescence, which is an intrinsic mechanism by which the telomere lengths are inherently shortened to prevent tumorigenesis. Mutations in the *TPP1* gene are known to play a key role in telomere elongation in melanoma (Yang et al. 2013).
- Tumor cells encounter selection pressures in the form of both intrinsic and extrinsic cell death pathways. Mutations in tumor suppressor genes like *TP53* can result in tumor cells developing *cell death resistance* (Kastan et al. 1995). For example, elephants have a lower risk of cancer because their genomes possess multiple copies of this gene (Abegglen et al. 2015).

Greater genomic and metabolomic diversity

- Introducing *genome instability* arises due to loss of function mutations in DNA repair genes like *BRCA1* or *BRCA2* (Yoshida and Miki 2004) or as a byproduct of cellular stress (Galhardo et al. 2007). This lends the cell more chances to gain advantageous mutations. However, we note that mutations are not always adaptive, as will be seen in our discussion of passenger mutations.
- Cancer cells *deregulate cellular metabolism* by switching their metabolism from mitochondrial oxidative phosphorylation to glycolysis. The intermediate metabolites generated in the glycolytic pathway result in faster energy release, facilitating accelerated tissue growth. Several clones often symbiotically evolve complementary metabolomic pathways such that each cell expends less energy to produce ATP. Mutations in the *IDH1* gene are known to deregulate cellular metabolism in gliomas (Zhang et al. 2013).

3.2.4 EXTRINSIC SELECTION PRESSURES

As the cancer cell population evolves, its local environment is altered, and several novel dynamics arise. An abnormal and sustained increase in size affects the space and tissues around the tumor, leading to an increased need for resources. In addition, the drastic increase in size and abnormality of tissue composition can attract more attention from the immune system. Cancer cells also develop non-genetic mechanisms of adaptations to stay robust to changing environments. This is observed in cells exhibiting epigenetic regulation and phenotypic plasticity that help alter gene expression. Interestingly, the size and heterogeneity of the tumor provide not just a challenge but also a potential benefit to its growth, as abnormal pathway regulation can help the tumor tissue mold its environment to accommodate requirements that arise from continued growth. For example, tumor cells trap immune cells into their microenvironment and use their innate proliferative signals to support their own growth (Gonzalez et al. 2018). The ensuant inflammation attracts more immune cells that further accelerate tumor growth. The spatial heterogeneity of tumors creates localized environmental differences, which can enable different clones to develop a symbiotic relationship with polymorphic microbial communities. This leads to the robust survival of multiple clone types. Extrinsic selection pressures yield multiple ways to increase fitness (resource acquisition, shaping the environment, evading predation, and non-genetic adaptation), each of which map to at least one hallmark of cancer (Figure 3.1).

Resource acquisition

- Tumor cells increase vasculature within the tissue through *angiogenesis* to support the increased resource requirement. This reduces the competition between cell populations that arises due to the spatial heterogeneity of resource availability. Mutations in the *VEGF* gene can be proangiogenic (Shibuya 2011).

Shaping the tumor microenvironment (TME)

- The immune system has an innate response of inflammation for pathogen detection. When immune cells detect a threat near tumors, they enter the TME from local lymphoid tissues or bone marrows and proliferate to intensify their response. But this results in redirecting proliferative signals and metabolites to replenish the TME. Mutations in the *MYC* gene are known to recruit T lymphocytes to the TME in osteosarcomas, resulting in *tumor-promoting inflammation* (Jiang et al. 2022).
- Within the altered microenvironment created by tumors, various kinds of microbes thrive and disturb the balance of the existing host-microbial dynamics. This results in a *polymorphic microbiome* that can promote the further development of tumors by reinforcing

tumor-promoting inflammation and malignant progression. A common example is the bacterium *Bacteroides fragilis*, which is said to remodel the colonic microbiota to promote colorectal cancer (Sheflin et al. 2014).

- Senescence is a property that is encoded in cells to prevent tumorigenesis by arresting proliferation as a response to excess mutation accumulation. Yet the *accumulation of senescent cells* promotes tumor growth through the gain of a senescence-associated secretory phenotype (SASP), which conveys hallmark capabilities through chemokine and cytokine signaling to the surrounding cancer cells in the TME (Campisi 2000).

Evading "predation"

- Loss of function mutations in HLA genes, which are responsible for antigen presentation, help cells *evade immune destruction* (Grasso et al. 2018).

Non-genetic adaptation

- To stay robust to extreme environmental changes, cancer cells *unlock phenotypic plasticity*, which generates non-genetic heterogeneity within the tumor. This phenomenon is described in more detail in a later section. The *HNF1A* gene is commonly mutated in hepatocellular carcinomas that exhibit phenotypic plasticity (Zheng et al. 2022).
- Another non-genetic alteration that results in differential expression of genes is *non-mutational epigenetic regulation*. This also plays a key role in coping with the host's defense mechanisms by changing the regulation of genes. Several mutations in genes like *PTEN*, *APC*, and *CDH1* have been hyper-methylated in non-melanoma skin cancers (Kashyap et al. 2022).

3.2.5 SELECTION PRESSURES IN NOVEL ENVIRONMENTS

Cancer cells can migrate from their initial location as an additional strategy to increase their fitness. While stem cells possess mobility, they lose this capability when they differentiate during tissue formation (cells making up a healthy tissue benefit from staying together). This is a transition between a mesenchymal to an epithelial state. However, since tumor cells can invoke plasticity, they are able to regulate pathways that change their status back to a mesenchymal state, enabling them to detach and move around within the body. Tumor cells are also able to regulate this plastic behavior and transition back into an epithelial cell and establish new colonies in a different location. This ability also maps to one of the hallmarks of cancer (Figure 3.1).

Migration

- By *activating invasion and metastasis* tumor cells can migrate to different locations with higher potential for their current fitness levels. A mutation in the gene *MLL3*, commonly observed in breast cancer is reported to increase the propensity for cells to switch states that enable them to detach and migrate away from the tissue of origin (Cui et al. 2023).

3.2.6 EVIDENCE OF SOMATIC MUTATIONS THAT HAVE DIFFERENT FITNESS EFFECTS

How common are beneficial or deleterious mutations in tumors? Several studies have claimed that while positive selection pressure is much stronger than negative selection, the majority of somatic mutations appear to be neutral (Martincorena et al. 2017). One reason for this is that the vast majority (>98%) of the human genome is noncoding DNA. In addition, many genes are nonessential in somatic cell lines. Two recent studies analyzing the same dataset and the same metric (dN/dS ratios) have yielded opposite conclusions about the relative importance of negative selection in tumors. Analyzing bulk tissue data from 29 different types of tumors from TCGA, Martincorena et al. claimed that the strength of negative selection was negligible (Martincorena et al. 2017). By contrast, Zapata et al. posit that negative selection in tumors is often missed because of the high mutation rates in cancer genomes (Zapata et al. 2018). When mutation rates are as high, as observed in tumors, it is difficult to identify deleterious mutations on a genomic background that has already undergone strong negative selection (Lakatos et al. 2020). We note that one limitation of the dN/dS metric is not ideal when it comes to bulk sequencing data, as different lineages of cells can have different selective histories. In addition, bulk tumor sequencing often fails to capture rare polymorphisms, which can skew one's understanding of the distribution of fitness effects (Kryazhimskiy and Plotkin 2008). Going forward, single-cell sequencing is likely to yield novel insights regarding tumor evolution.

3.2.7 CHANCE IN TUMOR EVOLUTION

Cancer evolution is also marked by a significant amount of stochasticity, not the least of which involves somatic mutations (Lipinski et al. 2016). Mutations are random; the specific oncogenes and tumor suppressor genes that are mutated in each patient tend to be tumor specific (Greif et al. 2011; Pezzuto et al. 2016). Furthermore, the temporal order of different mutations varies from tumor to tumor. The spatial expansion of tumors can also lead to stochastic effects, including the phenomenon known as allele surfing (Fusco et al. 2016). Another stochastic aspect of tumor evolution arises during metastasis. Strong founder effects can arise when cancer cells colonize new locations in the body, and different metastases in the same individual are often comprised of different cellular lineages (Birkbak and McGranahan 2020). Although natural selection is generally considered to be a deterministic phenomenon, the initial context of each somatic mutation matters. Mutations occur in different genetic backgrounds and microenvironments (Wu et al. 2016). Because fitness effects are often context-sensitive, this can impact the evolution of individual tumors.

3.3 LINKAGE AND NATURAL SELECTION IN TUMORS

3.3.1 GENETIC HITCHHIKING, BACKGROUND SELECTION, AND MULLER'S RATCHET

It is important to acknowledge that natural selection does not act in isolation. Mutations occur adjacent to other closely linked alleles. The combination of genetic linkage and selection results in two major evolutionary phenomena. The first of which is genetic hitchhiking, which occurs when alleles increase in frequency due to being tightly linked to another allele that has a selective advantage. The second of which is background selection (BGS), which refers to the loss of genetic diversity that arises from negative selection against linked deleterious alleles. In practice, BGS is harder to detect than genetic hitchhiking (Lakatos et al. 2020). As there is a lack of recombination during mitosis, both of these phenomena are particularly relevant to cancer cells. In effect, this means that even unlinked mutations can "hitchhike" during tumor evolution. This means that the evolutionary fates of new mutations in tumors are often context dependent.

The perpetual accumulation of deleterious mutations in populations of non-recombining individuals is known as Muller's ratchet (Muller 1964). This stochastic process is primarily influenced by population size, mutation rate, and selection strength. If mutation rates exceed a certain threshold, cells begin accumulating deleterious mutations that are unable to be successfully purged from a population, leading to what is known as a mutational meltdown or extinction vortex. An additional factor involves competition between clonal lineages of cells. This can further influence the context-dependent fitness of mutations, thus impacting probabilities of clonal expansion. Because of this, even once a neoplasm has developed, its progression to a malignant tumor need not be deterministic.

3.3.2 Driver and Passenger Mutations

In the context of tumor progression, alleles that increase in frequency due to their advantageous effects are called *driver mutations*, while alleles that increase in frequency due to linkage to drivers are called *passenger mutations* (Figure 3.2). Drivers are often found in genes that regulate cell division, DNA repair, and other essential cellular processes. Accurate annotations of drivers and passengers can help identify biomarkers for early detection and tracking the pace of tumor progression. Currently, about 97% of the commonly observed mutations in tumors are classified as passenger mutations (McFarland et al. 2017), though many of these mutations may be latent drivers.

Although many driver mutations involve structural variants or copy number alterations, single nucleotide variants can increase the fitness of cancer cells. *BRCA1* and *BRCA2* mutations in breast and ovarian cancer and *APC* mutations in familial adenomatous polyposis are prime examples of this. Note that the presence of a driver mutation does not immediately correlate with the formation of neoplasms. Many drivers require the support of additional drivers to act in a tumor-promoting manner (Pon and Marra 2015). The timing of specific driver acquisition is also key, as illustrated by mutations in TGF-β signaling. If a mutation upregulates TGF-β signaling, it can induce angiogenesis and invasion, which are beneficial in later stages of tumor growth, whereas if a mutation downregulates TGF-β signaling, it can prevent growth arrest and be beneficial in the earlier stages of tumor growth.

Mutations are classified as drivers if at least one of two criteria are met. First, drivers have a higher allele frequency in tumor tissues when compared to normal tissues. However, due to genetic hitchhiking some passenger mutations may be incorrectly classified as drivers. Second, driver mutations either alter a previously recognized oncogene or are involved in increased cell proliferation in functional assays. We note that several curated databases contain driver/passenger annotations specific to different cancer types (Khan et al. 2021).

There are several limitations to the existing methods of classification. First, it is hard to capture the fitness of mutations in a changing selection landscape using bulk tissue sequencing. For instance, a mutation that promotes angiogenesis might not be the most important mutation for tumor proliferation when it is still in a young neoplasm. This is further proved by studies that show that beneficial driver mutations occur even in the later stages of tumor progression (Gomez et al. 2018). Ideally, allele frequency trajectories from time-series sequence data can shed light on the relative fitnesses of different cancer genotypes (Schraiber et al. 2016). Second, as tumor progression also involves interactions with the local environment, it is important to track changes in the TME over time. Third, phenotypic plasticity also contributes to intra-tumoral heterogeneity. This means that in an unstable environment, identical genotypes can have different phenotypes due to differential gene expression, leading to misclassifications of driver and passenger mutations based on the stage of the tumor.

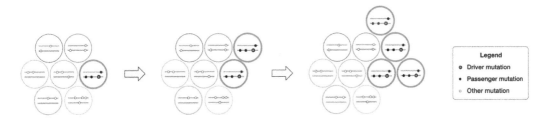

FIGURE 3.2 Genetic hitchhiking on a cellular level. Cellular lineages are represented by different colors, and there is a clonal expansion of cancer cells (red) over time. Due to a lack of recombination during mitosis, passenger mutations in tumors can be either linked or unlinked to driver mutations.

3.3.3 CLONAL EVOLUTION OF TUMORS

Since recombination does not occur during mitosis, one of the primary sources of new variation in tumor progression is mutations. Cell lines accumulate mutations and pass them onto their daughter cells. Over time, different ancestrally related cell lines accumulate enough mutations to diverge into distinct clones (Figure 3.2). This phenomenon is known as clonal divergence. These clones then interact with each other, much like different species would in an ecosystem. They compete for oxygen, energy, and space. Some clones may have evolved the ability to be resource efficient. For example, cells often switch their metabolic pathways to an inefficient but quicker pathway—aerobic glycolysis—known as the Warburg Effect (Warburg et al. 1927). These cells may be able to outcompete other cells in a resource-poor environment. Cancer cells often form symbiotic relationships as well. Some clones can evolve the ability to cooperate with each other by sharing metabolites. These clonal dynamics play a big part in the evolution of the tissue as a whole.

Clonal expansion can result in intra-tumoral heterogeneity in the progression of colorectal cancers. Here, the initial tumorigenic mutation typically affects cell division pathways to increase proliferation, such as mutations in the *APC* gene. A *Big Bang model* has been proposed by Sottoriva et al. to explain the evolution of colorectal cancer (Sottoriva et al. 2015). According to their model, colorectal cancers arise from a single mutant cell developing into a clonal population and branching off at various points. As a result, mutations that arise early on are spatially diffused across the tissue, while those that arise later are locally confined. Thus, mutations that occur early on are predominantly seen in the entire tissue. An additional pattern arising from the Big Bang model is that clonal expansion is more common during the initial stages of tumor growth. Finally, Sottoriva et al. note that some aggressive subclones might be preexisting in the earlier stages of tumor growth before progression and expand during the later stages of tumor progression (Sottoriva et al. 2015).

3.4 TUMOR EVOLUTION IN CHANGING ENVIRONMENTS

As populations expand within an ecosystem, they interact with other entities that influence their growth. These interactions include both competition and cooperation. Here, we describe how tumor progression is shaped by its environment, particularly through constantly evolving interactions between tumor cells and the TME landscape.

Tumor cells gradually co-evolve with their environment. As cancer tissues progress to malignancy, they create a supportive network of cells around themselves. They do so by recruiting various immune-related cells like macrophages and stem cells from the surrounding lymphoid and bone marrow tissues. Although immune genes have evolved to recognize and eliminate abnormal cells to protect the body, they are also capable of promoting the growth of pathogenic cancer cells (Gonzalez et al. 2018). This can be viewed as an example of antagonistic pleiotropy whereby alleles in immune genes that are strongly beneficial early in an organism's life cycle can be detrimental (from an organismal perspective) later in life (Lambeth 2007). Here, we describe different components of the TME and how their interactions with each other and tumor cells dictate the course of disease progression.

3.4.1 CHARACTERISTICS OF THE TUMOR MICROENVIRONMENT

The tumor microenvironment is a rich ecosystem that consists of cancer cells of different tumor-propagating potential, stromal cells, immune cells, and microbial communities (Figure 3.3). These cell types have specific functions in healthy tissues and often exhibit altered behavior when associated with tumors. Solid tumors can often contain subpopulations of cells with properties of stem-like behavior, with capacity to recapitulate the diverse phenotypes observed in the tumor and re-establish tumor growth at limiting dilution (Vessoni et al. 2020). Immune cells regularly target and kill pathogenic cells, but they exhibit pro-tumorigenic behavior when in persistent

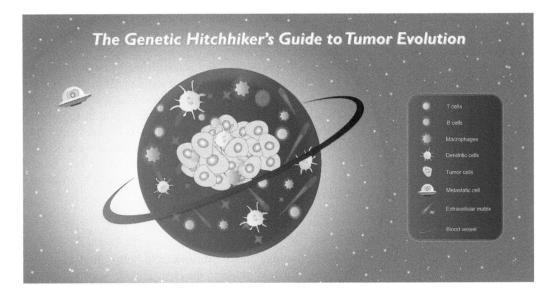

FIGURE 3.3 The tumor microenvironment is made up of several different cellular components. A metastasizing cell is off on an adventure where it will experience a new environment. This figure was created by Mimi Holness.

inflammatory conditions, like what is often present around tumor tissues (Wang et al. 2019). Among tumor-promoting immune cells, the most important cell type is macrophages. M1 macrophages (classically activated) have a pro-inflammatory response associated with defense against pathogens, while M2 macrophages (alternatively activated) are associated with functions related to wound healing such as tissue regeneration, angiogenesis, and extracellular matrix remodeling. In the TME, the excessive availability of cytokines, growth factors, damage-associated molecular patterns, and hypoxia-inducible factors trigger the repolarization of M1 macrophages to M2 macrophages, which have a pro-tumorigenesis effect (Noy and Pollard 2014). Similarly, an extensive repertoire of different immune cells promotes tumor initiation, progression, and conversion to malignancy by exhibiting an altered behavior in the TME (Gonzalez et al. 2018).

Tumor-promoting cells in the TME are located within the extracellular matrix surrounding a tumor. Fibroblasts make up a crucial component of the TME, as they secrete precursors for connective tissue formation (Figure 3.3). These precursors include matrix-degrading enzymes, inflammatory cytokines, and epithelial growth factors that can signal position and growth requirement-related information to neighboring cells (Shay and Roninson 2004). When aging fibroblasts senesce, they secrete their cellular contents into the extracellular matrix as they undergo apoptosis. The growth factors that are released stimulate proliferation in neighboring cells (Shay and Roninson 2004), and the release of inflammatory cytokines redirects immune cells to the tumor tissue, furthering cancer progression. In addition, there is evidence that senescent fibroblasts induce plastic behavior in epithelial cells and drive them toward pro-metastatic phenotypes (Parrinello et al. 2005; Tinaburri et al. 2021). Immune evasion can further reinforce fibroblast accumulation (De Blander et al. 2021).

Another key aspect in the TME is the microbial community. Not only are there tumor-specific microbes (Nejman et al. 2020), but there is also evidence that specific sub-niches around the tumor tissue host different microbial communities. These microbial communities can promote progression by modulating proliferation and cell death. Most importantly, they can induce metastasis and drug resistance in tumor cells by inducing an inflammatory response and promoting T-cell exclusion (Galeano Nino et al. 2022).

Cells at the core of a tumor experience a different microenvironment than cells at the periphery of a tumor. Key differences include nutrient availability and vasculature surrounding cells. The resulting spatial heterogeneity in resource availability contributes to genetic and transcriptomic heterogeneity within the tumor tissue. This means that resource heterogeneity affects clonal expansion of different cellular lineages (Lloyd et al. 2016). In addition, higher vasculature near the peripheral parts of a tumor increases selection pressures for immune resistance. There is also evidence that ER-positive and ER-negative cells in breast cancer can be distributed according to estrogen availability in the tissue (Lloyd et al. 2014). Multiple studies have shown that the presence and the density of tumor-infiltrating immune cells, cancer-associated fibroblasts, and vascular invasion are highly correlated with prognosis and treatment response in different types of tumors (Anderberg et al. 2009; Nakasone et al. 2012). Diagnostic and therapeutic research can benefit immensely by studying interactions between the tumor and its environment (Galon et al. 2006).

3.4.2 PHENOTYPIC PLASTICITY IN CANCER EVOLUTION

Phenotypic plasticity is an adaptive response that individuals often adopt to buffer from selection pressures. On an individual level, phenotypic plasticity refers to the change in the phenotype as a response to fluctuating environmental effects (Pigliucci 2005). This is accomplished by differential regulation of genes, and switching between different gene expression states can be facilitated by heritable epigenetic regulation. These epigenetic modifiers can also be altered by the local environment. Furthermore, plastic responses can sometimes become genetically encoded. This phenomenon, which is known as genetic assimilation, can drastically change a population's genetic composition over multiple generations.

One of the most lethal consequences (from a patient's perspective) of plasticity exhibited by tumor cells is transient drug-induced tolerance. Drug resistance is a lethal phenomenon that gives rise to cancer recurrence after chemotherapy. It can be developed as a result of either selection of resistant clones or plasticity induced in drug tolerant cells (Hata et al. 2016). This means that targeting specific drug resistant clones during chemotherapy is not guaranteed to kill the tumor tissue. Moreover, cells can develop multidrug tolerance using plasticity in metabolic rewiring. Recent studies have also shown that plastic behavior can simultaneously elicit both metastasis and drug resistance (Jolly et al. 2019).

3.5 METASTASIS

3.5.1 FOUNDER EFFECTS AND METASTASIS

Individuals often migrate to new ecosystems in search of suitable environments to propagate. As they migrate in small batches, there is an ensuing loss in genetic diversity compared to the original population. As a result, once these individuals propagate in a new location, the genetic makeup of the new colony is vastly different from the parent colony. This is known as a founder effect, and it often can be detrimental to individuals that find themselves in a new environment (James 1970). This is because lower genetic diversity often brings a higher risk of extinction. As tumors evolve, some cells can migrate to novel locations. This phenomenon is called metastasis, and it occurs in the later stages of cancer progression.

3.5.2 MECHANISMS OF METASTASIS

Cells within solid tissues tend to be epithelial and hence are anchored to the extracellular matrix surrounding the tissue. Under normal conditions, when cells lose this adhesion, they undergo a type of apoptosis named anoikis. However, when cells become metastatic, they develop resistance to anoikis (Valastyan and Weinberg 2011). They can easily penetrate neovasculature generated by tumor cells since this type of vascularization is more permeable than normal tissues. These meta-static cells can form emboli by interacting with platelets. As they leave the blood vessels and invade other organs, they permeate the luminal tissue in a process named extravasation. During this process, the tissue of destination determines the success of metastasis. Some tissues like the brain and pulmonary tissues have barriers surrounding them, which may make them more difficult to colonize as sites of secondary metastasis expansion. In many instances, primary tumors overcome this by secreting proteins like Angptl4 and COX-2 that increase the tissue permeability at distant locations (Valastyan and Weinberg 2011).

So how do specialized cancer cells successfully start the tumor progression stage all over again? Metastatic cells that successfully survive the journey in the blood vessels are faced with multiple challenges. Though these challenges are often tissue-specific, primary tumors are thought to secrete systemic signals to make these secondary sites more hospitable. These signals generally result in immunosuppression, stromal tissue remodeling, increased vascular permeability, and mobilization of bone marrow derived cells (de Visser and Joyce 2023). This ancillary system is established before the circulating metastatic cells settle into the pre-metastatic niche. Even with this support, most metastatic cells require cell-autonomous mechanisms to persist in the new environment, and once they persist, they require further supportive factors to proliferate. While gene expression changes invoked by plasticity contribute largely to bridging this gap, most circulating tumor cells are not successful in generating secondary tumors. Many of the changes needed for successful metastasis are either evolved along the way or induced by plasticity (Erler et al. 2009).

3.5.3 Plasticity and Metastasis

Phenotypic plasticity in tumor cells can also facilitate metastasis. This can occur through epithelial-mesenchymal plasticity (EMP), which enables a transition between epithelial and mesenchymal cellular states (Cook and Wrana 2022). Epithelial cells tend to have a higher proliferative potential and lower mobility compared to mesenchymal cells. EMP allows epithelial cells to lose cell-cell adhesion and mesenchymal cells to gain the ability. Switching permanently to either state does not allow the cell to go through the entire process of migration, invasion, and proliferation (Jolly et al. 2017). Hence the flexibility to exhibit both behaviors in the same cell is a big advantage as they can switch between the two states to enable both migration and propagation at different points in time. This potential is retained by cells that have intermediate epithelial and mesenchymal properties (Jolly et al. 2017).

3.5.4 Cell Fusions and Metastasis

Another interesting phenomenon that occurs during metastasis is cell-cell fusion, where two cells fuse to form a hybrid cell that retains advantages of both cells. When the fusing cells possess migratory and tumorigenic properties individually, the resulting hybrid cell has a high metastatic potential that enables cell detachment and establishment of new colonies. In a healthy physiological state, cells are either undifferentiated and mobile or differentiated and fixed in a tissue. But during cancer progression, myoblasts that are immortal can fuse with transformed fibroblasts, and the resulting metastatic hybrid cell possesses both genomic instability and plasticity and hence can disseminate into vasculature and regenerate in new locations (Delespaul et al. 2020).

3.5.5 Always Metastasizing—Hematologic Cancers

While most of the concepts explained in this chapter are geared toward solid tumors, other cancer types, such as hematological malignancies, also have similar evolutionary dynamics at play, albeit with a few subtle variations. Solid tumors and hematologic cancers differ based on the cell of origin—the former develop in specific organs like the stomach, prostate, and breast, while the latter usually develop in blood-forming tissues like bone marrow, white blood cells, red blood cells, and platelets. Hematologic cancers include leukemia, lymphoma, and myeloma. Myelomas tend to originate in plasma cells and localize in the bone marrow. Leukemia arises from hematopoietic cells and lymphoma arises from abnormal white blood cells, both of which circulate throughout the body.

Since hematologic cancers originate within the circulatory system, they have a higher chance of spreading throughout the body and forming tumors in other organs. They benefit from lacking encapsulation and possessing a high replicative capacity since they arise in stem cells. This plays a large role in how rapidly they progress toward malignancy. Two broad types of leukemias exist—acute and chronic. Chronic leukemia is characterized by abnormal function of matured blood cells, slower progression, and hence better prognosis, while acute leukemia is characterized by the inability of blood cells to reach maturation, faster progression, and hence a worse prognosis in patients. Both types of leukemias are initiated with driver mutations in hematopoietic stem/progenitor cells (HSPCs). As HSPCs with higher proliferative capacity are selected, they experience further driver mutations with an additional slew of passenger mutation accumulation (Welch et al. 2012). Despite this, the mutation load in hematologic cancers tends to be less than in solid tumors, and as a result, they tend to harbor fewer neoantigens. This contributes to the subversion of immune surveillance and promotes progression in most lymphomas (Curran et al. 2017). The continual dispersion of these cancer cells lowers the concentration of neoantigens, thus preventing a robust immune response. Both leukemias and lymphomas have unique pathways of immune evasion and immune-assisted progression. For example, in classic Hodgkin's lymphoma lymph nodes are densely populated with immunosuppressive cells, whereas in Burkitt lymphoma, the nodes are populated with malignant cells. Thus, TME niche construction is also different between solid tumors and hemtaologic cancers as well as within different types of hematologic cancers (Yan and Jurasz 2016). For this reason, these cancers respond differently to various immunotherapies, despite developing and progressing in the same tissue.

3.6   DON'T PANIC!

Adopting an evolutionary perspective can lead to improved cancer therapeutics. One promising approach involves what is known as adaptive therapy (see Chapter 2) (Gatenby et al. 2009; Thomas et al. 2018). Regular chemotherapy involves treating several different clones within a tumor with different drugs to decrease the size of the tumor in a step-by-step manner. This can give rise to the expansion of drug resistant clones that were preexisting within the tumor, resulting in the growth and expansion of an untreatable tumor within the body. Adaptive therapy advises using a combination of drugs in a planned treatment schedule such that each drug reduces the growth of each clone without fully eliminating them (Gatenby et al. 2009). This leads to the preservation of clonal competition for space and resources, thus preserving the heterogeneity of the tumor while avoiding extremely strong selective pressures that would otherwise force the tumor to become more aggressive.

In conclusion, we note that a full understanding of cancer progression requires far more than just sequencing tumor genomes or tracking biomarkers over time. It requires an understanding of the environmental contexts experienced by individual cells as well as an awareness of how specific mutations may (or may not) lead to increased fitness. To paraphrase Dobzhansky, nothing in cancer biology makes sense except in the light of evolution (Dobzhansky 1973; Seyfried 2012).

3.6.1   So Long and Thanks for All the Fish

We would like to thank Norman A. Johnson, Jason A. Somarelli, and members of the Center for Integrative Genomics at Georgia Institute of Technology for their assistance in writing this manuscript. We also thank Mimi Holness for generating Figure 3.3. This work was funded in part by a grant from the National Cancer Institute (R01CA259200). This chapter is dedicated to the memory of Douglas Adams who, aside from his stories of Arthur Dent and Zaphod Beeblebrox, was a well-known advocate of ecology and evolutionary biology (Adams and Carwardine 2013).

REFERENCES

Abegglen, L. M., A. F. Caulin, A. Chan, K. Lee, R. Robinson et al., 2015 Potential mechanisms for cancer resistance in elephants and comparative cellular response to DNA damage in humans. JAMA 314: 1850–1860.

Adams, D. and M. Carwardine, 2013 *Last chance to see*. Random House.

Amin, A. R., P. A. Karpowicz, T. E. Carey, J. Arbiser, R. Nahta et al., 2015 Evasion of anti-growth signaling: A key step in tumorigenesis and potential target for treatment and prophylaxis by natural compounds. In *Seminars in cancer biology*. Elsevier, pp. S55–S77.

Anderberg, C., H. Li, L. Fredriksson, J. Andrae, C. Betsholtz et al., 2009 Paracrine signaling by platelet-derived growth factor-CC promotes tumor growth by recruitment of cancer-associated fibroblasts. Cancer Res 69: 369–378.

Birkbak, N. J. and N. McGranahan, 2020 Cancer genome evolutionary trajectories in metastasis. Cancer Cell 37: 8–19.

Campisi, J., 2000 Cancer, aging and cellular senescence. In Vivo (Athens, Greece) 14: 183–188.

Cook, D. P. and J. L. Wrana, 2022 A specialist-generalist framework for epithelial-mesenchymal plasticity in cancer. Trends Cancer 8: 358–368.

Cui, J., C. Zhang, J. E. Lee, B. A. Bartholdy, D. Yang et al., 2023 MLL3 loss drives metastasis by promoting a hybrid epithelial-mesenchymal transition state. Nat Cell Biol 25: 145–158.

Curran, E. K., J. Godfrey and J. Kline, 2017 Mechanisms of immune tolerance in leukemia and lymphoma. Trends Immunol 38: 513–525.

De Blander, H., A. P. Morel, A. P. Senaratne, M. Ouzounova and A. Puisieux, 2021 Cellular plasticity: A route to senescence exit and tumorigenesis. Cancers (Basel) 13.

Delespaul, L., C. Gelabert, T. Lesluyes, S. Le Guellec, G. Perot et al., 2020 Cell-cell fusion of mesenchymal cells with distinct differentiations triggers genomic and transcriptomic remodelling toward tumour aggressiveness. Sci Rep 10: 21634.

de Visser, K. E. and J. A. Joyce, 2023 The evolving tumor microenvironment: From cancer initiation to metastatic outgrowth. Cancer Cell 41: 374–403.

Dewhurst, S. M., N. McGranahan, R. A. Burrell, A. J. Rowan, E. Grönroos et al., 2014 Tolerance of whole-genome doubling propagates chromosomal instability and accelerates cancer genome evolution. Cancer Discov 4: 175–185.

Dobzhansky, T., 1973 Nothing in biology makes sense except in the light of evolution. Am Biol Teach 35: 125–129.

Erler, J. T., K. L. Bennewith, T. R. Cox, G. Lang, D. Bird et al., 2009 Hypoxia-induced lysyl oxidase is a critical mediator of bone marrow cell recruitment to form the premetastatic niche. Cancer Cell 15: 35–44.

Eyre-Walker, A. and P. D. Keightley, 2007 The distribution of fitness effects of new mutations. Nat Rev Genet 8: 610–618.

Friedman, A. D., 2009 Cell cycle and developmental control of hematopoiesis by Runx1. J Cell Physiol 219: 520–524.

Fusco, D., M. Gralka, J. Kayser, A. Anderson and O. Hallatschek, 2016 Excess of mutational jackpot events in expanding populations revealed by spatial Luria-Delbruck experiments. Nat Commun 7: 12760.

Galeano Nino, J. L., H. Wu, K. D. LaCourse, A. G. Kempchinsky, A. Baryiames et al., 2022 Effect of the intratumoral microbiota on spatial and cellular heterogeneity in cancer. Nature 611: 810–817.

Galhardo, R. S., P. J. Hastings and S. M. Rosenberg, 2007 Mutation as a stress response and the regulation of evolvability. Crit Rev Biochem Mol Biol 42: 399–435.

Galon, J., A. Costes, F. Sanchez-Cabo, A. Kirilovsky, B. Mlecnik et al., 2006 Type, density, and location of immune cells within human colorectal tumors predict clinical outcome. Science 313: 1960–1964.

Gatenby, R. A., A. S. Silva, R. J. Gillies and B. R. Frieden, 2009 Adaptive therapy. Cancer Res 69: 4894–4903.

Gomez, K., S. Miura, L. A. Huuki, B. S. Spell, J. P. Townsend et al., 2018 Somatic evolutionary timings of driver mutations. BMC Cancer 18: 85.

Gonzalez, H., C. Hagerling and Z. Werb, 2018 Roles of the immune system in cancer: From tumor initiation to metastatic progression. Genes Dev 32: 1267–1284.

Grasso, C. S., M. Giannakis, D. K. Wells, T. Hamada, X. J. Mu et al., 2018 Genetic mechanisms of immune evasion in colorectal cancer. Cancer Discov 8: 730–749.

Greif, P. A., S. H. Eck, N. P. Konstandin, A. Benet-Pages, B. Ksienzyk et al., 2011 Identification of recurring tumor-specific somatic mutations in acute myeloid leukemia by transcriptome sequencing. Leukemia 25: 821–827.

Hanahan, D., 2022 Hallmarks of cancer: New dimensions. Cancer Disc 12: 31–46.

Hata, A. N., M. J. Niederst, H. L. Archibald, M. Gomez-Caraballo, F. M. Siddiqui et al., 2016 Tumor cells can follow distinct evolutionary paths to become resistant to epidermal growth factor receptor inhibition. Nat Med 22: 262–269.

Huminiecki, L. and G. C. Conant, 2012 Polyploidy and the evolution of complex traits. Int J Evol Biol 2012: 292068.

James, J., 1970 The founder effect and response to artificial selection. Genet Res. 16: 241–250.

Jiang, K., Q. Zhang, Y. Fan, J. Li, J. Zhang et al., 2022 MYC inhibition reprograms tumor immune microenvironment by recruiting T lymphocytes and activating the CD40/CD40L system in osteosarcoma. Cell Death Discov 8: 117.

Jolly, M. K., J. A. Somarelli, M. Sheth, A. Biddle, S. C. Tripathi et al., 2019 Hybrid epithelial/mesenchymal phenotypes promote metastasis and therapy resistance across carcinomas. Pharmacol Ther 194: 161–184.

Jolly, M. K., K. E. Ware, S. Gilja, J. A. Somarelli and H. Levine, 2017 EMT and MET: necessary or permissive for metastasis? Mol Oncol 11: 755–769.

Kashyap, M. P., R. Sinha, M. S. Mukhtar and M. Athar, 2022 Epigenetic regulation in the pathogenesis of non-melanoma skin cancer. Semin Cancer Biol 83: 36–56.

Kastan, M. B., C. E. Canman and C. J. Leonard, 1995 P53, cell cycle control and apoptosis: Implications for cancer. Cancer Metastasis Rev 14: 3–15.

Khan, A., T. Khan, S. N. Nasir, S. S. Ali, M. Suleman et al., 2021 BC-TFdb: A database of transcription factor drivers in breast cancer. Database (Oxford) 2021: baab018.

Kryazhimskiy, S. and J. B. Plotkin, 2008 The population genetics of dN/dS. PLoS Genet 4: e1000304.

Lakatos, E., M. J. Williams, R. O. Schenck, W. C. H. Cross, J. Househam et al., 2020 Evolutionary dynamics of neoantigens in growing tumors. Nat Genet 52: 1057–1066.

Lambeth, J. D., 2007 Nox enzymes, ROS, and chronic disease: An example of antagonistic pleiotropy. Free Radic Biol Med 43: 332–347.

Lean, C. and A. Plutynski, 2016 The evolution of failure: Explaining cancer as an evolutionary process. Biol Philos 31: 39–57.

Ling, S., Z. Hu, Z. Yang, F. Yang, Y. Li et al., 2015 Extremely high genetic diversity in a single tumor points to prevalence of non-Darwinian cell evolution. Proc Natl Acad Sci USA 112: E6496–E6505.

Lipinski, K. A., L. J. Barber, M. N. Davies, M. Ashenden, A. Sottoriva et al., 2016 Cancer evolution and the limits of predictability in precision cancer medicine. Trends Cancer 2: 49–63.

Lloyd, M. C., K. O. Alfarouk, D. Verduzco, M. M. Bui, R. J. Gillies et al., 2014 Vascular measurements correlate with estrogen receptor status. BMC Cancer 14: 279.

Lloyd, M. C., J. J. Cunningham, M. M. Bui, R. J. Gillies, J. S. Brown et al., 2016 Darwinian dynamics of intratumoral heterogeneity: Not solely random mutations but also variable environmental selection forces. Cancer Res 76: 3136–3144.

Lynch, M., 2010 Evolution of the mutation rate. Trends Genet 26: 345–352.

Martin, T. D., R. S. Patel, D. R. Cook, M. Y. Choi, A. Patil et al., 2021 The adaptive immune system is a major driver of selection for tumor suppressor gene inactivation. Science 373: 1327–1335.

Martincorena, I., K. M. Raine, M. Gerstung, K. J. Dawson, K. Haase et al., 2017 Universal Patterns of selection in cancer and somatic tissues. Cell 171: 1029–1041, e1021.

McFarland, C. D., J. A. Yaglom, J. W. Wojtkowiak, J. G. Scott, D. L. Morse et al., 2017 The damaging effect of passenger mutations on cancer progression. Cancer Res 77: 4763–4772.

Michod, R. E., 2005 On the transfer of fitness from the cell to the multicellular organism. Biol Philos 20: 967–987.

Muller, H. J., 1964 The relation of recombination to mutational advance. Mutat Res 106: 2–9.

Nakasone, E. S., H. A. Askautrud, T. Kees, J. H. Park, V. Plaks et al., 2012 Imaging tumor-stroma interactions during chemotherapy reveals contributions of the microenvironment to resistance. Cancer Cell 21: 488–503.

Nathanson, S. D., 2003 Insights into the mechanisms of lymph node metastasis. Cancer 98: 413–423.

Nejman, D., I. Livyatan, G. Fuks, N. Gavert, Y. Zwang et al., 2020 The human tumor microbiome is composed of tumor type–specific intracellular bacteria. Science 368: 973–980.

Noy, R. and J. W. Pollard, 2014 Tumor-associated macrophages: From mechanisms to therapy. Immunity 41: 49–61.

Parrinello, S., J. P. Coppe, A. Krtolica and J. Campisi, 2005 Stromal-epithelial interactions in aging and cancer: Senescent fibroblasts alter epithelial cell differentiation. J Cell Sci 118: 485–496.

Patel, A., 2020 Benign vs malignant tumors. JAMA Oncol 6: 1488.

Pezzuto, F., F. Izzo, L. Buonaguro, C. Annunziata, F. Tatangelo et al., 2016 Tumor specific mutations in TERT promoter and CTNNB1 gene in hepatitis B and hepatitis C related hepatocellular carcinoma. Oncotarget 7: 54253–54262.

Pigliucci, M., 2005 Evolution of phenotypic plasticity: Where are we going now? Trends Ecol Evol 20: 481–486.

Pon, J. R. and M. A. Marra, 2015 Driver and passenger mutations in cancer. Annu Rev Pathol 10: 25–50.

Schraiber, J. G., S. N. Evans and M. Slatkin, 2016 Bayesian inference of natural selection from allele frequency time series. Genetics 203: 493–511.

Seyfried, T., 2012 Nothing in cancer biology makes sense except in the light of evolution. In *Cancer as a metabolic disease: On the origin, management, and prevention of cancer.* John Wiley & Sons, pp. 261–275.

Shay, J. W. and I. B. Roninson, 2004 Hallmarks of senescence in carcinogenesis and cancer therapy. Oncogene 23: 2919–2933.

Sheflin, A. M., A. K. Whitney and T. L. Weir, 2014 Cancer-promoting effects of microbial dysbiosis. Curr Oncol Rep 16: 406.

Shibuya, M., 2011 Vascular endothelial growth factor (VEGF) and its receptor (VEGFR) signaling in angiogenesis: A crucial target for anti- and pro-angiogenic therapies. Genes Cancer 2: 1097–1105.

Somarelli, J. A., H. Gardner, V. L. Cannataro, E. F. Gunady, A. M. Boddy et al., 2020 Molecular biology and evolution of cancer: From discovery to action. Mol Biol Evol 37: 320–326.

Sottoriva, A., H. Kang, Z. Ma, T. A. Graham, M. P. Salomon et al., 2015 A big bang model of human colorectal tumor growth. Nat Genet 47: 209–216.

Stadler, P. F., 2002 Fitness landscapes. In *Biological evolution and statistical physics.* Springer, pp. 183–204.

Thomas, F., E. Donnadieu, G. M. Charriere, C. Jacqueline, A. Tasiemski et al., 2018 Is adaptive therapy natural? PLoS Biol 16: e2007066.

Tinaburri, L., C. Valente, M. Teson, Y. A. Minafo, S. Cordisco et al., 2021 The secretome of aged fibroblasts promotes EMT-like phenotype in primary keratinocytes from elderly donors through BDNF-TrkB axis. J Invest Dermatol 141: 1052–1062, e1012.

Tomasetti, C. and B. Vogelstein, 2015 Variation in cancer risk among tissues can be explained by the number of stem cell divisions. Science 347: 78–81.

Valastyan, S. and R. A. Weinberg, 2011 Tumor metastasis: Molecular insights and evolving paradigms. Cell 147: 275–292.

Vessoni, A. T., E. C. Filippi-Chiela, G. Lenz and L. F. Z. Batista, 2020 Tumor propagating cells: Drivers of tumor plasticity, heterogeneity, and recurrence. Oncogene 39: 2055–2068.

Wang, J., D. Li, H. Cang and B. Guo, 2019 Crosstalk between cancer and immune cells: Role of tumor-associated macrophages in the tumor microenvironment. Cancer Med 8: 4709–4721.

Warburg, O., F. Wind and E. Negelein, 1927 The metabolism of tumors in the body. J Gen Physiol 8: 519–530.

Welch, J. S., T. J. Ley, D. C. Link, C. A. Miller, D. E. Larson et al., 2012 The origin and evolution of mutations in acute myeloid leukemia. Cell 150: 264–278.

Werner, B. and A. Sottoriva, 2018 Variation of mutational burden in healthy human tissues suggests non-random strand segregation and allows measuring somatic mutation rates. PLoS Comput Biol 14: e1006233.

Wright, S., 1932 The roles of mutation, inbreeding, crossbreeding, and selection in evolution. Proc VI Intl Cong Genet 1: 356–366.

Wu, C.-I., H.-Y. Wang, S. Ling and X. Lu, 2016 The ecology and evolution of cancer: The ultra-microevolutionary process. Annu Rev Genet 50: 347–369.

Yan, M. and P. Jurasz, 2016 The role of platelets in the tumor microenvironment: From solid tumors to leukemia. Biochim Biophys Acta 1863: 392–400.

Yang, L., W. Wang, L. Hu, X. Yang, J. Zhong et al., 2013 Telomere-binding protein TPP1 modulates telomere homeostasis and confers radioresistance to human colorectal cancer cells. PLoS One 8: e81034.

Yoshida, K. and Y. Miki, 2004 Role of BRCA1 and BRCA2 as regulators of DNA repair, transcription, and cell cycle in response to DNA damage. Cancer Sci 95: 866–871.

Zapata, L., O. Pich, L. Serrano, F. A. Kondrashov, S. Ossowski et al., 2018 Negative selection in tumor genome evolution acts on essential cellular functions and the immunopeptidome. Genome Biol 19: 67.

Zhang, C., L. M. Moore, X. Li, W. K. Yung and W. Zhang, 2013 IDH1/2 mutations target a key hallmark of cancer by deregulating cellular metabolism in glioma. Neuro Oncol 15: 1114–1126.

Zheng, S., H. Bian, J. Li, Y. Shen, Y. Yang et al., 2022 Differentiation therapy: Unlocking phenotypic plasticity of hepatocellular carcinoma. Crit Rev Oncol Hematol: 103854.

4 Multicellularity, Phenotypic Heterogeneity, and Cancer

Christopher Helenek, Jason A. Somarelli, and Gábor Balázsi

4.1 INTRODUCTION: CANCER AND MULTICELLULARITY

Although the evolution of the first multicellular organisms is a major evolutionary transition, many aspects of this process are still unclear, such as when or how exactly this transformation first took place (Knoll, Javaux et al. 2006; Abedin and King 2010; Erwin, Laflamme et al. 2011). Similarly, multiple cell types from very different species have undergone independent transitions to multicellularity on our planet (Grosberg and Strathmann 2007). Likewise, examples of reverse transitions to unicellularity are also known (Rebolleda-Gomez and Travisano 2019; Urrejola, von Dassow et al. 2020). Green algae (Kirk 2005) and some fungi (Wohlbach, Thompson et al. 2009; Nagy, Varga et al. 2020) present a spectrum of related unicellular to multicellular species. The evolutionary transition to multicellularity (Maynard Smith and Szathmáry 1998) requires strong cooperation between the participating cellular units, such as kin recognition, propagules, and cheater exclusion. Ultimately, the multicellular form can become a new unit under higher levels of selection. Nonetheless, the constituent single cells can still acquire mutations that allow them to break free from the strict rules of cooperation, causing the collapse of the multicellular entity (Hammerschmidt, Rose et al. 2014; Kuzdzal-Fick, Chen et al. 2019; Rebolleda-Gomez and Travisano 2019). Unicellularity and multicellularity can even alternate in time as a result (Hammerschmidt, Rose et al. 2014).

The bodies of adult animals, including those of humans, might appear relatively unchanged. However, in most animals, tissues are continuously undergoing cell turnover through a balance of cell death and division, giving the impression of tissue and body constancy. This balance indicates the existence of size sensing and feedback response (Durant, Lobo et al. 2016; Kunche, Yan et al. 2016; Karin, Raz et al. 2020) as mechanisms of cooperation, whereby cells proliferate only when they receive signals to do so, e.g., during wound healing, upon cell death, or due to developmental programs. The "disobedient cells" that do not "follow the rules" of multicellularity are usually detected and eliminated by the immune system or by neighboring epithelial cells (Tanimura and Fujita 2020). Various mechanisms such as cell-cell competition (Parker, Madan et al. 2020; Parker, Gupta et al. 2021) can select for disobedient, abnormal cells that eventually reach a tipping point, escape elimination, and multiply to form a tumor, which can be benign, with well-defined boundaries and little to no metastatic capacity. Other tumors are malignant and consist of cells capable of continuous growth and metastatic dissemination. Thus, cancer is a manifestation of single cells breaking the rules of multicellularity (Aktipis, Boddy et al. 2015; Aktipis 2019).

Some authors have claimed that hallmarks of cancer are similar to dysfunction of multicellularity (Nedelcu 2020) and that cancer is a reversion of multicellularity to a "unicellular state" (Metazoa 1.0) (Davies and Lineweaver 2011). These thought-provoking ideas may further benefit from additional refinements that consider the unique features of cancer. For example, while the unicellular ancestors of animal cells were free-living, cancer cells are obligate inhabitants of animal bodies. Therefore, cancer cells approach a form of unicellularity that is not equivalent to naturally unicellular metazoans since they adapt to environments that did not exist before large animal bodies emerged. For example, cancer cells are surrounded by other cells or cell-derived materials, whereas in natural habitats cells interact with non-living substrates and environmental bacteria; the body temperature inside warm-blooded animals is relatively constant, whereas the temperature typically

DOI: 10.1201/9781003307921-4

fluctuates in natural environments; and visible light hardly penetrates multicellular bodies whereas most natural environments are exposed to sunlight, and so on. The similarities and differences of the transition from multicellularity to unicellularity in nature and in cancers require more investigation to understand if key differences exist between the sources of variation and the selective forces acting within the body and those for unicellular organisms in nature. In the following sections, we discuss environmental and stochastic sources of non-genetic variation and their role in unicellular and multicellular organisms, as well as in cancer.

4.2 BENEFITS AND DISADVANTAGES OF MULTICELLULARITY: ENVIRONMENTAL VARIATION

Multicellularity relies on strong cooperation between cells as well as mechanisms (e.g., kin recognition, propagules, and cheater exclusion) to ensure evolutionary stability. These measures require a large amount of energy overall and may contribute to the observed loss of reproduction for most differentiated, somatic cells compared to germ cells. Such task allocation in lower multicellular organisms might be reminiscent of normal animal tissues, where stem or progenitor cells take the task to divide as needed, while differentiated cells tend to remain quiescent until they die. Additionally, resources must be split among the individual cells in the group, which may lead to fewer net resources per cell. To offset these costs, multicellularity must provide decisive advantages in certain conditions to be evolutionarily stable.

To investigate this claim, several studies have utilized various yeast strains, some of which can adopt either constitutive or facultative multicellular and unicellular phenotypes (Barrere, Nanda et al. 2023), to test the advantages and disadvantages of multicellularity. The results indicated that multicellularity may be beneficial in stressful conditions, such as nutrient scarcity (Koschwanez, Foster et al. 2011; Colizzi, Vroomans et al. 2020; Blackstone and Gutterman 2021) or environmental stress (Smukalla, Caldara et al. 2008; Kuzdzal-Fick, Chen et al. 2019; Rebolleda-Gomez and Travisano 2019; Guinn, Lo et al. 2022), while being disadvantageous in nutrient-rich, stress-free conditions. Multicellularity can be beneficial by creating controlled environmental variation, i.e., cell-cell heterogeneity due to spatial positioning of cells in multicellular groups (Figure 4.1). Cells

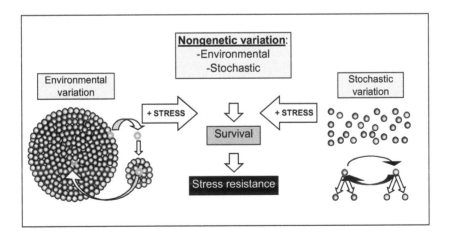

FIGURE 4.1 Comparison of multicellular/environmental (left) and stochastic (right) variation from the perspective of cellular fitness and stress resistance. Gray represents cells of lower relative fitness in the absence of stress but higher relative fitness in the presence of stress. Green cells are propagules. Both stochastic and environmental variation due to multicellularity can cause gray cells to survive for sufficient time during stress exposure to rescue the cell population and enable the evolution of genetic stress resistance.

in different positional contexts within a group are exposed to different microenvironments, resulting from cell-cell interactions and interactions with gradients of nutrient density, oxygen, pressure, cytokine and growth factor signals, and the extracellular matrix. Other cells can act as protective "shields" (Smukalla, Caldara et al. 2008; Kuzdzal-Fick, Chen et al. 2019; Rebolleda-Gomez and Travisano 2019; Guinn, Lo et al. 2022), which can be beneficial by reducing stress felt by individual cells but disadvantageous by simultaneously reducing access to nutrients.

Thus, multicellular entities naturally develop phenotypic heterogeneity due to microenvironmental cell-cell differences. This raises the question, "What are the similarities and differences between the effects of microenvironmentally-mediated heterogeneity and stochastic heterogeneity?"

4.3   NOISE, SURVIVAL, AND EVOLUTION

Like environmental variation in multicellular organisms, stochastic heterogeneity (noise) can drive cells in a genetically-identical group to adopt different fates (Balazsi, van Oudenaarden et al. 2011). This stems from the stochastic variability (noise) in the amounts of various cellular components (Blake, Kærn et al. 2003) that can fluctuate randomly in time, causing different phenotypes to emerge and in some cases, fluctuate in the population (Kaern, Elston et al. 2005; Balazsi, van Oudenaarden et al. 2011). Thus, it seems logical to assume that noise is also beneficial in stressful environments, since it can generate multiple phenotypes that may be helpful in this situation. However, while noise can aid survival in high stress, it can also hinder survival in low stress (Blake, Balazsi et al. 2006; Farquhar, Charlebois et al. 2019; Guinn, Wan et al. 2020), which is reminiscent of the nutrient-related costs and stress-related benefits of multicellularity.

Noise is omnipresent and unavoidable in all cell types, but cells can evolve to control noise according to profit from its benefits or to lower its disadvantages. In unicellular organisms, noise can aid survival during recurrent, high stress through a process of bet-hedging mediated by random phenotypic switching (Figure 4.1). A high variability in cellular phenotypes means that at least some cells in the population will survive during times of high stress (Kussell and Leibler 2005; Belete and Balazsi 2015). In more consistent environments, phenotypic variability is disadvantageous, and cells with initially-heterogenous expression evolve toward a more uniform expression profile that is optimal in the given environment (Gonzalez, Ray et al. 2015; Farquhar, Charlebois et al. 2019). The return of stress can re-evoke heterogeneity in these populations, leading to the reemergence of bet-hedging as a viable strategy (Beaumont, Gallie et al. 2009; Kheir Gouda, Manhart et al. 2019).

Similarly to microbial populations, cancer cells can also benefit from heterogeneity. Noise can drive differentiation and adaptability in isogenic populations (Aronson, Ricci-Tam et al. 2020), contributing to the versatility of cancers (Shen and Clairambault 2020), including metastatic spread (Lee, Lee et al. 2014) and survival during chemotherapy (Sharma, Lee et al. 2010; Shaffer, Dunagin et al. 2017). For example, in the process of metastasis, cancer cells from most solid tumors need to remove themselves from the original tissue, travel through the extracellular space toward a blood vessel, squeeze through the endothelial layer to enter the bloodstream, travel in the blood stream, exit the blood vessel, and then establish themselves at a secondary site. Thus, noise may be beneficial in generating cancer cells that can survive and complete diverse tasks in these multiple stressful and unique microenvironments. In addition, noise may be beneficial in giving cells plasticity to move between different phenotypes. Instead of being locked into one cell-fate, noise may allow single cells to move between different stable fates, due to varying amounts of differentiation and cell-fate cues. Thus, this enables cells to survive in an even wider range of conditions. However, it is worth noting that heterogeneity can be beneficial or disadvantageous depending on the nature of selection, e.g., the threshold that cells need to cross to survive (Farquhar, Charlebois et al. 2019; Guinn, Wan et al. 2020).

Importantly, tumors also share some features with free-living multicellular entities: they consist of cells that diversify due to their positions relative to oxygen and nutrient availability. This adds to the heterogeneity of cancer cell populations, boosting their ability to survive in various stresses and to metastasize. Nonetheless, more work is needed to precisely define the effects of noise and spatial heterogeneity within a population of cancer cells.

One area that is contributing to our improved understanding of these processes is in the realm of single cell omics technologies. These techniques have provided an unprecedented view into the heterogeneity of cancer cell populations as well as the dynamics of phenotypic switching during the evolution of therapy resistance and metastasis. A well-studied hallmark of resistance and metastasis is epithelial plasticity, a spectrum of phenotypic transitions along the epithelial to mesenchymal axis that, although with a controversial role in full metastasis (Fischer, Durrans et al. 2015; Zheng, Carstens et al. 2015), is known to promote cell invasion, survival, and therapy resistance (Yang, Antin et al. 2020). The reversible processes of epithelial-mesenchymal transition (EMT) and mesenchymal-epithelial transition (MET) contribute to various aspects of metastasis and chemoresistance biology. Single-cell gene expression and lineage tracing studies are illuminating the complex and heterogeneous processes by which cell populations transition between these phenotypes. Cells can take diverse and context-specific trajectories toward EMT, and the reversion (MET) is likewise along differing trajectories (Cook and Vanderhyden 2020). Heterogeneity also exists in the time it takes for individual cells to transition between states, even for the same growth factor in a cell culture system. Clustering of cell populations based on their timing of EMT induction is associated with distinct microRNAs and gene regulatory networks, and clusters of hybrid E/M states are prognostic for poorer clinical outcomes (Deshmukh, Vasaikar et al. 2021). Heterogeneity in phenotypic states promotes diversity in function within the population, which enhances the potential for survival (Grasset, Dunworth et al. 2022). For example, in some cases, hybrid E/M cells are more capable of disseminating than their more epithelial (E) counterparts, while the most mesenchymal (M) subpopulations alter the microenvironmental niche toward a pro-angiogenic and pro-tumor immune environment (Pastushenko, Brisebarre et al. 2018). Moreover, hybrid E/M cells are more plastic than both fully E and fully M cells (Brown, Abdollahi et al. 2022; Hari, Ullanat et al. 2022), suggesting a possible task allocation where hybrid E/M cells quickly respond and shift tumor composition according to environmental fluctuations, whereas stable E and M cells ensure the continuous presence of diverse cell types within tumors. While the exact "division of labor" within populations is likely to be cancer type and lineage-specific, this non-genetic heterogeneity within the cancer cell population contributes to population robustness and survival.

Thus, stochastic and environmental heterogeneity can diversify a cell population even without genetic differences. Instead of relying solely on mutations to mediate adaptation, environmental heterogeneity and gene expression noise can allow a cell population to rapidly adapt to its environment (or many diverse microenvironments) by generating non-genetic differences within the population (Figure 4.1). Consequently, properly probing non-genetic variation and its effects on cancer progression requires single cell measurements of gene expression and corresponding phenotypes. While this can give information regarding naturally-existing heterogeneity, more precise and thorough characterization of the phenotypic role of non-genetic variation requires that it be controlled in some way. In the following section we discuss the control of both environmental and stochastic heterogeneity.

4.4   SYNTHETIC BIOLOGICAL CONTROL OF HETEROGENEITY AND MULTICELLULARITY

Synthetic biology is a growing field focused on building new biological systems for predefined purposes, such as cellular sensing and control, chemical production, or environmental remediation. Synthetic gene circuits are genetic devices that control cells by altering the levels of proteins or RNA. Two interesting applications of synthetic gene circuits are the control of environmental heterogeneity arising from multicellularity and the control of stochastic heterogeneity.

Controlling environmental heterogeneity due to multicellularity requires controlling genes responsible for these phenotypes. For example, various strains of yeast can have a multitude of multicellular phenotypes, such as biofilms, flocs, chains, and clumps (Guinn, Lo et al. 2022) as opposed to most common lab strains of yeast that have unicellular phenotypes. A synthetic gene circuit controlling the expression of the *FLO1* flocculation gene could adjust the timing of yeast sedimentation (Ellis, Wang et al. 2009). Interestingly, yeast can evolve into or out of multicellular phenotypes, such as clumping, by changes in particular genes, such as *AMN1*, a mediator of cell-cell separation during exit from mitosis (Kuzdzal-Fick, Chen et al. 2019). Therefore, controlling the expression of genes like *AMN1* could confer adjustable degrees of multicellularity and thereby environmental heterogeneity in certain yeast strains. Larger multicellular groups should have larger environmental cell-cell differences. This would enable investigating the role of multicellularity in numerous processes, such as drug resistance (Guinn, Lo et al. 2022).

Similarly, environmental heterogeneity due to multicellularity could be controlled in mammalian cells by manipulating cellular adhesion. For example, expression of different cadherins can be controlled to produce self-organizing multicellular structures from cells that have a low propensity to cluster (i.e., are mainly unicellular) (Toda, Blauch et al. 2018). Similar effects can arise from artificial proteins, allowing for more specific control of multicellularity. For example, spatial and temporal control of cell-cell adhesion can be accomplished by artificial proteins that dimerize in response to light (Rasoulinejad, Mueller et al. 2020).

In addition to environmental variation, synthetic gene circuits can also be used to control stochastic cell-to-cell heterogeneity, driving cell populations to become more or less noisy. For example, early work involved mutating the TATA region of the GAL1 promoter in yeast. This created yeast strains with highly variable promoter kinetics, which caused higher or lower differences (noise) in protein levels (Blake, Balazsi et al. 2006). Thus, these yeast strains could reveal the effect of heterogeneity on drug resistance (Blake, Balazsi et al. 2006). This idea has been expanded for more advanced control of heterogeneity through the use of negative feedback (Nevozhay, Adams et al. 2009) and positive feedback (Nevozhay, Adams et al. 2012) gene circuits. Using such constructs, the expression noise of practically any protein can be adjusted independently of its mean expression in yeast cells, allowing for a deeper exploration of the role of noise in various cellular processes.

In addition to yeast cells, synthetic gene circuits can be similarly applied to control stochastic gene expression heterogeneity in mammalian cells (Nevozhay, Zal et al. 2013), which allows for the understanding of how stochastic heterogeneity contributes to drug resistance (Farquhar, Charlebois et al. 2019), metastasis (Lee, Lee et al. 2014), and cell-fate (Zañudo, Guinn et al. 2019). Overall, the control offered by synthetic biology can elucidate the role of environmental and stochastic heterogeneity in these and many other fundamental cell activities.

4.5 CONCLUSIONS

Currently, it is well appreciated that stochastic and environmental heterogeneity (multicellularity) contribute to the emergence and progression of cancer. However, research is just beginning to describe the underlying mechanisms and demonstrate how they manifest into observable phenotypic landscapes. Synthetic biology is uniquely positioned to help with understanding the role of these processes in cancer. By allowing for unprecedented control of single cells and cell populations, they can be precisely perturbed, which can generate hypotheses, inform models, or validate predictions. Furthermore, synthetic biology allows for studying the effects of noise alone, independent from population expression mean. This may enable studying the role of environmental variation due to multicellularity, which can uniquely provide insight into cancer and other related processes.

REFERENCES

Abedin, M. and N. King (2010). "Diverse evolutionary paths to cell adhesion." *Trends Cell Biol* **20**(12): 734–742.

Aktipis, C. A. (2019). *The cheating cell: how evolution helps us understand and treat cancer*. Princeton University Press.

Aktipis, C. A., A. M. Boddy, G. Jansen, U. Hibner, M. E. Hochberg, C. C. Maley and G. S. Wilkinson (2015). "Cancer across the tree of life: cooperation and cheating in multicellularity." *Philos Trans R Soc Lond B Biol Sci* **370**(1673): 20140219.

Aronson, M. S., C. Ricci-Tam, X. Zhu and A. E. Sgro (2020). "Exploiting noise to engineer adaptability in synthetic multicellular systems." *Curr Opin Biomed Eng* **16**: 52–60.

Balázsi, G., A. van Oudenaarden and J. J. Collins (2011). "Cellular decision making and biological noise: from microbes to mammals." *Cell* **144**(6): 910–925.

Barrere, J., P. Nanda and A. W. Murray (2023). "Alternating selection for dispersal and multicellularity favors regulated life cycles." *Curr Biol* **33**(9): 1809–1817 e1803.

Beaumont, H. J., J. Gallie, C. Kost, G. C. Ferguson and P. B. Rainey (2009). "Experimental evolution of bet hedging." *Nature* **462**(7269): 90–93.

Belete, M. K. and G. Balázsi (2015). "Optimality and adaptation of phenotypically switching cells in fluctuating environments." *Phys Rev E Stat Nonlin Soft Matter Phys* **92**(6): 062716.

Blackstone, N. W. and J. U. Gutterman (2021). "Can natural selection and druggable targets synergize? Of nutrient scarcity, cancer, and the evolution of cooperation." *Bioessays* **43**(2): e2000160.

Blake, W. J., G. Balazsi, M. A. Kohanski, F. J. Isaacs, K. F. Murphy, Y. Kuang, C. R. Cantor, D. R. Walt and J. J. Collins (2006). "Phenotypic consequences of promoter-mediated transcriptional noise." *Mol Cell* **24**(6): 853–865.

Blake, W. J., M. Kærn, C. R. Cantor and J. J. Collins (2003). "Noise in eukaryotic gene expression." *Nature* **422**(6932): 633–637.

Brown, M. S., B. Abdollahi, O. M. Wilkins, H. Lu, P. Chakraborty, N. B. Ognjenovic, K. E. Muller, M. K. Jolly, B. C. Christensen, S. Hassanpour and D. R. Pattabiraman (2022). "Phenotypic heterogeneity driven by plasticity of the intermediate EMT state governs disease progression and metastasis in breast cancer." *Sci Adv* **8**(31): eabj8002.

Colizzi, E. S., R. M. Vroomans and R. M. Merks (2020). "Evolution of multicellularity by collective integration of spatial information." *Elife* **9**.

Cook, D. P. and B. C. Vanderhyden (2020). "Context specificity of the EMT transcriptional response." *Nature Commun* **11**(1): 2142.

Davies, P. C. and C. H. Lineweaver (2011). "Cancer tumors as Metazoa 1.0: tapping genes of ancient ancestors." *Phys Biol* **8**(1): 015001.

Deshmukh, A. P., S. V. Vasaikar, K. Tomczak, S. Tripathi, P. den Hollander, E. Arslan, P. Chakraborty, R. Soundararajan, M. K. Jolly, K. Rai, H. Levine and S. A. Mani (2021). "Identification of EMT signaling cross-talk and gene regulatory networks by single-cell RNA sequencing." *Proc Natl Acad Sci USA* **118**(19).

Durant, F., D. Lobo, J. Hammelman and M. Levin (2016). "Physiological controls of large-scale patterning in planarian regeneration: a molecular and computational perspective on growth and form." *Regeneration (Oxford)* **3**(2): 78–102.

Ellis, T., X. Wang and J. J. Collins (2009). "Diversity-based, model-guided construction of synthetic gene networks with predicted functions." *Nature Biotech* **27**(5): 465–471.

Erwin, D. H., M. Laflamme, S. M. Tweedt, E. A. Sperling, D. Pisani and K. J. Peterson (2011). "The Cambrian conundrum: early divergence and later ecological success in the early history of animals." *Science* **334**(6059): 1091–1097.

Farquhar, K. S., D. A. Charlebois, M. Szenk, J. Cohen, D. Nevozhay and G. Balázsi (2019). "Role of network-mediated stochasticity in mammalian drug resistance." *Nat Commun* **10**(1): 2766.

Fischer, K. R., A. Durrans, S. Lee, J. Sheng, F. Li, S. T. Wong, H. Choi, T. El Rayes, S. Ryu and J. Troeger (2015). "Epithelial-to-mesenchymal transition is not required for lung metastasis but contributes to chemoresistance." *Nature* **527**(7579): 472–476.

Gonzalez, C., J. C. Ray, M. Manhart, R. M. Adams, D. Nevozhay, A. V. Morozov and G. Balázsi (2015). "Stress-response balance drives the evolution of a network module and its host genome." *Mol Syst Biol* **11**(8): 827.

Grasset, E. M., M. Dunworth, G. Sharma, M. Loth, J. Tandurella, A. Cimino-Mathews, M. Gentz, S. Bracht, M. Haynes, E. J. Fertig and A. J. Ewald (2022). "Triple-negative breast cancer metastasis involves complex epithelial-mesenchymal transition dynamics and requires vimentin." *Sci Transl Med* **14**(656): eabn7571.

Grosberg, R. K. and R. R. Strathmann (2007). "The evolution of multicellularity: a minor major transition?" *Annu Rev Ecol Evol Syst* **38**: 621–654.

Guinn, L., E. Lo and G. Balázsi (2022). "Drug-dependent growth curve reshaping reveals mechanisms of antifungal resistance in Saccharomyces cerevisiae." *Communications Biology* **5**(1): 292.

Guinn, M. T., Y. Wan, S. Levovitz, D. Yang, M. R. Rosner and G. Balázsi (2020). "Observation and control of gene expression noise: barrier crossing analogies between drug resistance and metastasis." *Front Genet* **11**: 586726.

Hammerschmidt, K., C. J. Rose, B. Kerr and P. B. Rainey (2014). "Life cycles, fitness decoupling and the evolution of multicellularity." *Nature* **515**(7525): 75–79.

Hari, K., V. Ullanat, A. Balasubramanian, A. Gopalan and M. K. Jolly (2022). "Landscape of epithelial-mesenchymal plasticity as an emergent property of coordinated teams in regulatory networks." *Elife* **11**.

Kaern, M., T. C. Elston, W. J. Blake and J. J. Collins (2005). "Stochasticity in gene expression: from theories to phenotypes." *Nat Rev Genet* **6**(6): 451–464.

Karin, O., M. Raz, A. Tendler, A. Bar, Y. Korem Kohanim, T. Milo and U. Alon (2020). "A new model for the HPA axis explains dysregulation of stress hormones on the timescale of weeks." *Mol Syst Biol* **16**(7): e9510.

Kheir Gouda, M., M. Manhart and G. Balázsi (2019). "Evolutionary regain of lost gene circuit function." *Proc Natl Acad Sci USA* **116**(50): 25162–25171.

Kirk, D. L. (2005). "A twelve-step program for evolving multicellularity and a division of labor." *Bioessays* **27**(3): 299–310.

Knoll, A. H., E. J. Javaux, D. Hewitt and P. Cohen (2006). "Eukaryotic organisms in Proterozoic oceans." *Philos Trans R Soc Lond B Biol Sci* **361**(1470): 1023–1038.

Koschwanez, J. H., K. R. Foster and A. W. Murray (2011). "Sucrose utilization in budding yeast as a model for the origin of undifferentiated multicellularity." *PLoS Biol* **9**(8): e1001122.

Kunche, S., H. Yan, A. L. Calof, J. S. Lowengrub and A. D. Lander (2016). "Feedback, lineages and self-organizing morphogenesis." *PLoS Comput Biol* **12**(3): e1004814.

Kussell, E. and S. Leibler (2005). "Phenotypic diversity, population growth, and information in fluctuating environments." *Science* **309**(5743): 2075–2078.

Kuzdzal-Fick, J. J., L. Chen and G. Balázsi (2019). "Disadvantages and benefits of evolved unicellularity versus multicellularity in budding yeast." *Ecol Evol* **9**(15): 8509–8523.

Lee, J., J. Lee, K. S. Farquhar, J. Yun, C. A. Frankenberger, E. Bevilacqua, K. Yeung, E.-J. Kim, G. Balázsi and M. R. Rosner (2014). "Network of mutually repressive metastasis regulators can promote cell heterogeneity and metastatic transitions." *Proc Natl Acad Sci* **111**(3): E364–E373.

Maynard Smith, J. and E. Szathmáry (1998). *The major transitions in evolution.* Oxford University Press.

Nagy, L. G., T. Varga, A. Csernetics and M. Viragh (2020). "Fungi took a unique evolutionary route to multicellularity: seven key challenges for fungal multicellular life." *Fungal Biol Rev* **34**(4): 151–169.

Nedelcu, A. M. (2020). "The evolution of multicellularity and cancer: views and paradigms." *Biochem Soc Trans* **48**(4): 1505–1518.

Nevozhay, D., R. M. Adams, K. F. Murphy, K. I. Josić and G. Balázsi (2009). "Negative autoregulation linearizes the dose–response and suppresses the heterogeneity of gene expression." *Proc Natl Acad Sci* **106**(13): 5123–5128.

Nevozhay, D., R. M. Adams, E. Van Itallie, M. R. Bennett and G. Balázsi (2012). "Mapping the environmental fitness landscape of a synthetic gene circuit." *PLoS Comput Biol* **8**(4): e1002480.

Nevozhay, D., T. Zal and G. Balázsi (2013). "Transferring a synthetic gene circuit from yeast to mammalian cells." *Nat Commun* **4**: 1451.

Parker, T. M., K. Gupta, A. M. Palma, M. Yekelchyk, P. B. Fisher, S. R. Grossman, K. J. Won, E. Madan, E. Moreno and R. Gogna (2021). "Cell competition in intratumoral and tumor microenvironment interactions." *EMBO J* **40**(17): e107271.

Parker, T. M., E. Madan, K. Gupta, E. Moreno and R. Gogna (2020). "Cell competition spurs selection of aggressive cancer cells." *Trends Cancer* **6**(9): 732–736.

Pastushenko, I., A. Brisebarre, A. Sifrim, M. Fioramonti, T. Revenco, S. Boumahdi, A. Van Keymeulen, D. Brown, V. Moers, S. Lemaire, S. De Clercq, E. Minguijón, C. Balsat, Y. Sokolow, C. Dubois, F. De Cock, S. Scozzaro, F. Sopena, A. Lanas, N. D'Haene, I. Salmon, J. C. Marine, T. Voet, P. A. Sotiropoulou and C. Blanpain (2018). "Identification of the tumour transition states occurring during EMT." *Nature* **556**(7702): 463–468.

Rasoulinejad, S., M. Mueller, B. Nzigou Mombo and S. V. Wegner (2020). "Orthogonal blue and red light controlled cell–cell adhesions enable sorting-out in multicellular structures." *ACS Synth Biol* **9**(8): 2076–2086.

Rebolleda-Gomez, M. and M. Travisano (2019). "Adaptation, chance, and history in experimental evolution reversals to unicellularity." *Evolution* **73**(1): 73–83.

Shaffer, S. M., M. C. Dunagin, S. R. Torborg, E. A. Torre, B. Emert, C. Krepler, M. Beqiri, K. Sproesser, P. A. Brafford, M. Xiao, E. Eggan, I. N. Anastopoulos, C. A. Vargas-Garcia, A. Singh, K. L. Nathanson, M. Herlyn and A. Raj (2017). "Rare cell variability and drug-induced reprogramming as a mode of cancer drug resistance." *Nature* **546**(7658): 431–435.

Sharma, S. V., D. Y. Lee, B. Li, M. P. Quinlan, F. Takahashi, S. Maheswaran, U. McDermott, N. Azizian, L. Zou, M. A. Fischbach, K. K. Wong, K. Brandstetter, B. Wittner, S. Ramaswamy, M. Classon and J. Settleman (2010). "A chromatin-mediated reversible drug-tolerant state in cancer cell subpopulations." *Cell* **141**(1): 69–80.

Shen, S. and J. Clairambault (2020). "Cell plasticity in cancer cell populations." *F1000Res* **9**.

Smukalla, S., M. Caldara, N. Pochet, A. Beauvais, S. Guadagnini, C. Yan, M. D. Vinces, A. Jansen, M. C. Prevost, J. P. Latge, G. R. Fink, K. R. Foster and K. J. Verstrepen (2008). "FLO1 is a variable green beard gene that drives biofilm-like cooperation in budding yeast." *Cell* **135**(4): 726–737.

Tanimura, N. and Y. Fujita (2020). "Epithelial defense against cancer (EDAC)." *Semin Cancer Biol* **63**: 44–48.

Toda, S., L. R. Blauch, S. K. Tang, L. Morsut and W. A. Lim (2018). "Programming self-organizing multicellular structures with synthetic cell-cell signaling." *Science* **361**(6398): 156–162.

Urrejola, C., P. von Dassow, G. van den Engh, L. Salas, C. W. Mullineaux, R. Vicuna and P. Sanchez-Baracaldo (2020). "Loss of filamentous multicellularity in cyanobacteria: the extremophile gloeocapsopsis sp. strain UTEX B3054 retained multicellular features at the genomic and behavioral levels." *J Bacteriol* **202**(12).

Wohlbach, D. J., D. A. Thompson, A. P. Gasch and A. Regev (2009). "From elements to modules: regulatory evolution in Ascomycota fungi." *Curr Opin Genet Dev* **19**(6): 571–578.

Yang, J., P. Antin, G. Berx, C. Blanpain, T. Brabletz, M. Bronner, K. Campbell, A. Cano, J. Casanova, G. Christofori, S. Dedhar, R. Derynck, H. L. Ford, J. Fuxe, A. García de Herreros, G. J. Goodall, A. K. Hadjantonakis, R. Y. J. Huang, C. Kalcheim, R. Kalluri, Y. Kang, Y. Khew-Goodall, H. Levine, J. Liu, G. D. Longmore, S. A. Mani, J. Massagué, R. Mayor, D. McClay, K. E. Mostov, D. F. Newgreen, M. A. Nieto, R. Puisieux, R. Runyan, P. Savagner, B. Stanger, M. P. Stemmler, Y. Takahashi, M. Takeichi, E. Theveneau, J. P. Thiery, E. W. Thompson, R. A. Weinberg, E. D. Williams, J. Xing, B. P. Zhou and G. Sheng (2020). "Guidelines and definitions for research on epithelial-mesenchymal transition." *Nat Rev Mol Cell Biol* **21**(6): 341–352.

Zañudo, J. G. T., M. T. Guinn, K. Farquhar, M. Szenk, S. N. Steinway, G. Balázsi and R. Albert (2019). "Towards control of cellular decision-making networks in the epithelial-to-mesenchymal transition." *Phys Biol* **16**(3): 031002.

Zheng, X., J. L. Carstens, J. Kim, M. Scheible, J. Kaye, H. Sugimoto, C.-C. Wu, V. S. LeBleu and R. Kalluri (2015). "Epithelial-to-mesenchymal transition is dispensable for metastasis but induces chemoresistance in pancreatic cancer." *Nature* **527**(7579): 525–530.

5 Feedback Loops in Gene Regulatory Networks and Cell-Cell Communication Networks

Drivers of Cancer Cell Plasticity

Yeshwanth Mahesh, Subbalakshmi Ayalur Raghu, and Mohit Kumar Jolly

5.1 INTRODUCTION

Cellular plasticity is defined as a reversible transition of cells from one state (phenotype) to another, often in response to changing environment (Gupta et al., 2019). In nature, various organisms can switch their behavior as per environmental conditions, for instance, change in leaf shape in plants, and the emergence of queen (reproductive) and worker (non-reproductive) populations in insects, despite the same genetic makeup (G. Li et al., 2019; Patalano et al., 2015). It is implicated in cancer metastasis and in evading various therapeutic attacks and is considered a hallmark of cancer (Hanahan, 2022). It manifests along various phenotypic axes, such as cancer stem cell (CSC) plasticity, epithelial-mesenchymal plasticity (EMP), metabolic reprogramming, and reversible drug resistance. CSC plasticity refers to switching among different subpopulations of CSCs and/ or between CSCs and non-CSCs (Gupta et al., 2011; Vipparthi et al., 2022). EMP includes reversible transitions among epithelial (E), mesenchymal (M) and hybrid E/M states through partial or full epithelial-mesenchymal transition (EMT) and the reverse process of mesenchymal-epithelial transition (MET) (Brown et al., 2022; Lourenco et al., 2020). Metabolic reprogramming includes switching between a more glycolytic and a more oxidative phosphorylation (OXPHOS)-dependent metabolic state (Jia et al., 2019).

These multiple axes of plasticity are often interconnected. For example, EMP enables estrogen-receptor positive breast cancer cells to evade tamoxifen, while tamoxifen resistance enables EMP (Sahoo et al., 2021). Similarly, EMP and metabolic reprogramming are linked such that epithelial cells are more dependent on oxidative phosphorylation, while partially or fully mesenchymal ones are relatively more glycolytic (Muralidharan et al., 2022; Schwager et al., 2022). This cross-linking of decision-making can enable metastasizing cells to navigate multiple bottlenecks simultaneously and is often driven by feedback loops among the set of molecules driving plasticity along each of these axes individually. For instance, ZEB1, an EMT-inducing transcription factor, represses estrogen receptor (ERα) and *vice versa* (Sahoo et al., 2021). Thus, such cross-linked mechanisms of cellular plasticity can impact selection of specific cellular traits providing higher fitness during metastasis. Similar feedback loops are implicated in reversible cell state switching along individual axis of plasticity as well, for instance, GRHL2—an MET-inducing transcription factor—and ZEB1 inhibit each other in multiple cancer types (Chung et al., 2016; Mooney et al., 2017; Somarelli et al., 2016).

Feedback loops need not be intracellular; direct or indirect interaction of cancer cells with elements of the tumor microenvironment (TME) such as extracellular matrix (ECM) and other stromal

DOI: 10.1201/9781003307921-5

and cancer cells can also form feedback loops that result in cell plasticity. For instance, cells under-going EMT secrete the collagen cross-linking enzyme LOXL2 that can stiffen the ECM, which, in turn, can promote EMT (Deng et al., 2021; Peng et al., 2017). Together, these feedback loops oper-ating at multiple levels—both intra-cellularly and inter-cellularly—shape the emergent dynamics of cell plasticity in cancer progression. Here, we discuss three main axes of cancer cell plasticity—EMP, CSC-like and metabolic plasticity—to illustrate some key feedback loops and identify com-mon principles in their topological structures.

5.2   EPITHELIAL-MESENCHYMAL PLASTICITY (EMP)

The first intracellular feedback loop reported in EMP was a combination of the microRNA-200 family and ZEB transcription factor family, which promote epithelial and mesenchymal cell states, respectively (Brabletz and Brabletz, 2010). ZEB1 and ZEB2 transcriptionally repress miR-200 fam-ily members and are inhibited at the post-transcriptional level. Mathematical modeling of this feed-back loop revealed that it could allow for multistability, i.e., the coexistence of multiple cell states together than can reversibly switch among one another (Lu et al., 2013; Tian et al., 2013). These models also predicted that EMP is not a binary process (i.e., cells do not directly switch between epithelial and mesenchymal phenotypes but can stably acquire one or more hybrid epithelial/mes-enchymal states), a prediction that has been validated experimentally in the past decade (Bocci, Tripathi et al., 2019; Jolly et al., 2016; Kröger et al., 2019; Pastushenko et al., 2018; Schliekelman et al., 2015; Selvaggio et al., 2020). Further investigations into this network suggested two key features. First, the feedback loops identified had a similar structure—EMT-inducing player(s) were engaged in such mutually inhibitory interactions with MET-inducing player(s), for instance, SNAIL and miR-34 (Siemens et al., 2011), TWIST and miR-129 (Silveira et al., 2020), ZEB1 and GRHL2 (Chung et al., 2016). Second, gene regulatory networks (GRNs) constructed for EMT/MET allowed for one or more hybrid E/M cell states depending on possible combinations of co-expressing EMT-specific and MET-specific players, thus highlighting that multistability is a common trait of net-works involved in shaping EMP dynamics (Font-Clos et al., 2018; Steinway et al., 2015). A recent analysis of topology of many such GRNs highlighted that they consist of "teams" of nodes such that members within a team activate each other effectively and those across teams inhibit each other, thus leading to a "toggle switch" between the two teams, one driving EMT and the other pushing MET (Hari et al., 2022). Interestingly, no such "team" existed for hybrid E/M phenotypes, which explained the relatively higher plasticity of hybrid E/M states as compared to E and M states noted experimentally (Pastushenko et al., 2018; Ruscetti et al., 2016). The higher the "team strength", the more bimodal the underlying phenotypic landscape was, with the deeper valleys corresponding to epithelial and mesenchymal phenotypes and the shallower ones corresponding to hybrid E/M ones (Hari et al., 2022) (Figure 5.1A).

EMP dynamics is also governed by feedback loops engaging components of the TME. For exam-ple, epithelial cancer cells can contribute to macrophage polarization into M1-like macrophages, while mesenchymal cells shift macrophages toward the M2-like phenotype. M2 macrophages can drive MET, thus effectively leading to formation of two "teams" in a tumor-immune feedback loop—epithelial cells and M1 macrophages and mesenchymal cells and M2 macrophages. Computational modeling of this interaction network also revealed underlying multistability, suggesting reversible switching in cell population state-space (X. Li et al., 2019) (Figure 5.1B). Similarly, a group of mesenchymal cells can induce a neighboring cell to undergo EMT in a non-autonomous manner. They secrete LOXL2 that can increase ECM stiffness thus increasing ZEB1 levels. This positive feedback loop between ZEB1 and ECM stiffness can amplify the EMT status of a cell popula-tion (Deng et al., 2021). Initial transcriptomic data analysis in The Cancer Genome Atlas (TCGA) cohorts showed a positive correlation between LOXL2 and EMT and enrichment of the EMT and M2-macrophage signatures, providing support for the presence of the aforementioned feedback

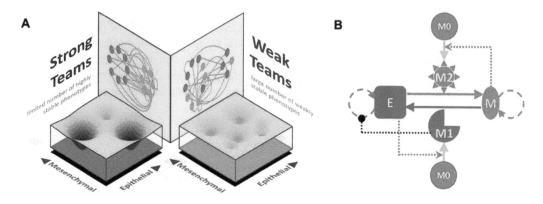

FIGURE 5.1 Intracellular regulatory networks and cell-cell communication networks engaged in cellular plasticity. A) Intracellular networks driving EMT/MET are comprised of "teams" of EMT-drivers and MET-drivers. The stronger the "teams", the deeper the valley (the more the stability) of epithelial and mesenchymal states relative to one or more hybrid states that are relatively less stable. Schematic shows the effect of team structure on the phenotypic stability landscape emergent from the network topology. B) Cell-cell interaction network among epithelial (E), mesenchymal (M) cancer cells and M1 and M2 polarized macrophages. It includes conversions between different cancer or immune cell states (solid lines), cell proliferation (dashed lines) and activation and inhibition (red and black dotted lines, respectively). (Adapted from Hari et al., 2022; X. Li et al., 2019).

loops in clinical specimens. However, the strength of these feedback loops, i.e., how strongly ZEB1 and ECM stiffness amplify each other, depends on factors such as spatial-temporal heterogeneity in LOXL2 concentration and diffusivity of cytokines secreted by macrophages to induce EMP, thus impacting the metastatic propensity of a cell population. This shows that the interaction networks among cancer cells can alter spatiotemporal dynamics of EMT and eventual functional traits of a heterogeneous cell population (Hapach et al., 2023; Neelakantan et al., 2017). For instance, Notch-Delta-Jagged signaling among cancer cells can stabilize cells in a hybrid E/M phenotype. The emergent dynamics of a coupled EMT-Notch circuit at a tissue level can lead to different spatial patterns of EMT-linked heterogeneity. When Notch-Jagged signaling dominates over Notch-Delta signaling, patches of hybrid E/M cells were observed, indicating possible stabilization of this cell state via cell-cell communication. Such cooperativity in terms of EMT induction can lead to formation of clusters of circulating tumor cells (CTCs) that are the primary drivers of cancer metastasis (Boareto et al., 2016; Jolly et al., 2017). A phenomenological dynamical model encapsulating a cell autonomous EMT induction rate and a non-cell autonomous cooperativity parameter in terms of cancer cells inducing EMT in their neighbors was sufficient to recapitulate the different size distributions of CTC clusters identified from patient samples (Bocci, Jolly et al., 2019). This observation highlights the bi-directionality in cause and effect between feedback loops formed by different cell-cell interactions in the TME and the corresponding tissue-level EMP dynamics.

5.3 CANCER STEM CELL (CSC) PLASTICITY

A long-standing conceptual framework in oncology has been that of a "hierarchical" cancer stem cell (CSC) model where cancer cell subpopulations are thought of as organized in hierarchical structure based on their differentiation capacity. At the apex of this hierarchy are CSCs that are the most stem-like and capable of regenerating the non-CSCs, repopulating tumor heterogeneity and enabling tumor recurrence (Cole et al., 2020). However, recent work across cancer types has highlighted that non-CSCs can give rise to CSCs, under varying environmental conditions (Andriani et al., 2016; Auffinger et al., 2014; Enderling, 2015; Gupta et al., 2011; Vipparthi et al., 2022), establishing the "plasticity" model (Thankamony et al., 2020). From the perspective of the plasticity model, "stemness" is a dynamic trait that can be gained or lost.

At an intracellular level, the GRNs underlying the control of stemness also engage mutually inhibitory feedback loops, such as the one between RNA-binding protein LIN28A/LIN28B and microRNA let-7 family, which controls the frequency of aldehyde dehydrogenase (ALDH1+) CSCs (Yang et al., 2010). Similarly, during matrix-detachment, while many cancer cells die (anoikis), a few of them switch to an pAMPK^{high}/pAKT^{low} state, which is critical for cell survival in suspension. Upon matrix reattachment, these cells switch to a pAMPK^{low}/pAKT^{high} cell state. Such reversible cell state switching is a direct consequence of multistability enabled by a mutually inhibitory loop between AMPK and Akt involving kinases and phosphatases (Chedere et al., 2021; Saha et al., 2018). Given the ability of CSCs to resist anoikis and form mammospheres in suspension (Kim et al., 2016), the dynamic switching between pAMPK^{high}/pAKT^{low} and pAMPK^{low}/pAKT^{high} is reminiscent of switching between CSCs and non-CSCs.

Feedback loops formed among CSCs and other components of the TME can also alter their plasticity. For instance, β-catenin-induced CCL2 levels can promote macrophage infiltration and M2 polarization. Co-culture of macrophages with breast cancer cells, in turn, increase CCL2 levels and enhanced CSC traits. This amplifying paracrine loop between macrophages and CSCs supports cancer metastasis (Zhang et al., 2021). Further, in hepatocellular carcinoma (HCC), tumor-associated neutrophils (TANs) secrete BMP2 and TGF-β that increase stemness in HCC cells through the miR-301b-3p/LSAMP axis. In turn, the TAN-induced HCC CSCs had enriched NF-kB signaling and CXCL5 levels, which promoted further TAN infiltration (Zhou et al., 2019). Thus, this positive feedback loop can drive CSC frequency and tumor progression in HCC. Another example of simultaneous modulation of TME and CSCs is the release of angiogenic factors such as VEGF and CXCL12 by CSCs to stimulate endothelial cell (EC) angiogenesis. ECs can, in turn, stimulate Notch signaling and Wnt/β-catenin pathway to activate CSC traits (Sipos and Műzes, 2023), setting up a self-sustaining feedback loop to aggravate tumor spread.

5.4   METABOLIC REPROGRAMMING/PLASTICITY

Metabolic reprogramming is a hallmark of cancer (Hanahan and Weinberg, 2011); cancer cells can alter their predominant mode of cellular energetics depending on their microenvironment—glycolysis or oxidative phosphorylation (OXPHOS)—and/or shortage of various intermediate metabolites inhibiting specific sub-pathways (Diehl et al., 2019; Faubert et al., 2020). Referred to as the Warburg Effect in cancer, such metabolic adaptability is reminiscent of Crabtree yeast observed in yeast, where cells use aerobic fermentation instead of the more efficient process of energy generation—respiration, thus allowing growth in high glucose situations (Malina et al., 2021).

Metabolic plasticity is seen across cancer types differently—in breast cancer, metastatic cells have higher OXPHOS activity, while in prostate and renal cancer metastasis, OXPHOS is suppressed (Fendt et al., 2020). Besides these two well-studied modes, a (high glycolysis, high OXPHOS) (Jia et al., 2019) and a (low glycolysis, low OXPHOS) (Jia et al., 2020) state has also been reported, indicating a hybrid metabolic state and a quiescent one. The (co-)existence of these four states has been demonstrated to be an emergent outcome of the underlying dynamics of a network involving feedback loops among HIF1, AMPK and other metabolic regulators.

Metabolic plasticity has been investigated in CSCs as well. For instance, two subsets of breast CSCs were isolated from multiple breast cancer cell lines (SUM149, MCF7 and T47D)—a) ALDH+ cells were more proliferative, epithelial-like and exhibited increased OXPHOS, and b) CD44+/CD24$^-$ cells were more quiescent, mesenchymal-like and glycolytic (Luo et al., 2018). Hypoxic or oxidative stress enabled reversible cell state transitions, showing metabolic adaptability according to external stimuli. Such instances of plasticity can help explain the repertoire of strategies metastatic breast cancer cells engage in, based on the metastatic site. For instance, breast cancer cells from bone or lung metastases preferentially use OXPHOS, but the liver-metastatic ones prioritize glycolysis (Lehúede et al., 2016).

Metabolic plasticity is also crucial for effective collective cell migration, where leader cells exhibit higher glucose uptake than the follower cells. As the leader cell depletes its energy, it is replaced by a follower cell. The timescale of retention of a leader state can be modulated by glucose starvation or AMPK activation (Zhang et al., 2019), demonstrating spatiotemporal metabolic coordination among cancer cells. Similar instances of metabolic symbiosis are witnessed through lactate shuttle, where cancer cells in hypoxic regions can release lactate through undergoing glycolysis, and this lactate can be used as fuel for TCA cycle in adjacent oxygenated tumor region (Nakajima and Van Houten, 2013).

Metabolic coordination or competition of cancer cells with stromal cells is often observed in primary tumor and metastatic TMEs, thus impacting the escape from immunosurveillance or other therapeutic interventions. For instance, ovarian cancer cells and stromal adipocytes can engage in metabolic symbiosis through metabolism of arginine into nitric oxide and citrulline by cancer cells and capture of citrulline by adipocytes to convert it back to arginine (Li and Simon, 2020). On the other hand, in the context of competition, glucose consumption by melanoma and sarcoma cells can metabolically restrict T-cells, thus limiting T-effector cell activity. Consequently, inhibition of glycolysis has been shown to enhance anti-tumor immunity mediated by T-cells (Cascone et al., 2018).

Overall, similar to EMP and CSC plasticity, both intracellular and intercellular feedback loops govern the dynamics of metabolic plasticity in the TME, with important implications for cancer cell competition, cancer-immune cross-talk, TME remodeling and therapeutic response.

5.5 CONCLUSION

The phenomenon of cellular plasticity is being increasingly recognized to play a crucial role in cancer progression; however, the salient dynamical and structural features of networks driving these processes remain poorly understood. Here, we highlight specific common traits observed in intracellular GRNs and cell-cell communication networks that exist along the different axes of plasticity—EMP, CSC plasticity and metabolic switching. These networks comprise multiple interconnected positive feedback loops that can reinforce a cell state and allow for the coexistence of multiple phenotypes that may interconvert. While intracellular GRNs incorporated mutually inhibitory loops among one or more master regulators, including forming the "teams" of nodes as well, cell-cell communication networks often had mutually amplifying structures. Both these network motifs (double negative or double positive) are effectively positive feedback loops enabling cellular plasticity. Such loops also regulate the coupling between different axes of plasticity. Mapping and investigating these interconnected feedback loops can provide unique insights into dynamics of cancer cell adaptation and eventually lead to new therapeutic targets as well.

REFERENCES

Andriani F, Bertolini G, Facchinetti F, Baldoli E, Moro M, Casalini P, et al. Conversion to stem-cell state in response to microenvironmental cues is regulated by balance between epithelial and mesenchymal features in lung cancer cells. Mol Oncol 2016;10:253–71. https://doi.org/10.1016/j.molonc.2015.10.002.

Auffinger B, Tobias AL, Han Y, Lee G, Guo D, Dey M, et al. Conversion of differentiated cancer cells into cancer stem-like cells in a glioblastoma model after primary chemotherapy. Cell Death Differ 2014;21:1119–31. https://doi.org/10.1038/CDD.2014.31.

Boareto M, Jolly MK, Goldman A, Pietilä M, Mani SA, Sengupta S, et al. Notch-Jagged signalling can give rise to clusters of cells exhibiting a hybrid epithelial/mesenchymal phenotype. J R Soc Interface 2016;13. https://doi.org/10.1098/RSIF.2015.1106.

Bocci F, Jolly MK, Onuchic JN. A biophysical model uncovers the size distribution of migrating cell clusters across cancer types. Cancer Res 2019;79:5527–35. https://doi.org/10.1158/0008-5472.CAN-19-1726.

Bocci F, Tripathi SC, Vilchez Mercedes SA, George JT, Casabar JP, Wong PK, et al. NRF2 activates a partial epithelial-mesenchymal transition and is maximally present in a hybrid epithelial/mesenchymal phenotype. Integrative Biology 2019;11:251. https://doi.org/10.1093/INTBIO/ZYZ021.

Brabletz S, Brabletz T. The ZEB/miR-200 feedback loop-a motor of cellular plasticity in development and cancer? EMBO Rep 2010;11. https://doi.org/10.1038/embor.2010.117.

Brown MS, Abdollahi B, Wilkins OM, Lu H, Chakraborty P, Ognjenovic NB, et al. Phenotypic heterogeneity driven by plasticity of the intermediate EMT state governs disease progression and metastasis in breast cancer. Sci Adv 2022;8. https://doi.org/10.1126/SCIADV.ABJ8002.

Cascone T, McKenzie JA, Mbofung RM, Punt S, Wang Zhe, Xu C, et al. Increased tumor glycolysis characterizes immune resistance to adoptive T cell therapy. Cell Metab 2018;27:977–987.e4. https://doi.org/10.1016/J.CMET.2018.02.024.

Chedere A, Hari K, Kumar S, Rangarajan A, Jolly MK. Multi-stability and consequent phenotypic plasticity in AMPK-Akt double negative feedback loop in cancer cells. J Clin Med 2021;10:1–16. https://doi.org/10.3390/JCM10030472.

Chung VY, Tan TZ, Tan M, Wong MK, Kuay KT, Yang Z, et al. GRHL2-miR-200-ZEB1 maintains the epithelial status of ovarian cancer through transcriptional regulation and histone modification. Sci Rep 2016;6. https://doi.org/10.1038/SREP19943.

Cole AJ, Fayomi AP, Anyaeche VI, Bai S, Buckanovich RJ. An evolving paradigm of cancer stem cell hierarchies: therapeutic implications. Theranostics 2020;10:3083–98. https://doi.org/10.7150/THNO.41647.

Deng Y, Chakraborty P, Jolly MK, Levine H. A theoretical approach to coupling the epithelial-mesenchymal transition (EMT) to extracellular matrix (ECM) stiffness via loxl2. Cancers (Basel) 2021;13:1609. https://doi.org/10.3390/CANCERS13071609/S1.

Diehl FF, Lewis CA, Fiske BP, Vander Heiden MG. Cellular redox state constrains serine synthesis and nucleotide production to impact cell proliferation. Nat Metab 2019;1:861–7. https://doi.org/10.1038/s42255-019-0108-x.

Enderling H. Cancer stem cells: small subpopulation or evolving fraction? Integr Biol (Camb) 2015;7:14–23. https://doi.org/10.1039/C4IB00191E.

Faubert B, Solmonson A, DeBerardinis RJ. Metabolic reprogramming and cancer progression. Science (1979) 2020;368. https://doi.org/10.1126/SCIENCE.AAW5473/ASSET/043E2E47-0328-47B9-927A-622370C71FB7/ASSETS/GRAPHIC/368_AAW5473_F4.JPEG.

Fendt SM, Frezza C, Erez A. Targeting metabolic plasticity and flexibility dynamics for cancer therapy. Cancer Discov 2020;10:1797. https://doi.org/10.1158/2159-8290.CD-20-0844.

Font-Clos F, Zapperi S, La Porta CAM. Topography of epithelial–mesenchymal plasticity. Proc Natl Acad Sci USA 2018;115:5902–7. https://doi.org/10.1073/PNAS.1722609115/-/DCSUPPLEMENTAL.

Gupta PB, Fillmore CM, Jiang G, Shapira SD, Tao K, Kuperwasser C, et al. Stochastic state transitions give rise to phenotypic equilibrium in populations of cancer cells. Cell 2011;146:633–44. https://doi.org/10.1016/j.cell.2011.07.026.

Gupta PB, Pastushenko I, Skibinski A, Blanpain C, Kuperwasser C. Phenotypic plasticity: Driver of cancer initiation, progression, and therapy resistance. Cell Stem Cell 2019;24:65–78. https://doi.org/10.1016/j.stem.2018.11.011.

Hanahan D. Hallmarks of cancer: New dimensions. Cancer Discov 2022;12:31–46. https://doi.org/10.1158/2159-8290.CD-21-1059.

Hanahan D, Weinberg RA. Hallmarks of cancer: The next generation. Cell 2011;144:646–74. https://doi.org/10.1016/j.cell.2011.02.013.

Hapach LA, Wang W, Schwager SC, Pokhriyal D, Fabiano ED, Reinhart-King CA. Phenotypically sorted highly and weakly migratory triple negative breast cancer cells exhibit migratory and metastatic commensalism. Breast Cancer Res 2023;25. https://doi.org/10.1186/S13058-023-01696-3.

Hari K, Ullanat V, Balasubramanian A, Gopalan A, Jolly MK. Landscape of epithelial mesenchymal plasticity as an emergent property of coordinated teams in regulatory networks. Elife 2022;11. https://doi.org/10.7554/ELIFE.76535.

Jia D, Lu M, Jung KH, Park JH, Yu L, Onuchic JN, et al. Elucidating cancer metabolic plasticity by coupling gene regulation with metabolic pathways. Proc Natl Acad Sci USA 2019;116:3909–18. https://doi.org/10.1073/PNAS.1816391116/SUPPL_FILE/PNAS.1816391116.SAPP.PDF.

Jia D, Paudel BB, Hayford CE, Hardeman KN, Levine H, Onuchic JN, et al. Drug-tolerant idling melanoma cells exhibit theory-predicted metabolic low-low phenotype. Front Oncol 2020;10:561215. https://doi.org/10.3389/FONC.2020.01426/BIBTEX.

Jolly MK, Boareto M, Debeb BG, Aceto N, Farach-Carson MC, Woodward WA, et al. Inflammatory breast cancer: A model for investigating cluster-based dissemination. NPJ Breast Cancer 2017;3:21. https://doi.org/https://doi.org/10.1101/119479.

Jolly MK, Tripathi SC, Jia D, Mooney SM, Celiktas M, Hanash SM, et al. Stability of the hybrid epithelial/mesenchymal phenotype. Oncotarget 2016;7:27067–84. https://doi.org/10.18632/oncotarget.8166.

Kim SY, Hong SH, Basse PH, Wu C, Bartlett DL, Kwon YT, et al. Cancer stem cells protect non-stem cells from anoikis: Bystander effects. J Cell Biochem 2016;117:2289–301. https://doi.org/10.1002/JCB.25527.

Kröger C, Afeyan A, Mraz J, Eaton EN, Reinhardt F, Khodor YL, et al. Acquisition of a hybrid E/M state is essential for tumorigenicity of basal breast cancer cells. Proc Natl Acad Sci USA 2019;116:7353–62. https://doi.org/10.1073/pnas.1812876116.

Lehúede C, Dupuy F, Rabinovitch R, Jones RG, Siegel PM. Metabolic plasticity as a determinant of tumor growth and metastasis. Cancer Res 2016;76:5201–8. https://doi.org/10.1158/0008-5472. CAN-16-0266/660578/P/metabolic-plasticity-as-a-determinant-of-tumor.

Li F, Simon MC. Cancer cells don't live alone: Metabolic communication within tumor microenvironments. Dev Cell 2020;54:183–95. https://doi.org/10.1016/J.DEVCEL.2020.06.018.

Li G, Hu S, Hou H, Kimura S. Heterophylly: Phenotypic plasticity of leaf shape in aquatic and amphibious plants. Plants 2019;8. https://doi.org/10.3390/PLANTS8100420.

Li X, Jolly MK, George JT, Pienta KJ, Levine H. Computational modeling of the crosstalk between macrophage polarization and tumor cell plasticity in the tumor microenvironment. Front Oncol 2019;9:1–12. https://doi.org/10.3389/fonc.2019.00010.

Lourenco AR, Ban Y, Crowley MJ, Lee SB, Ramchandani D, Du W, et al. Differential Contributions of pre- and post-EMT tumor cells in breast cancer metastasis. Cancer Res 2020;80:163–9. https://doi. org/10.1158/0008-5472.CAN-19-1427.

Lu M, Jolly MK, Levine H, Onuchic JN, Ben-Jacob E. MicroRNA-based regulation of epithelial-hybrid-mesenchymal fate determination. Proc Natl Acad Sci USA 2013;110:18144–9. https://doi.org/10.1073/ pnas.1318192110.

Luo M, Shang L, Brooks MD, Jiagge E, Zhu Y, Buschhaus JM, et al. Targeting breast cancer stem cell state equilibrium through modulation of redox signaling. Cell Metab 2018;28:69–86. https://doi.org/10.1016/j. cmet.2018.06.006.

Malina C, Yu R, Bjorkeroth J, Kerkhoven EJ, Nielsen J. Adaptations in metabolism and protein translation give rise to the Crabtree effect in yeast. Proc Natl Acad Sci USA 2021;118. https://doi.org/10.1073/ PNAS.2112836118/-/DCSUPPLEMENTAL.

Mooney SM, Talebian V, Jolly MK, Jia D, Gromala M, Levine H, et al. The GRHL2/ZEB feedback loop: A key axis in the regulation of EMT in breast cancer. J Cell Biochem 2017;118:2559–70. https://doi. org/10.1002/jcb.25974.

Muralidharan S, Sahoo S, Saha A, Chandran S, Majumdar SS, Mandal S, et al. Quantifying the patterns of metabolic plasticity and heterogeneity along the epithelial-hybrid-mesenchymal spectrum in cancer. Biomolecules 2022;12. https://doi.org/10.3390/BIOM12020297.

Nakajima EC, Van Houten B. Metabolic symbiosis in cancer: refocusing the Warburg lens. Mol Carcinog 2013;52:329–37. https://doi.org/10.1002/MC.21863.

Neelakantan D, Zhou H, Oliphant MUJ, Zhang X, Simon LM, Henke DM, et al. EMT cells increase breast cancer metastasis via paracrine GLI activation in neighbouring tumour cells. Nat Commun 2017;8:1–14. https://doi.org/10.1038/ncomms15773.

Pastushenko I, Brisebarre A, Sifrim A, Fioramonti M, Revenco T, Boumahdi S, et al. Identification of the tumour transition states occurring during EMT. Nature 2018;556:463–8. https://doi.org/10.1038/ s41586-018-0040-3.

Patalano S, Vlasova A, Wyatt C, Ewels P, Camara F, Ferreira PG, et al. Molecular signatures of plastic phenotypes in two eusocial insect species with simple societies. Proc Natl Acad Sci USA 2015;112:13970–5. https://doi.org/10.1073/PNAS.1515937112/SUPPL_FILE/PNAS.1515937112.SD05.XLSX.

Peng DH, Ungewiss C, Tong P, Byers LA, Wang J, Canales JR, et al. ZEB1 induces LOXL2-mediated collagen stabilization and deposition in the extracellular matrix to drive lung cancer invasion and metastasis. Oncogene 2017;36:1925–38. https://doi.org/10.1038/ONC.2016.358.

Ruscetti M, Dadashian EL, Guo W, Quach B, Mulholland DJ, Park JW, et al. HDAC inhibition impedes epithelial-mesenchymal plasticity and suppresses metastatic, castration-resistant prostate cancer. Oncogene 2016;35:3781–95. https://doi.org/10.1038/onc.2015.444.

Saha M, Kumar S, Bukhari S, Balaji SA, Kumar P, Hindupur SK, et al. AMPK-Akt double-negative feedback loop in breast cancer cells regulates their adaptation to matrix deprivation. Cancer Res 2018;78:1497–510. https://doi.org/10.1158/0008-5472.CAN-17-2090.

Sahoo S, Mishra A, Kaur H, Hari K, Muralidharan S, Mandal S, et al. A mechanistic model captures the emergence and implications of non-genetic heterogeneity and reversible drug resistance in ER+ breast cancer cells. NAR Cancer 2021;3. https://doi.org/10.1093/narcan/zcab027.

Schliekelman MJ, Taguchi A, Zhu J, Dai X, Rodriguez J, Celiktas M, et al. Molecular portraits of epithelial, mesenchymal and hybrid states in lung adenocarcinoma and their relevance to survival. Cancer Res 2015;75:1789–800. https://doi.org/10.1158/0008-5472.CAN-14-2535.

Schwager SC, Mosier JA, Padmanabhan RS, White A, Xing Q, Hapach LA, et al. Link between glucose metabolism and epithelial-to-mesenchymal transition drives triple-negative breast cancer migratory heterogeneity. IScience 2022;25. https://doi.org/10.1016/J.ISCI.2022.105190.

Selvaggio G, Canato S, Pawar A, Monteiro PT, Guerreiro PS, Brás MM, et al. Hybrid epithelial-mesenchymal phenotypes are controlled by microenvironmental factors. Cancer Res 2020;80:2407–20. https://doi.org/10.1158/0008-5472.CAN-19-3147.

Siemens H, Jackstadt R, Hünten S, Kaller M, Menssen A, Götz U, et al. miR-34 and SNAIL form a double-negative feedback loop to regulate epithelial-mesenchymal transitions. Cell Cycle 2011;10:4256–71. https://doi.org/10.4161/cc.10.24.18552.

Silveira DA, Gupta S, Mombach JCM. Systems biology approach suggests new miRNAs as phenotypic stability factors in the epithelial-mesenchymal transition. J R Soc Interface 2020;17. https://doi.org/10.1098/RSIF.2020.0693.

Sipos F, Műzes G. Cancer stem cell relationship with pro-tumoral inflammatory microenvironment. Biomedicines 2023;11. https://doi.org/10.3390/BIOMEDICINES11010189.

Somarelli JA, Shetler S, Jolly MK, Wang X, Bartholf Dewitt S, Hish AJ, et al. Mesenchymal-epithelial transition in sarcomas is controlled by the combinatorial expression of microRNA 200s and GRHL2. Mol Cell Biol 2016;36:2503–13. https://doi.org/10.1128/MCB.00373-16.

Steinway SN, Zañudo JGT, Michel PJ, Feith DJ, Loughran TP, Albert R. Combinatorial interventions inhibit TGFβ-driven epithelial-to-mesenchymal transition and support hybrid cellular phenotypes. NPJ Syst Biol Appl 2015;1:15014. https://doi.org/10.1038/npjsba.2015.14.

Thankamony AP, Saxena K, Murali R, Jolly MK, Nair R. Cancer stem cell plasticity: A deadly deal. Front Mol Biosci 2020;7:79. https://doi.org/10.3389/fmolb.2020.00079.

Tian X-J, Zhang H, Xing J. Coupled reversible and irreversible bistable switches underlying TGFβ-induced epithelial to mesenchymal transition. Biophys J 2013;105:1079–89. https://doi.org/10.1016/j.bpj.2013.07.011.

Vipparthi K, Hari K, Chakraborty P, Ghosh S, Patel AK, Ghosh A, et al. Emergence of hybrid states of stem-like cancer cells correlates with poor prognosis in oral cancer. IScience 2022;25. https://doi.org/10.1016/J.ISCI.2022.104317.

Yang X, Lin X, Zhong X, Kaur S, Li N, Liang S, et al. Double-negative feedback loop between reprogramming factor LIN28 and microRNA let-7 regulates aldehyde dehydrogenase 1-positive cancer stem cells. Cancer Res 2010;70:9463–72. https://doi.org/10.1158/0008-5472.CAN-10-2388.

Zhang F, Li P, Liu S, Yang M, Zeng S, Deng J, et al. β-Catenin-CCL2 feedback loop mediates crosstalk between cancer cells and macrophages that regulates breast cancer stem cells. Oncogene 2021;40:5854–65. https://doi.org/10.1038/s41388-021-01986-0.

Zhang J, Goliwas KF, Wang W, Taufalele P V., Bordeleau F, Reinhart-King CA. Energetic regulation of coordinated leader–follower dynamics during collective invasion of breast cancer cells. Proc Natl Acad Sci USA 2019;116:7867–72. https://doi.org/10.1073/pnas.1809964116.

Zhou SL, Yin D, Hu ZQ, Luo C Bin, Zhou ZJ, Xin HY, et al. A positive feedback loop between cancer stem-like cells and tumor-associated neutrophils controls hepatocellular carcinoma progression. Hepatology 2019;70:1214–30. https://doi.org/10.1002/HEP.30630.

6 Polygenic Evolution of Germline Variants in Cancer

Ujani Hazra and Joseph Lachance

6.1 INTRODUCTION

Cancer risks are influenced by a combination of inherited predisposition (Murff et al. 2004) and acquired factors such as exposure to carcinogens and old age (White et al. 2014). Over the years, familial and twin studies have established that many types of cancer have a significant heritable component (Mucci et al. 2016; Dai et al. 2017; Rashkin et al. 2020), with the risk of developing cancer increasing substantially when there is a family history of the disease (Goldgar et al. 1994; Ahlbom et al. 1997; Antoniou et al. 2003; Hemminki and Li 2004). Consistent with the appreciable heritability of many cancers, highly penetrant pathogenic germline variants in genes such as *BRCA1* and *BRCA2* have been associated with breast and prostate cancer (Levy-Lahad et al. 1997; Peto et al. 1999; Heramb et al. 2018; Messina et al. 2020), while germline variants in *EGFR* have been associated with non-small cell lung cancer (Bell et al. 2005). Similarly, Li Fraumeni syndrome, which results in the development of many tumors during childhood, is due to germline mutations in *TP53* (Li et al. 1988; Nichols et al. 2001; Guha and Malkin 2017). While candidate gene approaches have successfully mapped genes and pathways related to cancers, many of these variants are rare in populations, and they only account for a small fraction of the genetic predisposition to cancer (Easton 1999). Most cancers are not due to high penetrance germline variants (Zhang et al. 2015). Instead, there is a growing consensus that most cancers are highly polygenic (Zhang et al. 2020; Hassanin et al. 2022).

The emergence of high-throughput genotyping technologies and increased computational power has enabled genome-wide association studies (GWAS) to identify common germline variants associated with cancer (Escala-Garcia et al. 2019). Fueled by the common disease common variants hypothesis that states that when a heritable disease is common in the population, the genetic variants associated with it are also expected to be common in the population (Schork et al. 2009), GWASs have identified large numbers of germline variants that contribute to heritable disease risks (Uffelmann et al. 2021). To date, the NHGRI-EBI GWAS Catalog contains over 20,000 germline variants associated with various cancers (Sollis et al. 2023). These single nucleotide polymorphisms (SNPs) can be leveraged to generate polygenic risk scores (PRS), which quantify an individual's risk of developing cancer or other diseases (Kachuri et al. 2020; Mars et al. 2020; Patel et al. 2023).

This chapter explores the evolutionary genetics of germline variants that contribute to common cancers. It contains novel analyses that are motivated by the following questions: How much evolutionary constraint have cancer-associated loci experienced during recent history? Have these variants undergone recent polygenic selection, or is generic drift causing populations to differ in their cancer risk? First, we highlight examples of natural selection acting on complex traits. Next, we describe the population genetic properties of cancer-associated alleles. We then investigate signatures of selection acting on multiple spatial and temporal scales by examining sets of SNPs that are associated with lung, prostate, breast, ovarian, and colorectal cancer. Finally, we conclude with a discussion of why cancer-associated alleles are unlikely to be under strong natural selection.

DOI: 10.1201/9781003307921-6

6.2 SELECTION ACTING ON COMPLEX TRAITS

Phenotypic differences among human populations have been the subject of extensive study, with multiple pieces of evidence pointing to differential adaptation in various human traits. Several instances of adaptation can be found in genes associated with resistance to malaria and sickle cell anemia, such as the Duffy antigen protein (*DARC*) and glucose-6-phosphate dehydrogenase (*G6PD*) (de Carvalho and de Carvalho 2011). In this case, the signature of natural selection closely aligns with the prevalence and incidence of sickle cell anemia. Similar effects of natural selection have also seen in the case of the *LCT* gene, where populations of milk drinkers have evolved adaptive phenotypes of lactose tolerance (Gerbault et al. 2011).

Moving beyond candidate gene analyses, researchers have utilized SNP data to identify signals of ongoing or recently completed selective sweeps (Tang et al. 2007; Coop et al. 2009). These studies have revealed consistent patterns, including concentrated signals around genes rather than intergenic regions (Pritchard et al. 2010). Additional indicators of selection include large allele frequency differences between populations and reduced genetic diversity at nearby sites (Cai et al. 2009; McVicker et al. 2009). Along these lines, gene-rich regions of low recombination exhibit reduced genetic diversity (Barreiro et al. 2008). This observation can be explained by either positive selection or background selection against mildly deleterious alleles (Zeng et al. 2018; O'Connor et al. 2019). Despite recent advances in methods to detect signatures of natural selection (Pavlidis and Alachiotis 2017; Racimo et al. 2018; Kirsch-Gerweck et al. 2023), sets of cancer-associated SNPs have yet to be studied comprehensively from a polygenic perspective.

With this in mind, we note that cancer prevalence rates vary greatly across the globe (Sung et al. 2021). For example, men of African descent have much higher risks for prostate cancer than men of European descent (Rebbeck et al. 2013). Similarly, esophageal cancer affects Asian populations unequally compared to other populations (Zhang et al. 2012). Are differences in prevalence and mortality rates of cancer among human populations driven by natural selection?

6.3 SETS OF CANCER-ASSOCIATED ALLELES

Results from GWASs have now established that cancers are complex polygenic diseases. This means genetic predispositions to developing cancer are influenced by multiple germline variants with low effect sizes and/or low-penetrance mutations (Zhang et al. 2020). The prevailing notion is that these variants operate additively, where the cumulative effect sizes of the germline variants determine an individual's genetic predisposition to cancer (Polderman et al. 2015). This principle forms the basis for PRSs, which leverage findings from GWAS to determine an individual's predisposition to cancer by considering the additive effects of germline variants and accounting for other covariates, such as population structure (Lewis and Vassos 2020). While PRSs hold promise for future genetic risk determination and personalized medicine, their predictions are imperfect (Kim et al. 2022). Potential causes of subpar PRS performance include epistasis (Eichler et al. 2010), the limited ability of GWAS to detect rare variants of small effect (Yengo et al. 2022), and genotype-by-environment interactions (Araujo and Wheeler 2022). That said, PRS variants still capture our best understanding of the genetics of many complex diseases. The existence of sets of trait-associated SNPs offers a unique opportunity to test whether germline variants that are associated with cancer are governed more by natural selection or neutral evolution (Berg and Coop 2014).

6.3.1 GENOMIC DATASETS ANALYZED HERE

In this chapter we leverage findings from five different GWASs conducted in European populations, specifically lung, breast, prostate, colorectal, and gastric cancer (Table 6.1). The NHGRI-EBI

TABLE 6.1

Cancer Types Analyzed in This Chapter

Cancer	GWAS citation	# of SNPs analyzed
Breast cancer	(Mavaddat et al. 2019)	250
Colorectal cancer	(Law et al. 2019)	79
Lung cancer	(McKay et al. 2017)	552
Ovarian cancer	(Ahmed et al. 2022)	107
Prostate cancer	(Conti et al. 2021)	204

Note: Disease-associated germline variants are from corresponding genome-wide association studies. The third column lists the number of independent SNPs that were analyzed in each test of selection after applying a GWAS *p*-value filter of 10–5.

GWAS Catalog lists cancer-associated germline variants and their effect sizes and p-values (Sollis et al. 2023). After correcting for multiple tests, variants with a p-value of 5×10^{-8} or lower are considered to be statistically significant on a genome-wide scale (Pe'er et al. 2008). However, focusing on only the top trait-associated SNPs is known to reduce the effectiveness of polygenic scores (Choi et al. 2020). Because of this, our analyses focus on sets of independent (LD-pruned) SNPs with p-values < 5×10^{-5}. Our analyses focus on germline variants that are associated with breast (Mavaddat et al. 2019), colorectal (Law et al. 2019), lung (McKay et al. 2017), ovarian (Ahmed et al. 2022), and prostate (Conti et al. 2021) cancer.

The analyses described as follows utilize whole genome sequence data from Phase 3 of the 1000 Genomes Project (1000 Genomes Project et al. 2015). This dataset includes samples from five different continental super-populations: Europe, Africa, East Asia, South Asia, and the Americas. These super-populations can in turn be split into 26 different populations. Additional details about these populations can be found at https://www.internationalgenome.org. Each population in the 1000 Genomes Project has genotype data from approximately 90 to 100 healthy individuals.

6.3.2   EFFECT SIZES OF GERMLINE VARIANTS THAT ARE ASSOCIATED WITH CANCER

We first examine the effect sizes of cancer-associated variants. For each of the five types of cancer studied here we observe that most disease-associated alleles have small effect sizes (Figure 6.1a). Distributions with fatter tails indicate that some cancers have genetic architectures that includes a subset of loci with larger effect sizes. For example, rare alleles of large effect at 8q24.21 are known to contribute to prostate cancer risk (Matejcic et al. 2018). However, the general pattern that emerges is that each of these five cancers are highly polygenic diseases.

6.3.3   ALLELE FREQUENCIES OF GERMLINE VARIANTS THAT ARE ASSOCIATED WITH CANCER

Another crucial aspect to consider is how common disease-associated alleles are in different populations. Figure 6.1b–f shows the allele frequency distributions of cancer-associated variants in different continental super-populations. Broadly speaking, similar patterns are observed for breast, colorectal, lung, ovarian, and prostate cancer. In general, most cancer-associated alleles have intermediate allele frequencies. This trend is consistent with the limited statistical power of GWAS when alleles are rare. The lower minor allele frequencies observed for lung cancer (Figure 6.1d) are most likely due to relatively large sample sizes in the GWAS by McKay et al. 2017. We note that minor allele frequencies are elevated in Europe for the cancers shown in Figure 6.1b–f. This pattern is likely due to ascertainment bias, as individuals of European descent comprise the majority of GWAS study participants and thus have the ability to capture alleles that can be lost in smaller population studies as rare variants (Martin et al. 2017). Finally, we note that allele frequencies can vary across populations, either due to genetic drift or natural selection.

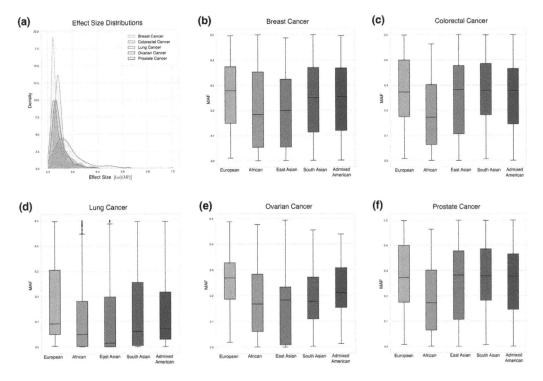

FIGURE 6.1 Genetic properties of cancer-associated variants. (a) Distributions of effect sizes of cancer-associated variants. Odds ratios were obtained from the GWASs described in Table 6.1. Most cancer-associated variants have small effect sizes. (b)–(f) Minor allele frequency (MAF) distributions of cancer-associated SNPs in the five super-populations from the 1000 Genomes Project: European, African, East Asian, South Asian, and Admixed Americans.

6.4 ARE CANCER-ASSOCIATED ALLELES UNDER STRONG SELECTION?

Following subsections, we analyze signals of natural selection of cancer-associated SNPs from a polygenic perspective. First, we test if sets of cancer-associated SNPs have undergone background selection on a species level. Second, we test for polygenic adaptation acting on a continental scale. Finally, we test for recent positive selection acting on a local (i.e., population) scale. Additional details about these tests of polygenic selection can be found elsewhere (Hazra and Lachance 2021).

6.4.1 TESTS OF BACKGROUND SELECTION

Background selection refers to reduced genetic diversity at a non-deleterious locus caused by negative selection against linked deleterious alleles (Charlesworth et al. 1993). This term emphasizes that a neutral mutation's genomic environment or genetic background significantly influences whether it will be preserved or eliminated from a population (McVicker et al. 2009). If cancers are subject to negative selection, then one would expect enrichment for signatures of background selection near disease-associated SNPs. B statistics can be used to measure the impact of background selection near individual genomic loci (Hudson and Kaplan 1995; Nordborg et al. 1996). These statistics quantify the expected fraction of neutral diversity present at a site. B values close to 0 indicate almost complete removal of diversity due to selection, and B values near 1 suggest minimal effect. Prior studies have revealed that a threshold B < 0.317 is indicative of SNPs that are under background selection.

We are interested in whether a given cancer has experienced background selection. Because of this, we adapted the B statistic framework to handle SNP sets. For each type of cancer, we obtained 1,000 control sets of SNPs that were matched to cancer-associated SNPs in terms of allele frequency, linkage disequilibrium, and distance to gene. As described elsewhere (Hazra and Lachance 2021), we computed the probability distributions of B statistics for each set of cancer-associated SNPs. We then identified the fraction of each distribution below 0.317 for each set of SNPs. Next, we obtained empirical percentile ranks for each disease by comparing each set of cancer-associated SNPs to their matched control sets of SNPs. Higher percentile ranks indicate that a set of cancer-associated SNPs have stronger signatures of background selection.

For each cancer studied here (breast, colorectal, lung, ovarian, and prostate), we find that disease-associated SNPs are enriched for signatures of background selection, i.e., they have a greater proportion of SNPs with B < 0.317 than matched control sets. This enrichment is strongest for breast, lung, and prostate cancer and weakest for ovarian cancer (Figure 6.2). Nevertheless, we find that each cancer has a percentile rank of at least 88% compared to null expectations (Torres et al. 2018). While 0.317 signifies the lowest 5 percentile of genome-wide B statistics values, our results are robust to a stricter threshold of 0.1. Taken together, our findings support the claim that polygenic diseases, including many cancers, are under weak purifying selection (O'Connor et al. 2019).

6.4.2 Tests of Polygenic Adaptation

Polygenic adaptation involves coordinated shifts in allele frequencies at numerous loci, each contributing to trait variation. Although individual allele frequency changes may be small, their collective impact on phenotype can be substantial, underscoring the significance of polygenic adaptation in shaping recent human evolution (Pritchard et al. 2010; Barghi et al. 2020). Allele frequencies at cancer-associated SNPs can differ among global populations. Are these differences due to polygenic adaptation or neutral evolution (including population bottlenecks and events like the out-of-Africa migration)?

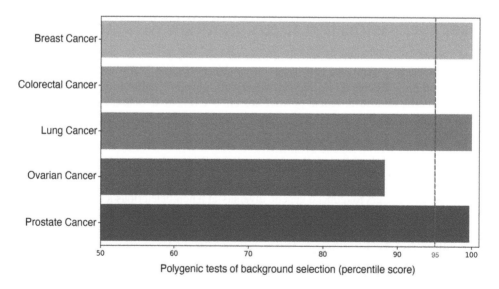

FIGURE 6.2 Tests of whether cancer-associated variants are enriched for signatures of background selection. Percentile ranks of cancer-associated SNPs were compared to matched control sets for signatures of background selection. The red dotted line corresponds to the 95th percentile of these control sets. In general, sets of cancer-associated SNPs have lower B statistics than matched sets of control SNPs.

Whether continental differences in cancer risk are due to polygenic adaptation, we first built a phylogenetic tree of the super-populations from the 1000 Genomes Project data using MixMapper (Lipson et al. 2013). This tree is technically an admixture graph as it allows for gene flow between different branches (Figure 6.3). As described elsewhere (Hazra and Lachance 2021), 1000 Genomes Project populations used to generate this demographic model include MSL, YRI, IBS, GBR, BEB, STU, CHB, JPT, and PEL (descriptions of each population are listed in the legend of Figure 6.3). We then used PolyGraph (Racimo et al. 2018) to detect for signatures of polygenic selection acting on sets of cancer-associated variants. Polygraph uses a Markov Chain Monte Carlo method that takes into account the ancestral and derived state of the trait-associated SNPs, their effect sizes, and allele frequencies in each population. Output from PolyGraph includes selection parameters for each branch of the tree, as well as a q-value summarizing whether a null-hypothesis of neutral evolution can be rejected.

Overall, we observe minimal signatures of polygenic adaptation for sets of SNPs that are associated with breast, colorectal, lung, ovarian, and prostate cancer (Figure 6.3). Despite this, we note that some branches have weak signatures of adaptation, most notably in branches leading to European populations in our analysis of lung cancer (Figure 6.3c). However, we note that none of the cancers analyzed here have q-values that pass a multiple testing threshold. This indicates that neutral evolution appears to be the primary cause of continental differences in allele frequencies at cancer-associated SNPs. Natural selection is not a major driver of germline differences in cancer risk across human populations.

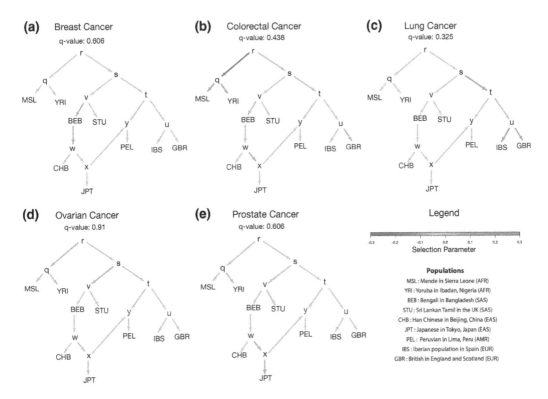

FIGURE 6.3 Tests of polygenic adaptation acting on germline variants that are associated with cancer risk. PolyGraph was used to test for signatures of selection (Racimo et al. 2018). Although weak signatures of polygenic adaptation are observed for some branches, a general pattern is that cancers have negligible signatures of polygenic adaptation (results do not pass the statistical significance threshold of q-value < 0.05 after multiple testing).

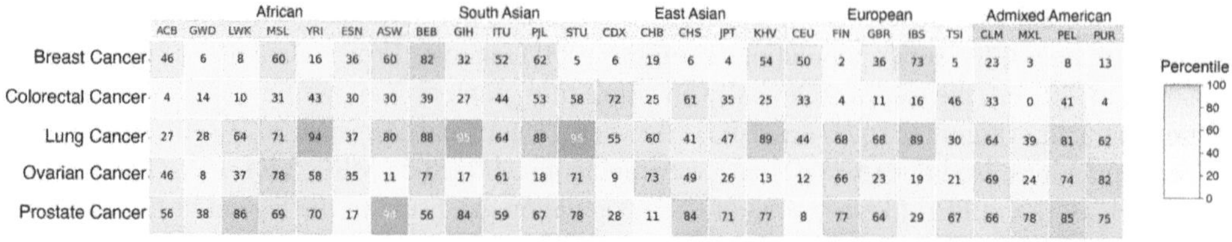

	African								South Asian					East Asian					European				Admixed American			
	ACB	GWD	LWK	MSL	YRI	ESN	ASW	BEB	GIH	ITU	PJL	STU	CDX	CHB	CHS	JPT	KHV	CEU	FIN	GBR	IBS	TSI	CLM	MXL	PEL	PUR
Breast Cancer	46	6	8	60	16	36	60	82	32	52	62	5	6	19	6	4	54	50	2	36	73	5	23	3	8	13
Colorectal Cancer	4	14	10	31	43	30	30	39	27	44	53	58	72	25	61	35	25	33	4	11	16	46	33	0	41	4
Lung Cancer	27	28	64	71	94	37	80	88	95	64	88	95	55	60	41	47	89	44	68	68	89	30	64	39	81	62
Ovarian Cancer	46	8	37	78	58	35	11	77	17	61	18	71	9	73	49	26	13	12	66	23	19	21	69	24	74	82
Prostate Cancer	56	38	86	69	70	17	94	56	84	59	67	78	28	11	84	71	77	8	77	64	29	67	66	78	85	75

Percentile: 100, 80, 60, 40, 20, 0

FIGURE 6.4 Tests of recent positive selection using iHS statistics. Percentile ranks of cancer-associated SNPs compared to matched control sets for signatures of recent positive selection are shown for 26 populations from the 1000 Genomes Project. Higher values indicate greater signatures of selection. The overall trend is that sets of cancer-associated variants are not enriched for recent positive selection.

6.4.3 Tests of Recent Positive Selection

In this section, we test for recent signatures of positive selection acting on cancer-associated SNPs using iHS (integrated Haplotype Score) statistics (Voight et al. 2006). These statistics identify alleles that are associated with extended haplotype homozygosity. One useful feature of iHS statistics is that they can detect partial selective sweeps, where favored alleles have risen in frequency but not yet reached fixation. Standardized iHS statistics use genome-wide empirical distributions to identify whether a given SNP is an outlier or not, i.e., they are z-scores. Accordingly, extreme negative or positive values of iHS scores are treated as putative indicators of recent positive selection (Voight et al. 2006; Johnson and Voight 2018).

We are interested in whether a given cancer has experienced recent positive selection. Because of this, we adapted the iHS framework to handle SNP sets. Like the tests of background selection described above, we obtained 1,000 sets of matched control SNPs for breast, colorectal, lung, ovarian, and prostate cancer. We then tested whether cancer-associated variants were enriched in outlier values of iHS statistics (iHS < −1.96 or > 1.96) compared to control SNPs (Hazra and Lachance 2021). Percentile ranks were generated for each combination of cancer type and 1000 Genomes Project population. Higher percentiles are indicative of sets of cancer-associated SNPs (as a whole) that are enriched for higher absolute values of iHS and hence recent positive selection, when compared to controls.

Figure 6.4 shows the percentile rank of each cancer in 26 global populations from the 1000 Genomes Project. We see that most cancers have low percentile values in all 26 populations. This indicates that there are minimal signatures of recent positive selection. Interestingly, lung cancer-associated SNPs are somewhat enriched for signatures of recent positive selection in two South Asian (GIH and STU) and one African (YRI) population. Prostate cancer-associated SNPs also exhibit weak enrichment for high iHS statistics in African Americans (ASW). However, we note that none of these percentile scores in Figure 6.4 are high enough to survive a Bonferroni correction. Because of this, any outliers should be treated with caution—especially since the general trend is that cancer-associated variants are not enriched for high iHS statistics.

6.5 DISCUSSION

Genome-wide association studies have established that cancers are polygenic diseases with multiple loci contributing to the disease with small effects. In this chapter, we studied the evolutionary dynamics of these germline variants associated with five common types of cancer (breast, colorectal, lung, ovarian, and prostate) across global populations. We observed that germline variants associated with cancer have undergone significant background or purifying selection when compared at a species level. Next, although cancer risks vary significantly among global populations, we did not observe significant signatures of polygenic adaptation. Similarly, tests of recent selection on germline variants were largely negative. These findings indicate that the inequities in cancer risk and prevalence are likely driven by neutral evolution and genetic drift, along with socioeconomic risk factors and lifestyle choices.

The observation that cancer-associated loci exhibit signatures of background selection raises intriguing questions about the underlying mechanisms. The prevalence of conserved cellular functions among many genes associated with cancer suggests that tumor cells often co-opt existing biological pathways for their advantage. This co-option may be one reason why we observe reduced genetic diversity and signatures of background selection near cancer-associated loci (Zapata et al. 2018).

Multiple factors may explain why minimal signatures of polygenic adaptation were observed for cancer-associated loci. First, the late age of onset for many cancers likely diminishes the strength of selection acting on these variants. Second, the polygenic nature of cancer, characterized by the cumulative effects of multiple low-risk variants, contributes to the observed lack of strong signatures of adaptation. Each variant makes only a modest contribution to disease risk, resulting in weak signatures of selection that may not be readily detectable. Finally, the decoupling of selection coefficients and disease risks further contributes to the absence of pronounced signals of adaptation (Agarwala et al. 2013). Even if a variant significantly impacts disease risk, its effect on fitness may be relatively minor due to factors such as incomplete penetrance.

Two potential explanations for large iHS scores warrant consideration. The first is pleiotropy, whereby genetic variants may influence multiple traits, some of which might be under selective pressures. Although it is beyond the scope for this chapter, one future research avenue involves studying the pleiotropic effects of germline variants that are associated with cancer risk. Second, genetic hitchhiking and linkage to advantageous alleles that are associated with other traits could potentially elevate iHS values for some cancer-associated SNPs.

In conclusion, this chapter comprehensively analyzes natural selection acting on cancer-associated germline variants and highlights the intricate dynamics that shape cancer risk across human populations. While background selection appears to shape the genetic landscape of cancer-associated loci, negligible signatures of polygenic adaptation emphasize the complex interplay between disease onset, effect sizes, and selective pressures. These findings underscore the need for nuanced approaches when considering the evolutionary forces influencing complex traits like cancer. As genetic datasets continue to grow in size and diversity, more sophisticated methods and refined analyses could provide deeper insights into the evolutionary dynamics of cancer-associated variants.

Note added in proof: We note that the exact q-values and percentile ranks arising from polygenic tests of selection depend on the specific sets of cancer-associated SNPs that are analyzed. Analyses in this chapter used a LD pruning threshold of $r^2 < 0.8$ to identify independent SNPs. Other thresholds would yield different summary statistics. However, our overarching conclusions are expected to be robust to how independent sets of cancer-associated SNPs are identified.

ACKNOWLEDGMENTS

We would like to thank Norman A. Johnson, Jason A. Somarelli, and members of the Center for Integrative Genomics at Georgia Institute of Technology for their assistance in writing this manuscript.

REFERENCES

1000 Genomes Project, A. Auton, L. D. Brooks, R. M. Durbin, E. P. Garrison et al., 2015 A global reference for human genetic variation. Nature 526: 68–74.

Agarwala, V., J. Flannick, S. Sunyaev, T. D. C. Go and D. Altshuler, 2013 Evaluating empirical bounds on complex disease genetic architecture. Nat Genet 45: 1418–1427.

Ahlbom, A., P. Lichtenstein, H. Malmstrom, M. Feychting, K. Hemminki et al., 1997 Cancer in twins: genetic and nongenetic familial risk factors. J Natl Cancer Inst 89: 287–293.

Ahmed, M., V. P. Makinen, A. Mulugeta, J. Shin, T. Boyle et al., 2022 Considering hormone-sensitive cancers as a single disease in the UK biobank reveals shared aetiology. Commun Biol 5: 614.

Antoniou, A., P. D. Pharoah, S. Narod, H. A. Risch, J. E. Eyfjord et al., 2003 Average risks of breast and ovarian cancer associated with BRCA1 or BRCA2 mutations detected in case Series unselected for family history: a combined analysis of 22 studies. Am J Hum Genet 72: 1117–1130.

Araujo, D. S. and H. E. Wheeler, 2022 Genetic and environmental variation impact transferability of polygenic risk scores. Cell Rep Med 3: 100687.

Barghi, N., J. Hermisson and C. Schlotterer, 2020 Author correction: polygenic adaptation: a unifying framework to understand positive selection. Nat Rev Genet 21: 782.

Barreiro, L. B., G. Laval, H. Quach, E. Patin and L. Quintana-Murci, 2008 Natural selection has driven population differentiation in modern humans. Nat Genet 40: 340–345.

Bell, D. W., I. Gore, R. A. Okimoto, N. Godin-Heymann, R. Sordella et al., 2005 Inherited susceptibility to lung cancer may be associated with the T790M drug resistance mutation in EGFR. Nat Genet 37: 1315–1316.

Berg, J. J. and G. Coop, 2014 A population genetic signal of polygenic adaptation. PLoS Genet 10: e1004412.

Cai, J. J., J. M. Macpherson, G. Sella and D. A. Petrov, 2009 Pervasive hitchhiking at coding and regulatory sites in humans. PLoS Genet 5: e1000336.

Charlesworth, B., M. T. Morgan and D. Charlesworth, 1993 The effect of deleterious mutations on neutral molecular variation. Genetics 134: 1289–1303.

Choi, S. W., T. S. Mak and P. F. O'Reilly, 2020 Tutorial: a guide to performing polygenic risk score analyses. Nat Protoc 15: 2759–2772.

Conti, D. V., B. F. Darst, L. C. Moss, E. J. Saunders, X. Sheng et al., 2021 Trans-ancestry genome-wide association meta-analysis of prostate cancer identifies new susceptibility loci and informs genetic risk prediction. Nat Genet 53: 65–75.

Coop, G., J. K. Pickrell, J. Novembre, S. Kudaravalli, J. Li et al., 2009 The role of geography in human adaptation. PLoS Genet 5: e1000500.

Dai, J., W. Shen, W. Wen, J. Chang, T. Wang et al., 2017 Estimation of heritability for nine common cancers using data from genome-wide association studies in Chinese population. Int J Cancer 140: 329–336.

de Carvalho, G. B., and G. B. de Carvalho, 2011 Duffy blood group system and the malaria adaptation process in humans. Rev Bras Hematol Hemoter 33: 55–64.

Easton, D. F., 1999 How many more breast cancer predisposition genes are there? Breast Cancer Res 1: 14–17.

Eichler, E. E., J. Flint, G. Gibson, A. Kong, S. M. Leal et al., 2010 Missing heritability and strategies for finding the underlying causes of complex disease. Nat Rev Genet 11: 446–450.

Escala-Garcia, M., Q. Guo, T. Dork, S. Canisius, R. Keeman et al., 2019 Genome-wide association study of germline variants and breast cancer-specific mortality. Br J Cancer 120: 647–657.

Gerbault, P., A. Liebert, Y. Itan, A. Powell, M. Currat et al., 2011 Evolution of lactase persistence: an example of human niche construction. Philos Trans R Soc Lond B Biol Sci 366: 863–877.

Goldgar, D. E., D. F. Easton, L. A. Cannon-Albright and M. H. Skolnick, 1994 Systematic population-based assessment of cancer risk in first-degree relatives of cancer probands. J Natl Cancer Inst 86: 1600–1608.

Guha, T. and D. Malkin, 2017 Inherited TP53 mutations and the Li-Fraumeni syndrome. Cold Spring Harb Perspect Med 7.

Hassanin, E., P. May, R. Aldisi, I. Spier, A. J. Forstner et al., 2022 Breast and prostate cancer risk: the interplay of polygenic risk, rare pathogenic germline variants, and family history. Genet Med 24: 576–585.

Hazra, U. and J. Lachance, 2021 Polygenic adaptation is not a major driver of disparities in disease mortality across global populations. medRxiv 2021: 2012. https://doi.org/10.1101/2021.12.10.21267630.

Hemminki, K. and X. Li, 2004 Familial risk in testicular cancer as a clue to a heritable and environmental aetiology. Br J Cancer 90: 1765–1770.

Heramb, C., T. Wangensteen, E. M. Grindedal, S. L. Ariansen, S. Lothe et al., 2018 BRCA1 and BRCA2 mutation spectrum—an update on mutation distribution in a large cancer genetics clinic in Norway. Hered Cancer Clin Pract 16: 3.

Hudson, R. R. and N. L. Kaplan, 1995 Deleterious background selection with recombination. Genetics 141: 1605–1617.

Johnson, K. E. and B. F. Voight, 2018 Patterns of shared signatures of recent positive selection across human populations. Nat Ecol Evol 2: 713–720.

Kachuri, L., R. E. Graff, K. Smith-Byrne, T. J. Meyers, S. R. Rashkin et al., 2020 Pan-cancer analysis demonstrates that integrating polygenic risk scores with modifiable risk factors improves risk prediction. Nat Commun 11: 6084.

Kim, M. S., D. Naidoo, U. Hazra, M. H. Quiver, W. C. Chen et al., 2022 Testing the generalizability of ancestry-specific polygenic risk scores to predict prostate cancer in sub-Saharan Africa. Genome Biol 23: 194.

Kirsch-Gerweck, B., L. Bohnenkamper, M. T. Henrichs, J. N. Alanko, H. Bannai et al., 2023 HaploBlocks: efficient detection of positive selection in large population genomic datasets. Mol Biol Evol 40.

Law, P. J., M. Timofeeva, C. Fernandez-Rozadilla, P. Broderick, J. Studd et al., 2019 Association analyses identify 31 new risk loci for colorectal cancer susceptibility. Nat Commun 10: 2154.

Levy-Lahad, E., R. Catane, S. Eisenberg, B. Kaufman, G. Hornreich et al., 1997 Founder BRCA1 and BRCA2 mutations in Ashkenazi Jews in Israel: frequency and differential penetrance in ovarian cancer and in breast-ovarian cancer families. Am J Hum Genet 60: 1059–1067.

Lewis, C. M. and E. Vassos, 2020 Polygenic risk scores: from research tools to clinical instruments. Genome Med 12: 44.

Li, F. P., J. F. Fraumeni, Jr., J. J. Mulvihill, W. A. Blattner, M. G. Dreyfus et al., 1988 A cancer family syndrome in twenty-four kindreds. Cancer Res 48: 5358–5362.

Lipson, M., P. R. Loh, A. Levin, D. Reich, N. Patterson et al., 2013 Efficient moment-based inference of admixture parameters and sources of gene flow. Mol Biol Evol 30: 1788–1802.

Mars, N., E. Widen, S. Kerminen, T. Meretoja, M. Pirinen et al., 2020 The role of polygenic risk and susceptibility genes in breast cancer over the course of life. Nat Commun 11: 6383.

Martin, A. R., C. R. Gignoux, R. K. Walters, G. L. Wojcik, B. M. Neale et al., 2017 Human demographic history impacts genetic risk prediction across diverse populations. Am J Hum Genet 100: 635–649.

Matejcic, M., E. J. Saunders, T. Dadaev, M. N. Brook, K. Wang et al., 2018 Germline variation at 8q24 and prostate cancer risk in men of European ancestry. Nat Commun 9: 4616.

Mavaddat, N., K. Michailidou, J. Dennis, M. Lush, L. Fachal et al., 2019 Polygenic risk scores for prediction of breast cancer and breast cancer subtypes. Am J Hum Genet 104: 21–34.

McKay, J. D., R. J. Hung, Y. Han, X. Zong, R. Carreras-Torres et al., 2017 Large-scale association analysis identifies new lung cancer susceptibility loci and heterogeneity in genetic susceptibility across histological subtypes. Nat Genet 49: 1126–1132.

McVicker, G., D. Gordon, C. Davis and P. Green, 2009 Widespread genomic signatures of natural selection in hominid evolution. PLoS Genet 5: e1000471.

Messina, C., C. Cattrini, D. Soldato, G. Vallome, O. Caffo et al., 2020 BRCA mutations in prostate cancer: prognostic and predictive implications. J Oncol 2020: 4986365.

Mucci, L. A., J. B. Hjelmborg, J. R. Harris, K. Czene, D. J. Havelick et al., 2016 Familial risk and heritability of cancer among twins in Nordic countries. JAMA 315: 68–76.

Murff, H. J., D. R. Spigel and S. Syngal, 2004 Does this patient have a family history of cancer? An evidence-based analysis of the accuracy of family cancer history. JAMA 292: 1480–1489.

Nichols, K. E., D. Malkin, J. E. Garber, J. F. Fraumeni, Jr. and F. P. Li, 2001 Germ-line p53 mutations predispose to a wide spectrum of early-onset cancers. Cancer Epidemiol Biomarkers Prev 10: 83–87.

Nordborg, M., B. Charlesworth and D. Charlesworth, 1996 The effect of recombination on background selection. Genet Res 67: 159–174.

O'Connor, L. J., A. P. Schoech, F. Hormozdiari, S. Gazal, N. Patterson et al., 2019 Extreme polygenicity of complex traits is explained by negative selection. Am J Hum Genet 105: 456–476.

Patel, A. P., M. Wang, Y. Ruan, S. Koyama, S. L. Clarke et al., 2023 A multi-ancestry polygenic risk score improves risk prediction for coronary artery disease. Nat Med 29: 1793–1803.

Pavlidis, P. and N. Alachiotis, 2017 A survey of methods and tools to detect recent and strong positive selection. J Biol Res (Thessalon) 24: 7.

Pe'er, I., R. Yelensky, D. Altshuler and M. J. Daly, 2008 Estimation of the multiple testing burden for genome-wide association studies of nearly all common variants. Genet Epidemiol 32: 381–385.

Peto, J., N. Collins, R. Barfoot, S. Seal, W. Warren et al., 1999 Prevalence of BRCA1 and BRCA2 gene mutations in patients with early-onset breast cancer. J Natl Cancer Inst 91: 943–949.

Polderman, T. J., B. Benyamin, C. A. de Leeuw, P. F. Sullivan, A. van Bochoven et al., 2015 Meta-analysis of the heritability of human traits based on fifty years of twin studies. Nat Genet 47: 702–709.

Pritchard, J. K., J. K. Pickrell and G. Coop, 2010 The genetics of human adaptation: hard sweeps, soft sweeps, and polygenic adaptation. Curr Biol 20: R208–R215.

Racimo, F., J. J. Berg and J. K. Pickrell, 2018 Detecting polygenic adaptation in admixture graphs. Genetics 208: 1565–1584.

Rashkin, S. R., R. E. Graff, L. Kachuri, K. K. Thai, S. E. Alexeeff et al., 2020 Pan-cancer study detects genetic risk variants and shared genetic basis in two large cohorts. Nat Commun 11: 4423.

Rebbeck, T. R., S. S. Devesa, B. L. Chang, C. H. Bunker, I. Cheng et al., 2013 Global patterns of prostate cancer incidence, aggressiveness, and mortality in men of African descent. Prostate Cancer 2013: 560857.

Schork, N. J., S. S. Murray, K. A. Frazer, E. J. Topol, 2009 Common vs. rare allele hypotheses for complex diseases. Curr Opin Genet Dev 19: 212–219.

Sollis, E., A. Mosaku, A. Abid, A. Buniello, M. Cerezo et al., 2023 The NHGRI-EBI GWAS Catalog: knowledgebase and deposition resource. Nucleic Acids Res 51: D977–D985.

Sung, H., J. Ferlay, R. L. Siegel, M. Laversanne, I. Soerjomataram et al., 2021 Global cancer statistics 2020: GLOBOCAN estimates of incidence and mortality worldwide for 36 cancers in 185 countries. CA Cancer J Clin 71: 209–249.

Tang, K., K. R. Thornton and M. Stoneking, 2007 A new approach for using genome scans to detect recent positive selection in the human genome. PLoS Biol 5: e171.

Torres, R., Z. A. Szpiech and R. D. Hernandez, 2018 Human demographic history has amplified the effects of background selection across the genome. PLoS Genet 14: e1007387.

Uffelmann, E., Q. Q. Huang, N. S. Munung, J. de Vries, Y. Okada et al., 2021 Genome-wide association studies. Nat Rev Methods Primers 1: 59.

Voight, B. F., S. Kudaravalli, X. Wen and J. K. Pritchard, 2006 A map of recent positive selection in the human genome. PLoS Biol 4: e72.

White, M. C., D. M. Holman, J. E. Boehm, L. A. Peipins, M. Grossman et al., 2014 Age and cancer risk: a potentially modifiable relationship. Am J Prev Med 46: S7–S15.

Yengo, L., S. Vedantam, E. Marouli, J. Sidorenko, E. Bartell et al., 2022 A saturated map of common genetic variants associated with human height. Nature 610: 704–712.

Zapata, L., O. Pich, L. Serrano, F. A. Kondrashov, S. Ossowski et al., 2018 Negative selection in tumor genome evolution acts on essential cellular functions and the immunopeptidome. Genome Biol 19: 67.

Zeng, J., R. de Vlaming, Y. Wu, M. R. Robinson, L. R. Lloyd-Jones et al., 2018 Signatures of negative selection in the genetic architecture of human complex traits. Nat Genet 50: 746–753.

Zhang, H. Z., G. F. Jin and H. B. Shen, 2012 Epidemiologic differences in esophageal cancer between Asian and Western populations. Chin J Cancer 31: 281–286.

Zhang, J., M. F. Walsh, G. Wu, M. N. Edmonson, T. A. Gruber et al., 2015 Germline mutations in predisposition genes in pediatric cancer. N Engl J Med 373: 2336–2346.

Zhang, Y. D., A. N. Hurson, H. Zhang, P. P. Choudhury, D. F. Easton et al., 2020 Assessment of polygenic architecture and risk prediction based on common variants across fourteen cancers. Nat Commun 11: 3353.

7 Two-Phased Cancer Evolution
The Pattern and Scale of Genomic and Non-Genomic Landscapes

Andrzej Kasperski and Henry H. Heng

7.1 INTRODUCTION

After waging war on cancer for over 50 years using various cutting-edge molecular platforms, cancer researchers are now starting to embrace a new frontier: cancer evolution (Heng and Heng, 2022a). Although the concept of cancer evolution is not new—it was formally proposed in the 1960s and 1970s (Cairns, 1975; Levan, 1967; Nowell, 1976)—it was largely overlooked until the Cancer Genome Project generated vast amounts of diverse data. Researchers then realized that the promised patterns of cancer gene mutations specific to cancer types and stages were not being delivered. As some leading molecular researchers discuss the theoretical challenges and potential for a new framework (Hayden, 2010; Levine, 2019; Weinberg, 2014), we believe that cancer evolution holds the key. Following decades of efforts to promote the importance of cancer evolution (Breivik, 2005; Crespi and Summers, 2005; Gatenby, 1991; Greaves, 2007; Heng et al, 2006c; Jablonka, 2006; Maley and Reid, 2005; Merlo et al, 2006; Shibata, 1997; Vincent and Gatenby, 2008; Vineis and Berwick, 2006; Weiss, 2006), the recent establishment of a cancer evolution working group in AACR (The American Association for Cancer Research) reflects a transformation in attitude among cancer researchers, who now consider cancer as an issue of evolution. With over 3,500 members in this working group in just over the first year of its existence, there is a high expectation of applying evolutionary principles to fight cancer. Many questions are discussed, including what is cancer evolution, what are the key similarities and differences between cancer and organismal evolution, and how can evolutionary principles guide essential cancer research and its clinical implications?

7.1.1 THE CONCEPT OF CANCER EVOLUTION

Cancer evolution refers to the dynamic process by which populations of proliferating cells are subject to natural selection. Fundamentally, cancers—which often display altered genomes—can be considered as emergent new cellular systems that derive from normal cells. During various physiological and pathological conditions, cells need to adapt to cellular stresses for both survival and function. Cellular adaptation can be achieved through processes such as energy consumption, changes in gene expression, modification of epigenetic status, alterations in cell signaling pathways, gene mutations, and reorganization of genomes. While altered genomic structures at different scales are vital for cellular adaptation under high level stresses, as an unavoidable trade-off, these alterations (from cellular to immune systems), some of which are reversible, can alter the cellular evolutionary landscape and play an essential role in cancer evolution (Heng, 2017, 2019; Horne et al, 2014, 2015; Kultz, 2005; Mojica and Kültz, 2022; Stevens et al, 2011). Cancer, manifested as a set of cancerous cells displaying population dynamics, fits all essential elements or conditions of Darwinian evolution, including inherited variations and accumulation of genomic changes; advantageous phenotypes, such as increased fitness following competition for nutrition, space, growth, and population dominance reflected by clonal expansion; as well as tissue and organ microenvironments. Therefore, it is logical to consider cancer formation as an evolutionary process, albeit at the

DOI: 10.1201/9781003307921-7

somatic level (Greaves and Maley, 2012; Heng, 2007a; Heng et al, 2011b; Kasperski and Kasperska, 2018; Merlo et al, 2006; Noble, 2021; Somarelli et al, 2020, 2022). If we accept the concept of cancer evolution, studying cancer seems straightforward: we just need to apply the already-known evolutionary principles to cancer studies. Among the many principles of neo-Darwinism, two concepts are fundamental: 1) microevolution is achieved by gene frequency changes within a population, and 2) the accumulation of small microevolution (small changes) leads to macroevolution (big changes) over a long period of time. Under these considerations, Peter Nowell proposed that the accumulation of mutations followed by stepwise selection, leading to the domination of different clonal expansions, represents the key process of cancer evolution (Nowell, 1976). These same principles (gene-based gradualism) are still commonly used in today's cancer evolutionary studies. For example, guided by these principles, cancer evolution has been classified into different types, mainly based on mutational profiles, to cover the dynamic process: linear, convergent, branched, and parallel (Heng, 2017; Vendramin et al, 2021; Venkatesan and Swanton, 2016). Despite the increasing number of publications that apply evolutionary analyses, the results are less certain, and clinical predictability remains low. The overwhelming heterogeneity discovered across different genomic and non-genomic scales challenges the stepwise evolutionary model and even raises questions about whether cancer actually follows typical Darwinian evolution (Comaills and Castellano-Pozo, 2023; Gerlinger et al, 2014; Heng, 2019, 2015, 2007a; Heng and Heng, 2022b, 2021a; Ling et al, 2015; Shapiro and Noble, 2021; Vendramin et al, 2021).

7.1.2 The Discovery of Genome-Based Evolution: Two-Phased Cancer Evolution and Its Validation

Cancer is often referred to as a "cellular growth problem" that leads to "out of control growth". In the literature, this process has often been believed by other researchers to be a microevolutionary process, where the accumulation of a few common cancer gene mutations is the key mechanism (Fearon and Vogelstein, 1990; Heng, 2015, 2007a). Therefore, identifying the pattern of stepwise evolution involves overgrowth, within which early genetic events hold the answer for cancer diagnosis and treatment. However, from the emergent framework, cancer is fundamentally an issue of macroevolution where new systems emerge from normal tissue by overcoming various constraints (from normal cell cycle, cellular regulation, and tissue architecture all the way to multiple systems homeostasis). To study this complex evolutionary process, we use experimental platforms to "watch evolution in action" by comparing the changes in the genetic makeup, gene expression, and physical characteristics of cells longitudinally before, during, and after key phase transitions, including immortalization, transformation, metastasis, and drug resistance.

Briefly, the profiles of karyotype, transcriptome, genome instability, cellular growth/death, and cell population dynamics, including the emergence of new populations under crisis, have been compared by focusing on each phase transition. These years-long experiments led to the following conclusions:

a) The evolutionary pattern of the karyotype landscape differs from that of gene mutation landscapes by dominating different phases of evolution.

b) DNA clones ≠ Karyotype clones. Karyotype represents a system-level of inheritance, while genes represent a parts-level of inheritance.

c) Karyotype reorganization through genome chaos is an important platform for creating new system-level information, while gene mutations contribute to information modification. Different phases of evolution depend on different types of information management. Punctuated Macroevolution ≠ Microevolution + Time.

d) The vast majority of genome-level changes are nonclonal (Heng et al, 2006a, 2006b, 2006c; Zhang et al, 2014), and the clonal aberrations represent only the tip of the iceberg, symbolizing the successful outcomes of macro- and microevolutionary selection. This process is highly dynamic and difficult to duplicate even under similar experimental conditions. The nonclonal nature of multiple levels of genomic and non-genomic variations has led to the concept of "fuzzy inheritance" which serves as the basis for cancer heterogeneity (Heng et al, 2013, 2019; Heppner, 1984; Heppner and Miller, 1989; Ye et al, 2018).

e) Phase transitions are driven by system destabilizations that lead to genome chaos and emergence of new stable systems. When new high-stress conditions occur, they can initiate a new phase transition.

f) Macro- and microevolution can be described by karyotype and gene mutation, respectively.

The entire evolutionary process should be divided into punctuated macroevolution and stepwise microevolution. Significantly, both genotypes and phenotypes display two-phased evolution. Figure 7.1 illustrates the cycle of two-phased evolution.

Significantly, the two-phased evolutionary process is validated by sequencing a large number of cancer samples. Initially led by the example of the "big bang" phenomenon in colon cancer (Sottoriva et al, 2015), many different types of cancers, including breast, prostate, and pancreatic cancers, display similar patterns. More interestingly, the similar pattern of two-phased evolution

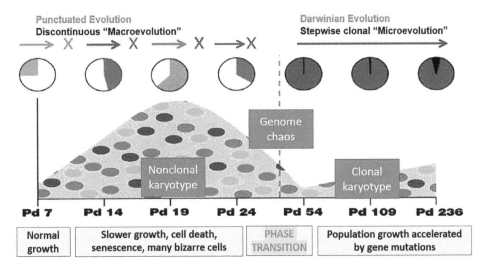

FIGURE 7.1 The model of two-phased cancer evolution illustrates the relationship between phenotype, genotype (karyotype), and evolutionary phases during cellular immortalization. The basic pattern is also observed in other phase transition models, such as metastasis and drug resistance. The x-axis represents the passage doubling (pd) numbers of the continuous cell culture, phenotypes, and the nonclonal and clonal statuses of karyotypes. The colors within each pie diagram (green, purple, orange, blue, and red) represent the fraction of the total population that exhibits a clonal pattern at each stage of the continuous culture. In the punctuated macroevolutionary phase, there is no stepwise pattern. From pd7 to pd24, each separate clonal event represents only a portion of the cell population. From pd54 to pd236, the majority of cell populations share the same clonal karyotype patterns. The y-axis represents the frequencies of nonclonal karyotypes, indicating overall chromosome instability. Higher frequencies of nonclonal karyotypes indicate greater genome instability within the corresponding population. Genome chaos encompasses different types of karyotype reorganizations, such as chromothripsis, massive translocations, polyploid giant cancer cells, micronuclei clusters, chromosome fragmentation, and extra chromosomal DNA. All of these represent unstable transitional genomic structures contributing to the formation of new cellular systems with new genomes. Clonal expansion is a signature of the microevolutionary phase (from pd54 to pd236). Phase transition is achieved by genome chaos and macroevolutionary selection. The entire process of cancer evolution often involves multiple cycles of two-phased evolution. For more information, please refer to Heng [2019] and Heng et al [2006a, 2006b, 2006c]. The image and legend have been modified from Heng and Heng [2021a].

can also be noticed in normal multicellular and unicellular organismal evolution (Kasperski, 2023, 2021; Kasperski and Kasperska, 2021, 2016; Shimizu, 2022; Sun et al, 2018). Similar patterns can be found in polyploidy-mediated speciation in animals and plants (Li et al, 2021; Mayrose et al, 2010). To illustrate the importance of the macroevolutionary phase and its new relationship with microevolution, as opposed to the traditional view that macroevolution is simply the continuous microevolution over a longer period of time, we studied the meaning of genome-level reorganization by examining the relationship between karyotype alterations and gene mutations with inheritance at the system and parts levels, respectively (Heng and Heng, 2022b, 2021a). Further definitions and discussions can be found in the glossary.

This chapter will explore the similarities and differences between the two phases of cancer evolution, as well as the different genomic and non-genomic scales of cancer, by comparing various landscapes. In particular, our discussions will focus on the changes of cellular states within the microevolution phase and genome organization-mediated macroevolution. As different phases and scales mainly involve different modes of evolutionary dynamics and genomic mechanisms, we need to consider the concept of information management to understand cancer, the cellular species, and their impact on cancer ecological research. This analysis is important not only for understanding cancer evolution and how to use genome and information-based principles to study cancer ecology, but it could provide a different perspective to current evolutionary theories, which is largely focused on gene-based microevolution.

7.2 THE PATTERN AND SCALE OF GENOMIC AND NON-GENOMIC LANDSCAPES IN CANCER

The complexity of cancer, encompassing various non-genomic landscapes, gene/genomic landscapes, and cellular and tissue landscapes, presents a significant challenge in understanding the molecular mechanism of cancer. There are numerous contributing factors that can directly or indirectly trigger cancer, ranging from internal genome instability and gene mutations, essential variables for cellular adaptation, genetic errors, to environmental stressors (chemicals, radiation, viral infections), metabolic stress, mental stress, medical treatment stress, biological stress responses, and aging. The combination of these factors is too vast to fully comprehend (Heng, 2015; Ye et al, 2007, 2019, 2020). As a result, the specific role of individual molecular mechanisms in cancer can be partially illustrated but remains highly uncertain. This complexity has led to confusion among researchers who often prioritize their preferred scale of research while disregarding other scales. Moreover, researchers may discover that their chosen research topic is linked to cancer, which can be advantageous for publishing and securing additional funding. This contributes to the diversity of cancer research, with numerous studies focusing on the same cancer gene in different contexts or different genes in the same context. However, it is crucial to acknowledge that the complexity of the system extends beyond individual factors, as many interconnected elements may not be the key features of the specific system under investigation.

Our endeavor to employ the evolutionary mechanism of cancer as a unifying framework for diverse molecular mechanisms is a timely action that will aid us in prioritizing research strategies. To accomplish this goal, it is necessary to provide a concise overview of the cancer genomic landscapes using the two-phased cancer evolution framework. We will utilize four case studies to illustrate this approach.

7.2.1 SCALES OF MOLECULAR EVOLUTION AND THE MULTIPLE CYCLES OF A TWO-PHASED MODEL

To gain a deeper comprehension of cancer's evolutionary dynamics, it is crucial to investigate the various scales of molecular evolution using the two-phased model. In the context of cancer's molecular evolution, the term "scales" refers to different dimensions or levels at which cancer evolution can manifest. These scales encompass epigenetic states, DNA/RNA sequences, chromosomal regions, karyotypes, individual cells, cellular populations, and tissues. Each scale signifies a distinct level of complexity and encompasses specific time points during which evolutionary dynamics occur. By studying these scales, we can obtain a comprehensive understanding of the evolutionary processes at play in cancer. Interestingly, when we classify the different scales within the two-phased evolutionary model, the logical relationship among these diverse factors becomes much simpler: they all fall under the category of variations and assume different dominant roles depending on the phase of evolution and the levels of stress response. Equally significant is the application of chaos theory in explaining the multiple cycles of two-phased cancer evolution, which play a crucial role in the continuous progression of cancer, including transformation, metastasis, and drug resistance. Each key phase transition can be characterized by one or a few cycles within this framework. Note that "chaos" in this context does not imply complete disorder, but rather a process that involves both certainty and uncertainty. Certainty reflects the fact that under high levels of stress, cells undergo genome chaos-mediated macroevolution, leading to generation of new genomes for survival. Gene mutations and epigenetic alterations promote population growth through stochastic associations within microevolution. Uncertainty arises from the stochastic nature of triggering factors and the diverse mechanisms used for genome reorganization, ranging from massive translocations to extreme forms of polyploidy. The resulting karyotypes and the associated cancer genes can also vary significantly. Despite all the uncertainty, cancer cells do indeed emerge, some manage to survive, and they can pose a significant threat to patients' lives. This can be seen as a typical example of chaos.

Our discussions will primarily focus on three specific examples: 1) the epigenetic landscapes, which represent non-genomic landscapes; 2) the karyotype landscapes, which represent genomic landscapes; and 3) the tissue landscapes. However, a similar analytical approach can be applied to other types of landscapes as well, including the gene mutation landscape, although they may not be extensively covered in this particular chapter.

7.2.1.1 Epigenetic Landscape, Cell States, and Microevolution

Despite the fact that many epigenetic features have been linked to cancer, such as abnormal DNA methylation, histone modifications, chromatin remodeling/compaction, RNA splicing, the function of noncoding RNA, the activity of endogenous retroviruses, specific types of cellular and environmental stresses, system instability, and chaotic behavior (Esteller, 2006; Feinberg et al, 2006; Heng et al, 2009, 2010, 2016, 2011b; Jaffe, 2005; Jones and Baylin, 2007; Ottaiano et al, 2023; Russo et al, 2021), the relationship between the epigenetic landscape and other types of genomic and non-genomic landscapes remains unclear. This lack of clarity hinders efforts to prioritize research platforms (Heng, 2019, 2015; Heng and Heng, 2021a). Nevertheless, epigenetic studies have gained popularity due to findings that some epigenetic factors can act as regulators of cancer genes, which helps explain the absence of driver gene mutations in cancers (Garraway and Lander, 2013).

> For example, the epigenetic silencing of key tumor suppressor, regulatory, and repair genes has been identified in various cancers, which are also associated with the interruption of some important cellular signaling pathways (DNA repair, RB1/CDK4 cell cycle regulation, Wnt/β-catenin, TGF-β, and cellular differentiation pathways), control of replication timing and nuclear architecture, and epigenetic regulation of repeat elements.
>
> **Heng (2017)**

Furthermore, the unexpected limitations of using the gene mutation landscape to explain the common mechanisms of cancer have prompted an urgent need for alternative explanations. Epigenetic abnormalities have emerged as a convenient choice in this regard. Moreover, certain types of cancers are clearly associated with cellular differentiation issues, and the developmental landscape bears similarities to the epigenetic landscape of cancer. Therefore, it is logical to integrate the developmental landscape with the evolutionary fitness landscape. Equally significant, the approach of systems biology has also contributed to our understanding of the epigenetic landscape. For example, the term "cancer attractors" has been introduced by Stuart Kauffman and experimentally supported by genomic technologies (Huang et al, 2009; Kauffman, 1971). Considering the cancer genome as a complex network of mutually regulating genes, Kauffman's idea explains how this network after destabilization can produce a variety of stable, discretely distinct cell phenotypes, termed as attractors (Greaves and Maley, 2012; Huang et al, 2009; Huang and Kauffman, 2012; Kauffman, 1993, 1969a, 1969b). This powerful model links network dynamics, cell states, microevolutionary selection, and cancer fitness. A similar concept has been applied to the study of intrinsically disordered proteins (IDPs) as well as endogenous networks (Ao et al, 2010; Kulkarni et al, 2013).

Not surprisingly, increased cancer-related factors have been linked to the cellular epigenetic landscape model initially proposed by Waddington (Waddington, 1957), including efforts to understand polyploid giant cancer cells (PGCCs) and life cycle (Liu, 2020, 2018). We anticipate that the epigenetic landscape model will continue to gain popularity. For example, in the light of unified cell bioenergetics, a reason of cancer transformation and then cancer progression is cell overenergization causing excessive disturbances in current cell-fate (Kasperski, 2008, 2022; Kasperski and Kasperska, 2018, 2013). In these conditions the overenergized normal cell can change to a cancerous cell-fate in order to reduce overenergization (Kasperski, 2022; Kasperski and Kasperska, 2021). A switch to a cancerous cell-fate is called cancer transformation (Kasperski, 2022; Kasperski and Kasperska, 2021). After this switch, the cell undergoes stabilization and attains stability in the cancerous cell-fate attractor (Kasperski, 2022; Kasperski and Kasperska, 2021). After attaining stability, new clones appear in the cell-fate attractor (i.e., new clones are trapped in the same cell-fate attractor as the cell of origin). These clones, stable in the cell-fate attractor, are subject to the stresses related to permanent over-energization (and also changes of environment), which can cause random DNA mutations (Kasperski, 2022; Kasperski and Kasperska, 2021). This stress can lead to subsequent steps of microevolution causing destabilization of current cell-fate of some clones, leaving the current cell-fate attractor and attaining stability in new cell-fate attractors as a result. This procedure is repeated because of the permanent cloning of cancerous cells that are subject to constant stresses.

Discovery of two-phased cancer evolution points out that many epigenetic models overlooked the concept of stress-triggered macroevolution. These models assumed that changes in cell states, similar to those observed in developmental processes, were the primary cause of the majority of cancers. In the light of two-phased cancer evolution, random DNA mutations can cause not only destabilizations of current cell-fates and subsequent steps of microevolution, but also subsequent steps of macroevolution. This is because an accumulation of random DNA mutations, similar to different types of stress response (Horne et al, 2014; Stevens et al, 2011; Stevens et al, 2007), can lead to destabilization of the genome (Heng, 2019; Ye et al, 2018). Genome destabilization can cause genome chaos and by auto-transformation to a genome attractor, attaining genome stability in the new genome attractor (Kasperski, 2022; Kasperski and Heng, 2024, 2023; Kasperski and Kasperska, 2021). In this case, subsequent steps of microevolution can be repeated for cells that are trapped in different genome attractors. Microevolution that occurs for a population of cancer cells trapped in different genome attractors greatly enhances the ability of cancer to adapt to an environment because in this case adaptation can occur in many different new ways depending on the "potential" of genome attractors (Kasperski and Heng, 2023).

7.2.1.2 Genome or Karyotype Landscape, Genome Chaos, and Macroevolution

The karyotype landscape of cancer refers to the overall pattern of chromosomal abnormalities and variations detected within cancer cells or populations of cancer cells. The karyotype encompasses the complete set of chromosomes in a cell, including their number (both polyploidy and aneuploidy), structure, and organization. In clinical cytogenetics, the karyotype profiles typically rely on clonal karyotypes. However, in cancer, the karyotype landscape can be highly heterogeneous and dynamic, often exhibiting a mixture of various chromosomal alterations, including gains, losses, and rearrangements, at different regional and whole genome levels. Identifying clonal karyotypes in solid cancer poses particular challenges (Heng et al, 2011a, 2016, 2024; Ye et al, 2007).

Our experimental studies on observing karyotype evolution in real time have provided significant insights into the karyotype landscape of cancer. First, we have observed multiple cycles of nonclonal chromosome aberrations (NCCA) transitioning to clonal chromosome aberrations (CCA) during the evolution of cancer. However, the crucial phase transition that marks important milestones such as transformation, metastasis, and drug resistance occurs through extreme high stress-induced genome chaos. Second, there exist various subtypes of chaotic genomes, including chromothripsis, massive translocations, extreme high polyploidy, and extremely fragmented chromosomes (Baca et al, 2013; Heng et al, 2006a; Lange et al, 2022; Liu et al, 2014; Pellestor and Gatinois, 2020; Stephens et al, 2011; Stevens et al, 2007; Ye et al, 2018, 2019; Zhang et al, 2014). Although these transitional structures may be unstable and often lead to dead-ends, they are vital for the survival of newly emerging cancer populations. Therefore, even though they may be less detectable in the later stages of the evolutionary process, they play a dominant role during the phase transition as information carriers. Third, analyzing the collective data on different types of abnormal karyotypes within a population can help assess the overall instability or evolutionary potential of that population. Karyotype dynamics serve as an excellent indicator for monitoring cancer evolution. Furthermore, recent cytogenetic data, which provides better clinical predictions compared to gene mutation profiles (Davoli et al, 2017; Frias et al, 2019; Jamal-Hanjani et al, 2017; Park et al, 2010; Vargas-Rondón et al, 2017), supports the importance of utilizing karyotype profiles for clinical prediction (Heng et al, 2023).

Now that we have a better understanding of both the epigenetic and karyotype landscapes and their roles in different phases of cancer evolution, it is important to prioritize studies that focus on specific scales. For instance, if a study is addressing the macroevolution phase, which involves treatment-induced genome chaos, key phase transitions, and cellular speciation, prioritizing the profiling of karyotype dynamics is crucial. On the other hand, for investigating the microevolution phase, which encompasses the regulation of population size, clonal expansion, and stress response within physiological ranges, gene and epigenetic profiles can be more effective.

Furthermore, we emphasize that macroevolution is not equivalent to microevolution over time. In the context of cancer evolution, the previously accepted strategy of using developmental landscape models to understand the cancer landscape requires significant modification. We have proposed that the survival landscape and adaptive landscape are distinct (Heng, 2017, 2015; Heng et al, 2011b). Survival involves genome chaos and macroevolution, which give rise to the formation of new cellular systems with different genomes. On the other hand, small adaptive steps involve gradual modifications of fitness through different cell states with the same genome but varying gene mutations and/or epigenetic profiles.

Upon accepting the two-phased cancer evolution and the distinct landscape models of survival and adaptation, we need to address a long-standing issue that has caused confusion in cancer research. When studying cancer evolution through epigenetic dynamics, many researchers often assume that there are no significant genomic changes at the genome or karyotype level. Consequently, they believe that epigenetic changes alone can alter or reverse cancer phenotypes. However, this belief is based on an incorrect conceptual understanding and a misunderstanding of experimental facts.

The molecular genetic tradition has largely disregarded changes at the karyotype or genome level, assuming that the genome, being a collection of genes, merely serves as a carrier of genes. Consequently, they prioritize gene mutations over chromosomal changes, believing that gene mutations alone can explain cancer. This viewpoint has led many to focus solely on epigenetic factors as the key drivers of cancer. However, it is crucial to correct this perspective and recognize the significance of the karyotype in cancer evolution.

While a few systems biologists have attempted to integrate genome alterations into their cellular states model (Huang, 2013), their efforts have not yet had a substantial impact on the field as a whole. For many researchers, cancer is still perceived primarily as a developmental abnormality. However, it is important to note that pathological observations and synthesis reveal two distinct types of tumors.

Type one tumors are differentiated and can be associated with developmental abnormalities. Their tissue architectures resemble the tissue of origin, and their genomes exhibit relative stability. On the other hand, type two tumors, which constitute the majority of common tumors, are undifferentiated. They often exhibit nuclear atypia and are dissimilar to the tissue of origin. Genome chaos is frequently involved in these tumors (Liu, 2023, 2020). While some embryogenic codes may play a role in genome reorganization and cellular growth, it is essential to recognize that this process is evolutionary in nature and not simply a part of normal development (Heng, 2024; Heng and Heng, 2022b).

Besides the conceptual misunderstandings, there are also biases in the factual explanations of certain classical experiments. For instance, some publications have claimed that epigenetic changes can reverse cancer phenotypes without considering the examination of the karyotype. While some papers have highlighted the significance of genomic changes, they are often overlooked by readers. To illustrate this point, here is a quote from Heng and Heng.

> There seems to be strong evidence that epigenetic or embryonic environment change (without genomic alteration) is sufficient for the formation or reversion of cancer. One often-mentioned case cited to support the idea that epigenetics drives cancer phenotypes is Mintz's classical experiment—or perhaps more accurately, a misinterpretation of it (Illmensee and Mintz, 1976). In the experiment, after isolated single teracarcinoma cells were injected into mouse blastocysts, restored totipotency and normal differentiation were observed. This result has been considered one of the most convincing examples that the embryonic environment determines cancer state, sidelining genomic factors (since the same genotype can display cancer and normal developmental phenotypes in different environments). However, as written in this paper's abstract, the genomic factor is critical. According to the authors, "The high modal frequency of euploidy in these individually tested cells thus tends to indicate that a near-normal chromosome complement is sufficient for total restoration of orderly gene expression in a normal embryonic environment; it may also be necessary for teratoma stem-cell proliferation to be terminated there." In other words, a normal or nearly-normal karyotype is required for the cell line's later reversion, and once that karyotypic barrier is surpassed, full "restoration" is no longer possible.
>
> **Heng and Heng (2022b)**

Certainly, integrating the epigenetic and gene mutation landscapes with the genomic landscape, both at the local and global levels, holds great significance. This integration has the potential to resolve many confusions and challenges in the field of cancer research. By considering the interplay between these different layers of information, we can gain a more comprehensive understanding of the complex nature of cancer and its evolution.

7.2.1.3 Cancer and Normal Tissue Landscape, Constraint, and Stress Response

The cancer tissue landscape refers to the dynamic composition of various cell types and their spatial organization within a tumor. It encompasses the intricate interactions between cancer cells and surrounding stromal cells, immune cells, blood vessels, and other factors, collectively forming the tumor microenvironment. This microenvironment is subject to the influence of multiple factors that can alter its characteristics. Additionally, the tumor microenvironment operates within the constraints and stresses imposed by the normal tissue landscape and normal physiological processes. Both the aging process and tissue damage can reduce the constraining influence of normal tissue, thereby increasing the evolutionary potential of cancer. Given the complexity and dynamic nature of the tumor microenvironment, it is logical to approach its study from a synthetic evolutionary perspective rather than focusing solely on individual factors in isolation. Unfortunately, despite the significance of the tumor microenvironment as a research topic, the majority of current studies remain focused on gene mutations and epigenetic alterations. However, by adopting an evolutionary approach, we can enhance our understanding of the complex interactions, adaptations, and changes that take place within the tumor microenvironment over time. Such an approach allows us to examine the broader context and dynamics of the tumor microenvironment, shedding light on its influence on tumor progression, treatment response, and overall cancer evolution.

With the new understanding of cancer evolution, it is crucial to address key questions related to the interplay between cancer cells and the surrounding tissue environment. These questions include how newly formed genomes and their populations can overcome the inherent constraints imposed by the normal tissue; how the normal tissue can suppress the growth of cancer cells, leading to dormancy; how cancer cells can manipulate and recruit immune cells to support their growth and survival; why certain cancer cells show a preference for specific organs; and how to effectively treat cancers and reduce their population without inducing genome chaos. To successfully address these questions, it is crucial to understand the average and outlier behavior of the tissue, as well as the tissue's stress response under both low and high stress conditions. Additionally, a comprehensive understanding of new system emergence is necessary (Heng, 2015; Horne et al, 2014).

Even though cancer can be referred to as a tissue disease, studying cancer evolution must include the genomic components at this level, as without heritable genomic variation, the somatic evolution cannot be seriously studied. Clearly, the changes of spatial organization and composition of various cell types in tumor tissue are achieved by cellular evolution. Recently, studies using organoid culture have demonstrated that polyploid giant cancer cells (PGCCs) can directly contribute to the formation of tumor tissue. Additionally, the formation of high-grade tumors exhibits striking similarities to pre-embryonic development, characterized by a continuous cycle between macroevolution through polyploidy and microevolution through diploid genomes (Li et al, 2023).

One classical example of the tissue's constraints is contact inhibition (Lipsick, 2019). This means that normal cells respect the boundaries of their neighbors and stop dividing once they touch. In contrast, cancer cells do not respect the boundaries of their neighbors, i.e., cancer cells grow over and on top of one another (Lipsick, 2019). In accordance with the layered model of evolution of cellular functionalities, functionalities of cells of multicellular organisms are located in the three layers: bioenergetic layer, unicellular layer, and multicellular layer (Kasperski, 2022). During the process of cancer evolution, disturbed (including disrupted) multicellular layer functionalities are responsible for creating cancerous tissue (Kasperski and Heng, 2023). Because of these disturbances (disruption) of multicellular layer functionalities, the created cancerous tissue is morbid and abnormal. Moreover, over time, due to progressive damage of multicellular layer functionalities, cancerous tissue may begin to disintegrate, which can lead to elevated risk of metastases. From this point of view the risk of metastases is especially high when macroevolution occurs during cancer progression because macroevolution (caused by genome destabilization followed by genome chaos) can cause huge and immediate destruction of multicellular layer functionalities.

7.2.2 CANCER AS NEW CELLULAR SPECIES

Although it remains a subject of controversy, the idea of considering cancer as a new species is not a recent concept. As early as the 1950s, the evolutionary thinker Julian Huxley proposed that the occurrence of cancer could be viewed as a speciation event. Huxley stated that once the neoplastic process achieves autonomy, the resulting tumor could be logically regarded as a distinct biological species (Huxley, 1956). Similarly, evolutionary biologist Van Valen suggested that HeLa cells could be considered a new species (van Valen and Mairorana, 1991). In the field of cancer research, Peter Duesberg and David Rasnick also regarded cancer as a species in its own right (Duesberg and Rasnick, 2000). Subsequently, an increasing number of researchers have supported this viewpoint, particularly due to the association of new karyotypes with cancer (Bloomfield and Duesberg, 2016; Heng, 2015, 2009; Vincent, 2010; Ye et al, 2007).

How to define species, and whether species is a tangible entity in evolution, has long been a subject of debate. If macroevolution is understood as the accumulation of microevolutionary changes over time, then it becomes challenging to delineate species as there is a difficulty in drawing a clear boundary among countless intermediate populations (Heng, 2019, 2009).

The establishment of the two-phased model and the introduction of the concept of karyotype coding have laid the conceptual foundation for defining species. Particularly, with the recognition that the primary role of sexual reproduction is to preserve the identity of a specific species, there exists a new basis for reevaluating the concept of species. By synthesizing literature on chromosome-mediated speciation, we have proposed the genome architecture theory, which asserts that the similarity of the karyotype, encompassing gene content and genomic topology, serves as the foundation for defining species (Heng, 2019, 2015, 2009, 2007b). The genome architecture theory presents both a challenge and an alteration to the current neo-Darwinian framework (Crkvenjakov and Heng, 2022; Heng, 2019).

Interestingly, our understanding of cancer evolution has a profound impact on how we perceive organismal evolution, broadening our comprehension of evolution as a comprehensive process. Likewise, insights derived from studying organismal evolution can also enhance our understanding of cancer evolution. These two fields of study mutually inform and influence each other. Consequently, it is justifiable to categorize cancer cells with distinct karyotypes as separate cellular species. The growing evidence of contagious cancers in certain cases further bolsters the notion of considering cancer as a new species. Additional supporting arguments can be found in our earlier publications (Heng, 2019, 2015).

There are some positive impacts of considering cancers as new species: first, it will help the needed shift of the research focus from genes to genome, as the karyotype changes and preservation is a key for cancer as well as speciation for the majority of plants and animals. Second, it demands new technical platforms for reducing treatment-induced cancer resistance, and third, it will attract more evolutionary biologists to study cancer. In fact, the cancer and evolution conferences organized in 2020 have led to the formation of a cancer evolutionary working group in AACR. Such an integrated platform will have profound mutual benefits for cancer research and organismal evolutionary studies (Bussey and Davies, 2021; Dasari et al, 2021; Furst, 2021; Heng and Heng, 2021a; Laukien, 2022, 2021; Levin, 2021; Marshall, 2021; Noble, 2021; Paul, 2021; Shapiro, 2022, 2021; Shapiro and Noble, 2021).

7.2.3   Information Management and Two-Phased Cancer Evolution

The framework of genome chaos in cancer evolution has provided us with a perspective on information management, particularly when considering the profound reorganization of the genome through massive cell death as a potent mechanism for macroevolutionary survival. Recognizing that genome reorganization is a process of system information creation, and that evolutionary selection aims to achieve information preservation, the study of cancer evolution has revealed that information lies at the heart of cancer's evolutionary dynamics.

During the development of the karyotype coding concept, we put forth the idea that information management can be categorized into four essential components: information creation, preservation, modification, and utilization (Heng and Heng, 2022b, 2021a, 2021b). The karyotype coding serves as a fundamental code of many organic codes, while macroevolution and microevolution play distinct roles in information management. Specifically, genome chaos-mediated information creation at the system level is of utmost importance for macroevolution, while evolutionary selection favors the preservation of stable genomes to maintain system information. During the microevolutionary phase, gene mutations serve as information modifications. Throughout the entire evolutionary process, the utilization of information can be achieved. Furthermore, it is crucial to acknowledge that genome chaos-mediated information creation is a process of self-creation of information. Alongside the mechanisms of information preservation, they are equally important, if not more important, than the process of self-organization (Heng and Heng, 2023a).

We propose that the history of evolution on Earth can be unified through multiple cycles of two-phased evolution, each employing distinct mechanisms. These cycles are driven by the emergence of new organic codes (cell cycle codes, epigenetic codes, codes for meiosis, development, and animal body plan), leading to a greater complexity of information (Heng, 2024). Emphasizing the ultimate importance of system information self-creation and preservation represents a new paradigm for understanding information management in biology. This framework departs from traditional approaches that solely focus on information storage and transmission from individual genes to proteins (relying on part-level information usage).

7.2.4 Cancer Ecology and Two-Phased Evolution

Ecology studies the relationships between organisms and their environment by focusing on how organisms interact with each other and their physical surroundings. This includes understanding the diversity, distribution, and dynamics of organisms, as well as the flow of nutrients and energy within ecosystems. Ecology and evolution are closely connected fields, often sharing analytical approaches.

Cancer ecology investigates various aspects related to tumor growth and its impact on the overall health and survival of the host organism. It explores the ecological factors that influence tumor development, such as the interactions between cancer cells and the immune system. Additionally, it examines the role of the tumor microenvironment in shaping cancer evolution and progression.

Interestingly, some basic characteristics of cancer, such as competitive growth and invasive behavior, have long been studied within the framework of ecology, albeit without considering genome types or assuming that gene-mediated clonal expansion is the sole basis for genotype (Anderson et al, 2006; Gatenby, 1991). Recently, the use of ecological principles and methods to study cancer evolution has gained wider acceptance in the field (Amend and Pienta, 2015; Dujon et al, 2021). Fortunately, the rapid advancement of spatial biology and genomic profiling for cellular populations has provided the necessary technical knowledge. What we need is an appreciation of the distinctive differences between the phases and scales of evolutionary patterns.

In conventional ecological conditions, complex interactions involve individuals from the same or different species. Similarly, in the context of cancer, it is important to consider different cancer cells with distinct genomes as separate cellular species, and evolutionary selection occurs at various scales and phases with different mechanisms. However, cancer ecology research has yet to systematically apply macroevolutionary principles. Typically, researchers simply follow the neo-Darwinian viewpoint, using a gene-based approach to study ecological questions without paying sufficient attention to the macroevolutionary aspects that require a genome-based platform.

Studying cancer ecology through different phases and scales of cancer evolution can provide valuable insights. By examining ecological questions within the context of cancer's evolutionary processes, researchers can gain a deeper understanding of the ecological consequences of cancer and potentially identify novel approaches for its management and treatment.

7.3 IMPLICATIONS

The discovery of two-phased cancer evolution offers a new perspective on the evolution of cancer. It illustrates that cancer evolution does not follow the typical gradual neo-Darwinian model, where accumulated gene mutations are the key. Instead, it emphasizes the significance of distinctive scales and phases of evolution. Clearly, the choice of research strategies that are best suited for specific scales depends on the phase of evolution. Despite the enormous implications of applying the two-phased evolution model in cancer research, diagnosis, and potential treatment, some immediate utilities can be briefly mentioned.

First, a systematic comparison is needed to identify which types of landscapes, at which scale, and during which phase(s) of cancer evolution, or what types of combinations of landscapes, should be prioritized. In recent years, a large amount of landscape data on gene mutations, epigenetic alterations, copy number variations (CNVs), and cellular populations has become available. However, little effort has been made to integrate data on the karyotype landscape. Unfortunately, without the karyotype landscape, which represents the highest genomic scale and the genotype responsible for systemic selection, there is a lack of comprehensive understanding. It is likely that the karyotype landscape can unify different landscapes below the genome level and provide a systemic context of information for all scales lower than the karyotype. This higher level of systemic constraints can explain many phenomena. For example, the karyotype serves as the selection unit for macroevolution, and the contributions of many gene mutations and CNVs may only be observable during microevolutionary phases. The reason why many gene mutations and CNVs display neutral selection is because researchers should focus on the karyotype level, which is not neutral. Even though clonal expansion can be observed in healthy tissue as well (Kakiuchi and Ogawa, 2021), it is possible (and usually visible in experiments) that the karyotype landscape differs significantly between normal tissue and cancer tissue, leading to distinct overall evolutionary dynamics. In normal tissue, the average profile is dominant, while in unstable tissues, the outliers have a greater chance of becoming dominant via macroevolutionary selection.

Second, when designing clinical diagnosis and treatment strategies, the two-phased evolution model needs to be carefully considered. Specifically, future efforts to search for evolutionary biomarkers should take into account the distinction between microevolution and macroevolution, as these different phases are associated with distinct genomic mechanisms. Solely focusing on profiling the gene landscape has limited power to study macroevolution, regardless of the number of samples studied.

Furthermore, given that aggressive treatment can induce rapid drug resistance through genome chaos (Heng, 2019, 2015; Heng et al, 2010, 2021; Ye et al, 2021), the commonly used maximal killing strategy of drug treatment in clinical practice needs to be reevaluated. Adaptive therapy (Gatenby et al, 2009), which proposes an alternating approach between drug treatment and drug-free periods to promote competition between sensitive and resistant cells and slow down cancer evolution after treatment (Gallaher et al, 2023), has gained attention recently. However, an alternative explanation could be that the treatment options actually reduce the triggering of genome chaos. With fewer newly generated and more aggressive cancer cell species, the cancer cell population can be better managed within the microevolutionary phase.

Based on this hypothesis, further research is needed to explore how to effectively manage the cancer cell population using a moderate strategy within the microevolutionary phase.

Third, information management in cancer is a chaotic evolutionary process that involves complex dynamics between different scales of genomic and non-genomic landscapes, genotype and phenotype interactions, and active evolution within the context of various environmental factors.

Interestingly, the four steps of bioinformation management can be effectively integrated into the chaotic evolutionary process. Specifically, system information creation takes place during the macroevolutionary phase through the occurrence of genome chaos. The preservation of system information occurs through macroevolutionary selection, which results in stable genomes. Information modification can be achieved through stochastic changes in gene and epigenetic landscapes within the microevolutionary phase, without altering the karyotypes (information package). Collectively, the dynamic information derived from different scales can be utilized by cellular populations for survival and adaptation purposes.

It is important to emphasize that the aforementioned process aligns with a typical chaotic phenomenon in biology: the dynamic interplay between order and disorder during evolution, resulting in an overall increase in information complexity alongside the presence of specific mechanisms that introduce uncertainty (Heng, 2024; Heng and Heng, 2023b). Despite the diverse array of high-stress conditions encountered (uncertainty), cancer cells possess the ability to activate the program of genome chaos (certainty), which facilitates the process of information self-creation. Although the mechanisms involved in genome reorganization can vary significantly, ranging from chromosome translocations to polyploidy (uncertainty), genome chaos often leads to the generation of new genomes (certainty). With massive cell death resulting in the emergence of winners (certainty), there is typically a different karyotype for each successive round of this process (uncertainty). While the specific genes associated with the microevolutionary phase may be stochastically determined (uncertainty), the growth of the cancer cell population and its clinical implications represent certain outcomes (certainty). In summary, both the environments and the mechanisms of evolution are highly uncertain. However, it is certain that bioinformation increases in complexity through the information and genome alteration mediated by the two-phased evolution (also see Box 7.1 for more discussions).

Fourth, the two-phased cancer evolution model has opened our eyes and forced us to reconsider the current theories of organismal evolution. If the gradual stepwise evolution driven by gene mutation accumulation does not apply to cancer evolution, and cancer is undoubtedly a biological system that follows the principles of evolution, it is logical to reexamine the patterns of organismal evolution. Many concepts in traditional evolutionary theories lack direct experimental validation. During the process of reexamination that has spanned over a decade, we have made connections between system information and karyotype coding, part-level information and gene codes, genome chaos and genome reorganization during times of crisis, the role of sex in preserving species' identity through karyotype preservation, the influence of environmental stress and active evolution, and the chaotic two-phased evolution and information management. Together, the genome architecture theory, also known as the genome theory, is proposed as a conceptual framework for understanding evolution (Heng, 2015, 2009).

Interestingly, the two-phased evolution pattern and its basis of information management provide an explanation for how punctuated equilibrium operates in evolution. It is important to note that the relationship between the punctuated phase and the equilibrium phase is not solely determined by a time factor, but by different types of evolution and the environmental conditions. This fundamental difference from the neo-Darwinian gradual model helps explain why some researchers undervalue the punctuated nature of macroevolution and why Steve Jay Gould was frustrated in his attempts to reconcile punctuated phenotypes in fossils with Darwinian gradualism. It is unfortunate that, without knowledge of the two-phased evolution model, he eventually abandoned the remarkable concept of evolution (Crkvenjakov and Heng, 2022; Gould, 2002; Heng, 2019, 2009).

In recent years, with the availability of genomic data from various animal and plant genome projects, the two-phased evolutionary model has been commonly observed in the majority of species studied (Christmas et al, 2023; Damas et al, 2022; Damas et al, 2021; Dodsworth et al, 2016; Hoang et al, 2022; Murat et al, 2015; Schultz et al, 2023; Simakov et al, 2022). One general pattern that emerges is the occurrence of whole genome duplication events, which result in an unstable and transitional genome, followed by karyotype reorganization that can be traced through various synteny relationships. This aligns precisely with the predictions of the genome theory, which suggests karyotype or chromosomal reorganization via genome chaos (Heng, 2009; Pellestor and Gatinois, 2020; Schubert, 2007). The prolonged stasis observed in some species can be explained by the constraints imposed by the karyotype during microevolution (Gorelick and Heng, 2011; Heng, 2007b; Wilkins and Holliday, 2009). The rapidly accumulating data calls for a new genome-based evolutionary paradigm, as the gene-centric evolutionary concept, which focuses on the search for new genes responsible for speciation, is no longer relevant.

Fortunately, chromosome-mediated macroevolution, which has long been overlooked, has been persistently observed and proposed by notable thinkers and researchers such as Barbara McClintock and Richard Goldschmidt (Goldschmidt, 1940; McClintock, 1984), representing a valuable tradition of cytogenetics. However, due to the dominance of molecular genetics over the past 70 years and the lack of a genome-based theory to compete with the gene-centric theory, the significance of the karyotype in macroevolution has remained invisible to the majority of evolutionary biologists. This knowledge gap has driven us to establish the genome theory as the foundation for evolutionary theory. By embracing the genome architecture theory, the evolutionary framework will incorporate new features, including an active and likely directional process that involves a variety of historically contingent mechanisms (Heng, 2024; Heng and Heng, 2023a; Laukien, 2022; Shapiro, 2022).

ACKNOWLEDGMENTS

This article is part of a series of studies entitled "The mechanisms of somatic cell and organismal evolution". Due to the focused scope of this review, we apologize to others in this diverse research area whose contributions should be also acknowledged.

BOX 7.1   GLOSSARY AND EXPLANATIONS

PHASE TRANSITION

A phase transition is a physical or chemical concept that describes a process in which a substance undergoes a change in its physical state or chemical composition. In evolutionary biology, it often refers to a radical shift in the evolutionary trajectory of a population or a species, reflecting significant alterations in genotype or phenotype. Here, in the two-phased evolution model, it describes the macroevolutionary transition through genome reorganization and immediately followed by microevolution to grow the population. From an information perspective, this transition reflects the creation of new system information followed by an information preservation mechanism to propagate the newly created and selected information package. As cancers consist of a cycle of two-phased evolution, including transformation, metastasis, and drug resistance, each of these key transitions needs to go through phase transitions. After the phase transition, the previously dominant clonal karyotypes (shown in blue in the pie chart in Figure 7.1), especially the massively chaotic ones, are replaced by new and stable ones (shown in red in the pie chart in Figure 7.1). Only functionally survivable NCCAs, in addition to stability under selection, can form dominant clonal chromosome aberrations or CCAs. For cancer to progress through all major stages, the new dominant karyotype becomes the vital platform as well as the end product. The formation of the new system (or new cellular species) relies on the creation and preservation of a stable karyotype, providing the ultimate benefit. Given that the karyotype encodes system information and is preserved through inheritance, this is no longer surprising.

THE SEQUENTIAL RELATIONSHIP BETWEEN MACRO- AND MICROEVOLUTION

The current evolutionary theory is based on a key assumption: that macroevolution results from the accumulation of microevolution over time. When illustrating this relationship, it is more appropriate for the first phase to be Darwinian microevolution. However, as shown in Figure 7.1, genome alteration-mediated cellular macroevolution (big changes) precedes gene mutation-mediated cellular microevolution (small changes). This pattern is particularly common in cancer evolution and has been observed in both *in vitro* and *in vivo* models of transformation, treatment-induced drug resistance, as well as giant cell formation and diploidization (Heng, 2019; Heng et al, 2006a, 2006b, 2006c; Niu et al, 2021; Ye et al., 2021; Zhang et al., 2013). Furthermore, chromosomal level changes followed by gene-mediated pathway activation have been frequently reported in major journals (Anderson et al., 2018; Bakhoum et al., 2018; Gao et al., 2016; Sottoriva et al., 2015; Wang et al., 2014). This sharp contradiction between the data and the assumption has rightfully prompted some to question the accepted basic relationship between micro- and macroevolution, which forms the backbone of evolution.

Once it is accepted that gene mutations represent inheritance at the parts level, and karyotype changes represent inheritance at the system level, there is no rationale to support the assumption that microevolution leads to macroevolution. The accumulation of gene mutations in a population will not lead to genome alteration.

Furthermore, when considering the informational nature of evolution, the order of macro-evolution first makes even more sense. New system creation must come first, and then such a new system can be copied and propagated, resulting in a large population. Of course, as the entire evolutionary process is comprised of a series of two-phased events, the order of micro-evolution can also be viewed as the initial phase, depending on one's perspective. However, the key lies in identifying the pair as an independent event, such as transformation, metasta-sis, or induced drug resistance.

The experiment illustrated in Figure 7.1 was combined with additional drug treatment to initiate a new cycle of a two-phased evolution. As expected, the newly emergent resistant cells displayed different karyotypes compared to previously treated cell populations. Within the immediately followed microevolutionary phase, the karyotype was preserved. Clearly, the independent pair of macro- and microevolution is triggered by the drug treatment, and the shared karyotype, representing the system information package, links the macroevolution and microevolution phases within one independent evolutionary event.

A better explanation is that macroevolution is not the direct consequence of forming a new system through microevolution. Instead, it occurs only when the old system can no lon-ger survive through microevolutionary adaptation (when faced with a crisis, microevolution alone is insufficient within the system). Macroevolution then takes place by reorganizing the genome of the old system, emerging as a new, survivable system. This is followed by a new round of microevolution to form a large population, resulting in a cell population with clonal new karyotypes. In this scenario, parental cells not only contribute genomic materials but also activate a survival strategy to reorganize the genome, preparing it for macroevolution-ary selection in another cycle of evolution. All parental cells are physically deceased, but not without meaning. They are not dead-ends. They are physically gone, but their information, in a new form, lives on, albeit in different carriers. This perhaps represents the ultimate self-sacrifice for the sake of information continuation (Heng, 2024).

Perhaps more significantly, the same holds true for organismal evolution. When many new species form, the creation of a new karyotype is typically the initial step, followed by gene-mediated system modifications and population growth; the stasis phase. This process is com-monly observed in polyploidy-mediated speciation in plants, hybridization in animals, and the spontaneous chromosomal formation of new karyotypes in rats, primates, and humans (Heng, 2019).

The involvement of karyotype changes in cancer is significant not only because it entails a large-scale arrangement of chromosomes, often containing hundreds of genes but also due to the organizational principles of the genome that govern gene function. Only when the karyotype is viable can potent genes assist in producing more copies. When the karyotype is not viable, regardless of how powerful individual genes or components are, they will be elimi-nated. This is also why most nonclonal chromosome aberrations, or NCCAs, cannot survive, despite containing some of the most powerful cancer genes. In contrast, a viable karyotype can receive assistance from diverse types of cancer genes. This is why so many cancer genes can be involved.

GENOME CHAOS

Chaos as a system behavior does not simply mean random disorder . . . . Genome behavior during crises shares common features with complex systems described by chaos theory. Chaos theory, a non-linear dynamics concept, states that "within the apparent randomness of chaotic complex systems, there are underlying patterns, interconnectedness, constant feedback loops, repetition, self-similarity, fractals, and self-organization" (Wikipedia: Chaos theory). Interestingly, noted chaos theory scholar Joseph Ford considered evolution "chaos with feedback." Similarly, some evolutionary biologists believe deterministic chaos is a major component of evolution (Koonin,

2011). It is important to note that, within the collective frameworks of chaos theory, apparently-random states of disorder and irregularities of dynamic systems are often governed by deterministic laws.

Heng and Heng (2022a)

A close examination of genome chaos reveals its underlying pattern: the deterministic two-phased evolution leading to the birth of new populations and the uncertainty of key stages. The triggering factors can be highly diverse as long as they produce a high level of stress, and the mechanisms of genome reorganization can vary, including translocation, chromothripsis, polyploidy, micronuclei clusters, and chromosome fragmentations. The final selected karyotypes are typically not the same, and neither are the employed cancer genes in microevolution. In other words, the deterministic two-phased evolution system displays stochastic behavior at each key stage, including triggers, mechanisms for genome reorganization, the newly selected karyotypes in macroevolution, and genes in microevolution. All of these are extremely sensitive to small changes in both initial and perturbation conditions. This illustrates the typical chaotic process. Even though cancer cells can choose between highly dynamic changes or constrained changes, the end products selected by evolution remain unpredictable to the organism. This explains why genome alteration and gene mutation in evolution are not random but chaotic. By the way, only this chaotic process can generate massive alterations in such a short time, rather than through a random process, which lacks the realistic probability for creating new systems (Heng, 2024).

REFERENCES

Amend, S.R., Pienta, K.J. (2015). Ecology meets cancer biology: the cancer swamp promotes the lethal cancer phenotype. *Oncotarget, 6(12)*, 9669–78. doi: https://doi.org/10.18632/oncotarget.3430

Anderson, A.R., Weaver, A.M., Cummings, P.T., Quaranta, V. (2006). Tumor morphology and phenotypic evolution driven by selective pressure from the microenvironment. *Cell*, Dec 1; *127(5)*, 905–15. doi: https://doi.org/10.1016/j.cell.2006.09.042

Anderson, N.D., de Borja, R., Young, M.D., Fuligni, F., Rosic, A., Roberts, N.D., Hajjar, S., Layeghifard, M., Novokmet, A., Kowalski, P.E., Anaka, M., Davidson, S., Zarrei, M., Id Said, B., Schreiner, L.C., Marchand, R., Sitter, J., Gokgoz, N., Brunga, L., Graham, G.T., Shlien, A., et al. (2018). Rearrangement bursts generate canonical gene fusions in bone and soft tissue tumors. *Science (New York, N.Y.), 361(6405)*, eaam8419. https://doi.org/10.1126/science.aam8419

Ao, P., Galas, D., Hood, L., Yin, L., Zhu, X.M. (2010). Towards predictive stochastic dynamical modeling of cancer genesis and progression. *Interdiscip Sci*, Jun; *2(2)*, 140–4. doi: https://doi.org/10.1007/s12539-010-0072-3

Baca, S.C., Prandi, D., Lawrence, M.S., Mosquera, J.M., Romanel, A., Drier, Y., Park, K., Kitabayashi, N., MacDonald, T.Y., Ghandi, M., et al. (2013). Punctuated evolution of prostate cancer genomes. *Cell, 153*, 666–77.

Bakhoum, S.F., Ngo, B., Laughney, A.M., Cavallo, J.A., Murphy, C.J., Ly, P., Shah, P., Sriram, R.K., Watkins, T.B.K., Taunk, N.K., Duran, M., Pauli, C., Shaw, C., Chadalavada, K., Rajasekhar, V.K., Genovese, G., Venkatesan, S., Birkbak, N.J., McGranahan, N., Lundquist, M., Cantley, L.C., et al. (2018). Chromosomal instability drives metastasis through a cytosolic DNA response. *Nature, 553(7689)*, 467–72. https://doi.org/10.1038/nature25432

Bloomfield, M., Duesberg, P. (2016). Inherent variability of cancer-specific aneuploidy generates metastases. *Mol Cytogenet, 9*, 90. doi: https://doi.org/10.1186/s13039-016-0297-x

Breivik, J. (2005). The evolutionary origin of genetic instability in cancer development. *Semin Cancer Biol.*, Feb; *15(1)*, 51–60. doi: https://doi.org/10.1016/j.semcancer.2004.09.008 PMID: 15613288.

Bussey, K.J., Davies, P.C.W. (2021). Reverting to single-cell biology: the predictions of the atavism theory of cancer. *Prog Biophys Mol Biol*, Oct; *165*, 49–55. doi: https://doi.org/10.1016/j.pbiomolbio.2021.08.002

Cairns, J. (1975). Mutation selection and the natural history of cancer. *Nature*, *255*, 197–200.

Christmas, M.J., Kaplow, I.M., Genereux, D.P., Dong, M.X., Hughes, G.M., Li, X., Sullivan, P.F., Hindle, A.G., Andrews, G., Armstrong, J.C., Bianchi, M., Breit, A.M., Diekhans, M., Fanter, C., Foley, N.M., Goodman, D.B., Goodman, L., Keough, K.C., Kirilenko, B., Kowalczyk, A., Lawless, C., Lind, A.L., Meadows, J.R.S., Moreira, L.R., Redlich, R.W., Ryan, L., Swofford, R., Valenzuela, A., Wagner, F., Wallerman, O., Brown, A.R., Damas, J., Fan, K., Gatesy, J., Grimshaw, J., Johnson, J., Kozyrev, S.V., Lawler, A.J., Marinescu, V.D., Morrill, K.M., Osmanski, A., Paulat, N.S., Phan, B.N., Reilly, S.K., Schäffer, D.E., Steiner, C., Supple, M.A., Wilder, A.P., Wirthlin, M.E., Xue, J.R., Zoonomia Consortium, Birren, B.W., Gazal, S., Hubley, R.M., Koepfli, K.P., Marques-Bonet, T., Meyer, W.K., Nweeia, M., Sabeti, P.C., Shapiro, B., Smit, A.F.A., Springer, M.S., Teeling, E.C., Weng, Z., Hiller, M., Levesque, D.L., Lewin, H.A., Murphy, W.J., Navarro, A., Paten, B., Pollard, K.S., Ray, D.A., Ruf, I., Ryder, O.A., Pfenning, A.R., Lindblad-Toh, K., Karlsson, E.K. (2023). Evolutionary constraint and innovation across hundreds of placental mammals. *Science*, Apr 28; *380(6643)*, eabn3943. doi: https://doi.org/10.1126/science.abn3943

Comaills, V., Castellano-Pozo, M. (2023). Chromosomal instability in genome evolution: from cancer to macroevolution. *Biology*, *12(5)*, 671. https://doi.org/10.3390/biology12050671

Crespi, B., Summers, K. (2005). Evolutionary biology of cancer. *Trends Ecol Evol*, *20*, 545–52.

Crkvenjakov, R., Heng, H.H. (2022). Further illusions: on key evolutionary mechanisms that could never fit with modern synthesis. *Prog Biophys Mol Biol*, Mar–May; *169–70*, 3–11. doi: https://doi.org/10.1016/j.pbiomolbio.2021.10.002

Damas, J., Corbo, M., Kim, J., Turner-Maier, J., Farré, M., Larkin, D.M., Ryder, O.A., Steiner, C., Houck, M.L., Hall, S., Shiue, L., Thomas, S., Swale, T., Daly, M., Korlach, J., Uliano-Silva, M., Mazzoni, C.J., Birren, B.W., Genereux, D.P., Johnson, J., Lindblad-Toh, K., Karlsson, E.K., Nweeia, M.T., Johnson, R.N., Zoonomia Consortium, Lewin, H.A. (2022). Evolution of the ancestral mammalian karyotype and syntenic regions. *Proc Natl Acad Sci USA*, Oct 4; *119(40)*, e2209139119. doi: https://doi.org/10.1073/pnas.2209139119

Damas, J., Corbo, M., Lewin, H.A. (2021). Vertebrate chromosome evolution. *Annu Rev Anim Biosci*, Feb 16; *9*, 1–27. doi: https://doi.org/10.1146/annurev-animal-020518-114924

Dasari, K., Somarelli, J.A., Kumar, S., Townsend, J.P. (2021). The somatic molecular evolution of cancer: mutation, selection, and epistasis. *Prog Biophys Mol Biol*, Oct; *165*, 56–65. doi: https://doi.org/10.1016/j.pbiomolbio.2021.08.003

Davoli, T., Uno, H., Wooten, E.C., Elledge, S.J. (2017). Tumor aneuploidy correlates with markers of immune evasion and with reduced response to immunotherapy. *Science*, *355*, eaaf8399. doi: https://doi.org/10.1126/science.aaf8399

Dodsworth, S., Chase, M.W., Leitch, A.R. (2016). Is post-polyploidization diploidization the key to the evolutionary success of angiosperms? *Bot J Linn Soc*, *180(1)*, 1e5. doi: https://doi.org/10.1111/boj.12357

Duesberg, P., Rasnick, D. (2000). Aneuploidy, the somatic mutation that makes cancer a species of its own. *Cell Motil Cytoskeleton*, Oct; *47(2)*, 81–107. doi: https://doi.org/10.1002/1097-0169(200010)47:2<81::AID-CM1>3.0.CO;2-#

Dujon, A.M., Aktipis, A., Alix-Panabières, C., Amend, S.R., Boddy, A.M., Brown, J.S., Capp, J.P., DeGregori, J., Ewald, P., Gatenby, R., Gerlinger, M., Giraudeau, M., Hamede, R.K., Hansen, E., Kareva, I., Maley, C.C., Marusyk, A., McGranahan, N., Metzger, M.J., Nedelcu, A.M., Noble, R., Nunney, L., Pienta, K.J., Polyak, K., Pujol, P., Read, A.F., Roche, B., Sebens, S., Solary, E., Staňková, K., Ewald, H.S., Thomas, F., Ujvari, B. (2021). Identifying key questions in the ecology and evolution of cancer. *Evol Appl.*, Feb 8; *14(4)*, 877–92. doi: https://doi.org/10.1111/eva.13190

Esteller, M. (2006). Epigenetics provides a new generation of oncogenes and tumor suppressor genes. *Br J Cancer*, *94*, 179–83.

Fearon, E.R., Vogelstein, B. (1990). A genetic model for colorectal tumorigenesis. *Cell*, Jun 1; *61(5)*, 759–67. doi: https://doi.org/10.1016/0092-8674(90)90186-i PMID: 2188735.

Feinberg, A.P., Ohlsson, R., Henikoff, S. (2006). The epigenetic progenitor origin of human cancer. *Nat Rev Genet*, *7*, 21–33. doi: https://doi.org/10.1038/nrg1748

Frias, S., Ramos, S., Salas, C., Molina, B., Sánchez, S., Rivera-Luna, R. (2019). Nonclonal chromosome aberrations and genome chaos in somatic and germ cells from patients and survivors of Hodgkin lymphoma. *Genes*, *10*, 37. doi: https://doi.org/10.3390/genes10010037

Furst, R. (2021). The importance of Henry H. Heng's genome architecture theory. *Prog Biophys Mol Biol*, Oct; *165*, 153–6. doi: https://doi.org/10.1016/j.pbiomolbio.2021.08.009

Gallaher, J., Strobl, M., West, J., Gatenby, R., Zhang, J., Robertson-Tessi, M., Anderson, A.R.A. (2023). Inter- and intra-metastatic heterogeneity shapes adaptive therapy cycling dynamics. *Cancer Res*, May 19; *83(16)*. doi: https://doi.org/10.1158/0008-5472.CAN-22-2558

Gao, C., Su, Y., Koeman, J., Haak, E., Dykema, K., Essenberg, C., Hudson, E., Petillo, D., Khoo, S.K., Vande Woude, G.F. (2016). Chromosome instability drives phenotypic switching to metastasis. *Proc Natl Acad Sci USA*, *113(51)*, 14793–8. https://doi.org/10.1073/pnas.1618215113

Garraway, L.A., Lander, E.S. (2013). Lessons from the cancer genome. *Cell*, Mar 28; *153(1)*, 17–37. https://doi.org/10.1016/j.cell.2013.03.002 PMID: 23540688.

Gatenby, R.A. (1991). Population ecology issues in tumor growth. *Cancer Res*, May 15; *51(10)*, 2542–7. PMID: 2021934.

Gatenby, R.A., Silva, A.S., Gillies, R.J., Frieden, B.R. (2009). Adaptive therapy. *Cancer Res*, Jun 1; *69(11)*, 4894–903. doi: https://doi.org/10.1158/0008-5472

Gerlinger, M., McGranahan, N., Dewhurst, S.M., Burrell, R.A., Tomlinson, I., Swanton, C. (2014). Cancer: evolution within a lifetime. *Annu Rev Genet*, *48*, 215–36. doi: https://doi.org/10.1146/annurev-genet-120213-092314

Goldschmidt, R.B. (1940). *The material basis of evolution*. Yale University Press, New Haven.

Gorelickz, R., Heng, H.H. (2011). Sex reduces genetic variation: a multidisciplinary review. *Evolution*, Apr; *65(4)*, 1088–98. doi: https://doi.org/10.1111/j.1558-5646.2010.01173.x

Gould, S.J. (2002). *The structure of evolutionary theory*. Harvard University Press, Cambridge.

Greaves, M. (2007). Darwinian medicine: a case for cancer. *Nat Rev Cancer*, Mar; *7(3)*, 213–21. doi: https://doi.org/10.1038/nrc2071 Epub 2007 Feb 15. PMID: 17301845.

Greaves, M., Maley, C.C. (2012). Clonal evolution in cancer. *Nature*, Jan 18; *481(7381)*, 306–13. doi: https://doi.org/10.1038/nature10762 PMID: 22258609; PMCID: PMC3367003.

Hayden, C.E. (2010). The human genome at ten: Life is complicated. *Nature*, *464*, 664–7. doi: https://doi.org/10.1038/464664a

Heng, E., Moy, A., Liu, G., Heng, H.H., Zhang, K. (2021). ER stress and micronuclei cluster: stress response contributes to genome chaos in cancer. *Front Cell Dev Biol*, Aug 4; *9*, 673188. doi: https://doi.org/10.3389/fcell.2021.673188 PMID: 34422803; PMCID: PMC8371933.

Heng, E., Thanedar, S., Heng, H.H. (2023). Challenges and opportunities for clinical cytogenetics in the 21st century. *Genes*, *14*, 493. doi: https://doi.org/10.3390/genes14020493

Heng, E., Thanedar, S., Heng, H.H. (2024). The importance of monitoring non-clonal chromosome aberrations (NCCAs) in cancer research. In: Heng, H.H., Ye, C.J. (eds) *Cancer cytogenetics and cytogenomics*. Springer, New York (in press).

Heng, H.H. (2007a). Cancer genome sequencing: the challenges ahead. *BioEssays News Rev Mol Cell Dev Biol*, *29(8)*, 783–94. doi: https://doi.org/10.1002/bies.20610

Heng, H.H. (2007b). Elimination of altered karyotypes by sexual reproduction preserves species identity. *Genome*, May; *50(5)*, 517–24. doi: https://doi.org/10.1139/g07-039

Heng, H.H. (2009). The genome-centric concept: resynthesis of evolutionary theory. *Bioessays*, May; *31(5)*, 512–25. doi: https://doi.org/10.1002/bies.200800182 PMID: 19334004.

Heng, H.H. (2015). *Debating cancer: the paradox in cancer research*. World Scientific Publishing Co., Singapore.

Heng, H.H. (2017). The genomic landscape of cancers (Chapter 5). In: Ujvari, B., Roche, B., Thomas, F. (eds) *Ecology and evolution of cancer*. Elseiver, London, pp. 69–83.

Heng, H.H. (2019). *Genome chaos: rethinking genetics, evolution, and molecular medicine*. Academic Press, San Diego, CA.

Heng, H.H. (2024). *Genome chaos: rethinking genetics, evolution, and molecular medicine*, 2nd edition, Academic Press (in press).

Heng, H.H., Bremer, S.W., Stevens, J.B., Horne, S.D., Liu, G., Abdallah, B.Y., Ye, K.J., Ye, C.J. (2013). Chromosomal instability (CIN): what it is and why it is crucial to cancer evolution. *Cancer Metastasis Rev*, *32(3–4)*, 325–40. doi: https://doi.org/10.1007/s10555-013-9427-7

Heng, H.H., Bremer, S.W., Stevens, J.B., Ye, K.J., Liu, G., Ye, C.J. (2009). Genetic and epigenetic heterogeneity in cancer: a genome-centric perspective. *J Cell Physiol*, Sep; *220(3)*, 538–47. doi: https://doi.org/10.1002/jcp.21799

Heng, H.H., Bremer, S.W., Stevens, J., Ye, K.J., Miller, F., Liu, G., Ye, C.J. (2006a). Cancer progression by non-clonal chromosome aberrations. *J Cell Biochem*, *98(6)*, 1424e1435. doi: https://doi.org/10.1002/jcb.20964

Heng, J., Heng, H.H. (2021a). Two-phased evolution: genome chaos-mediated information creation and maintenance. *Prog Biophys Mol Biol*, *165*, 29–42. doi: https://doi.org/10.1016/j.pbiomolbio.2021.04.003

Heng, J., Heng, H.H. (2021b). Karyotype coding: the creation and maintenance of system information for complexity and biodiversity. *Biosystems*, *208*, 104476. doi: https://doi.org/10.1016/j.biosystems.2021.104476

Heng, J., Heng, H.H. (2022a). Genome chaos, information creation, and cancer emergence: searching for new frameworks on the 50th anniversary of the "war on cancer". *Genes (Basel)*, Dec 31; *13(1)*, 101. doi: https://doi.org/10.3390/genes13010101 PMID: 35052441; PMCID: PMC8774498.

Heng, J., Heng, H.H. (2022b). Genome chaos: creating new genomic information essential for cancer macro-evolution. *Semin Cancer Biol, 81*, 160–75. doi: https://doi.org/10.1016/j.semcancer.2020.11.003

Heng, J., Heng, H.H. (2023a). Karyotype as code of codes: an inheritance platform to shape the pattern and scale of evolution. *BioSystems, 233*, 105016. Advance online publication. https://doi.org/10.1016/j.biosystems.2023.105016

Heng, J., Heng, H.H. (2023b). *Polyploidy: a transitional platform for creating and preserving newly emergent system information.* Polyploidy 2023: Across the Tree of Life, Palm Coast, FL, May 9–12.

Heng, H.H., Liu, G., Alemara, S., Regan, S., Zachary, A., Ye, C.J. (2019). The mechanisms of how genomic heterogeneity impacts bioemergent properties: the challenges for precision medicine. In: Sturmberg, J.P. (ed) *Embracing complexity in health.* Springer Nature, Cham, Switzerland, pp. 95–109.

Heng, H.H., Liu, G., Bremer, S., Ye, K.J., Stevens, J., Ye, C.J. (2006b). Clonal and nonclonal chromosome aberrations and genome variation and aberration. *Genome, 49(3)*, 195e204. doi: https://doi.org/10.1139/g06-023

Heng, H.H., Liu, G., Stevens, J.B., Bremer, S.W., Ye, K.J., Abdallah, B.Y., Horne, S.D., Ye, C.J. (2011a). Decoding the genome beyond sequencing: the new phase of genomic research. *Genomics, 98(4)*, 242–52. doi: https://doi.org/10.1016/j.ygeno.2011.05.008

Heng, H.H., Liu, G., Stevens, J.B., Bremer, S.W., Ye, K.J., Ye, C.J. (2010). Genetic and epigenetic heterogeneity in cancer: the ultimate challenge for drug therapy. *Curr Drug Targets*, Oct; *11(10)*, 1304–16. doi: https://doi.org/10.2174/138945011007011304

Heng, H.H., Regan, S.M., Liu, G., Ye, C.J. (2016). Why it is crucial to analyze non clonal chromosome aberrations or NCCAs? *Mol Cytogenet, 9*, 15. doi: https://doi.org/10.1186/s13039-016-0223-2

Heng, H.H., Stevens, J.B., Bremer, S.W., Liu, G., Abdallah, B.Y., Ye, C.J. (2011b). Evolutionary mechanisms and diversity in cancer. *Adv Cancer Res, 112*, 217–53. doi: https://doi.org/10.1016/B978-0-12-387688-1.00008-9

Heng, H.H., Stevens, J.B., Liu, G., Bremer, S.W., Ye, K.J., Reddy, P., Wu, G.S., Wang, Y.A., Tainsky, M.A., Ye, C.J. (2006c). Stochastic cancer progression driven by non-clonal chromosome aberrations. *J Cell Physiol, 208(2)*, 461–72. doi: https://doi.org/10.1002/jcp.20685

Heppner, G.H. (1984). Tumor heterogeneity. *Cancer Res, 44*, 2259–65.

Heppner, G.H., Miller, B.E. (1989). Therapeutic implications of tumor heterogeneity. *Semin Oncol, 16*, 91–105.

Hoang, P.T.N., Fuchs, J., Schubert, V., Tran, T.B.N., Schubert, I. (2022). Chromosome numbers and genome sizes of all 36 duckweed species (*Lemnaceae*). *Plants (Basel)*, Oct 11; *11(20)*, 2674. doi: https://doi.org/10.3390/plants11202674

Horne, S.D., Chowdhury, S.K., Heng, H.H. (2014). Stress, genomic adaptation, and the evolutionary trade-off. *Front Genet*, Apr 23; *5*, 92. doi: https://doi.org/10.3389/fgene.2014.00092 PMID: 24795754; PMCID: PMC4005935.

Horne, S.D., Pollick, S.A., Heng, H.H. (2015). Evolutionary mechanism unifies the hallmarks of cancer. *Int J Cancer*, May 1; *136(9)*, 2012–21. doi: https://doi.org/10.1002/ijc.29031 Epub 2014 Jun 30. PMID: 24957955.

Huang, S. (2013). Genetic and non-genetic instability in tumor progression: link between the fitness landscape and the epigenetic landscape of cancer cells. *Cancer Metastasis Rev*, Dec; *32(3–4)*, 423–48. doi: https://doi.org/10.1007/s10555-013-9435-7 PMID: 23640024.

Huang, S., Ernberg, I., Kauffman, S. (2009). Cancer attractors: a systems view of tumors from a gene network dynamics and developmental perspective. *Semin Cell Dev Biol, 20(7)*, 869–76.

Huang, S., Kauffman, S.A. (2012). Complex gene regulatory networks—from structure to biological observables: cell fate determination. In: Meyers, R.A. (ed) *Computational complexity: theory, techniques, and applications.* Springer, New York, pp. 527–60.

Huxley, J. (1956). Cancer biology: comparative and genetic. *Biol Rev, 31*, 474–514. doi: https://doi.org/10.1111/j.1469-185X.1956.tb01558.x

Illmensee, K., Mintz, B. (1976). Totipotency and normal differentiation of single teratocarcinoma cells cloned by injection into blastocysts. *Proc Natl Acad Sci USA*, Feb; *73(2)*, 549–53. doi: https://doi.org/10.1073/pnas.73.2.549 PMID: 1061157; PMCID: PMC335947.

Jablonka, E. (2006). Commentary: induction and selection of variations during cancer development. *Int J Epidemiol*, Oct; *35(5)*, 1163–5. doi: https://doi.org/10.1093/ije/dyl188 Epub 2006 Sep 20. PMID: 16990287.

Jaffe, L.F. (2005). A calcium-based theory of carcinogenesis. *Adv Cancer Res, 94*, 231–63. doi: https://doi.org/10.1016/S0065-230X(05)94006-2

Jamal-Hanjani, M., Wilson, G.A., McGranahan, N., Birkbak, N.J., Watkins, T.B.K., Veeriah, S., Shafi, S., Johnson, D.H., Mitter, R., Rosenthal, R., et al. (2017). Tracking the evolution of non–small-cell lung cancer. *N Engl J Med, 376*, 2109–21. doi: https://doi.org/10.1056/NEJMoa1616288

Jones, P.A, Baylin, S.B. (2007). The epigenomics of cancer. *Cell*, Feb 23; *128(4)*, 683–92. doi: https://doi.org/10.1016/j.cell.2007.01.029 PMID: 17320506; PMCID: PMC3894624.

Kasperski, A. (2008). Modelling of cells bioenergetics. *Acta Biotheor*, *56*, 233–47. doi: https://doi.org/10.1007/s10441-008-9050-0

Kasperski, A. (2021). Genome attractors as places of evolution and oases of life. *Processes*, *9(9)*, 1646. doi: https://doi.org/10.3390/pr9091646

Kasperski, A. (2022). Life entrapped in a network of atavistic attractors: how to find a rescue. *Int J Mol Sci*, *23(7)*, 4017. https://doi.org/10.3390/ijms23074017

Kasperski, A. (2023). Recognition of timestamps and reconstruction of the line of organism development. *Processes*, *11*, 1316. doi: https://doi.org/10.3390/pr11051316

Kasperski, A., Heng, H.H. (2023). Cancer formation as creation and penetration of unknown life spaces. In: Salgia, R., et al. (eds) *Cancer systems biology and translational mathematical oncology*. Oxford University Press, Oxford (in press).

Kasperski, A., Heng, H.H. (2024). Digital world of cytogenetic and cytogenomic web resources. In: Heng, H.H. and Ye, C.J. (eds) *Cancer cytogenetics and cytogenomics*. Springer, New York (in press).

Kasperski, A., Kasperska, R. (2016). A new approach to the automatic identification of organism evolution using neural networks. *Biosystems*, *142–3*, 32–42. doi: https://doi.org/10.1016/j.biosystems.2016.03.005

Kasperski, A., Kasperska, R. (2021). Study on attractors during organism evolution. *Sci Rep*, *11*, 9637. doi: https://doi.org/10.1038/s41598-021-89001-0

Kasperski, A., Kasperska, R. (2013). Selected disease fundamentals based on the unified cell bioenergetics. *J Investig Biochem*, *2(2)*, 93–100. doi: https://doi.org/10.5455/jib.20130227041230

Kasperski, A., Kasperska, R. (2018). Bioenergetics of life, disease and death phenomena. *Theory Biosci*, *137*, 155–68. doi: https://doi.org/10.1007/s12064-018-0266-5

Kultz, D. (2005) Molecular and evolutionary basis of the cellular stress response. *Annu Rev Physiol*, 67: 225–57. doi: https://doi.org/10.1146/annurev.physiol.67.040403.103635

Lange, J.T., Rose, J.C., Chen, C.Y., Pichugin, Y., Xie, L., Tang, J., Hung, K.L., Yost, K.E., Shi, Q., Erb, M.L., Rajkumar, U., Wu, S., Taschner-Mandl, S., Bernkopf, M., Swanton, C., Liu, Z., Huang, W., Chang, H.Y., Bafna, V., Henssen, A.G., Mischel, P.S., et al. (2022). The evolutionary dynamics of extrachromosomal DNA in human cancers. *Nat Genet*, *54*(10), 1527–33. https://doi.org/10.1038/s41588-022-01177-x

Levine, A.J. (2019). *Seventy years and two paradigm shifts: the changing faces of biology: what does one learn from all this reductionism without the organism?* https://www.ias.edu/ideas/seventy-years-and-two-paradigm-shifts-changing-faces-biology

Li, Z., McKibben, M.T.W., Finch, G.S., Blischak, P.D., Sutherland, B.L., Barker, M.S. (2021). Patterns and processes of diploidization in land plants. *Annu Rev Plant Biol*, Jun 17; *72*, 387–410. doi: https://doi.org/10.1146/annurev-arplant-050718-100344

Ling, S., Hu, Z., Yang, Z., Yang, F., Li, Y., Lin, P., Chen, K., Dong, L., Cao, L., Tao, Y., Hao, L., Chen, Q., Gong, Q., Wu, D., Li, W., Zhao, W., Tian, X., Hao, C., Hungate, E.A., Catenacci, D.V., Hudson, R.R., Li, W.H., Lu, X., Wu, C.I. (2015). Extremely high genetic diversity in a single tumor points to prevalence of non-Darwinian cell evolution. *Proc Natl Acad Sci USA*, Nov 24; *112(47)*, E6496–505. doi: https://doi.org/10.1073/pnas.1519556112

Maley, C.C., Reid, B.J. (2005). Natural selection in neoplastic progression of Barrett's esophagus. *Semin Cancer Biol*, Dec; *15(6)*, 474–83. doi: https://doi.org/10.1016/j.semcancer.2005.06.004

Mayrose, I., Barker, M.S., Otto, S.P. (2010). Probabilistic models of chromosome number evolution and the inference of polyploidy. *Syst Biol*, Mar; *59(2)*, 132–44. doi: https://doi.org/10.1093/sysbio/syp083

Merlo, L.M., Pepper, J.W., Reid, B.J., Maley, C.C. (2006). Cancer as an evolutionary and ecological process. *Nat Rev Cancer*, *6*, 924–35.

Mojica, E.A., Kültz, D. (2022) Physiological mechanisms of stress-induced evolution. *J Exp Biol*, *225(Suppl_1)*, jeb243264. https://doi.org/10.1242/jeb.243264

Niu, N., Yao, J., Bast, R.C., Sood, A.K., Liu, J. (2021). IL-6 promotes drug resistance through formation of polyploid giant cancer cells and stromal fibroblast reprogramming. *Oncogenesis*, *10(9)*, 65. https://doi.org/10.1038/s41389-021-00349-4

Noble, D. (2021). Cellular Darwinism: regulatory networks, stochasticity, and selection in cancer development. *Prog Biophys Mol Biol*, *165*, 66–71. doi: https://doi.org/10.1016/j.pbiomolbio.2021.06.007

Nowell, P.C. (1976). The clonal evolution of tumor cell populations. *Science*, *194*, 23–8.

Kakiuchi, N., Ogawa, S. (2021). Clonal expansion in non-cancer tissues. *Nat Rev Cancer*, *21*, 239–56. doi: https://doi.org/10.1038/s41568-021-00335-3

Kauffman, S.A. (1969a). Homeostasis and differentiation in random genetic control networks. *Nature*, *224*, 177–8.

Kauffman, S.A. (1969b). Metabolic stability and epigenesis in randomly constructed genetic nets. *J Theor Biol*, *22*, 437–67. doi: https://doi.org/10.1016/0022-5193(69)90015-0

Kauffman, S.A. (1971). Differentiation of malignant to benign cells. *J Theor Biol*, *31*, 429–51.

Kauffman, S.A. (1993). *The origins of order*. Oxford University Press, New York.

Koonin, E. (2011). Are there laws of genome evolution? *PLoS Comput Biol*, *7(8)*, e1002173. doi: https://doi.org/10.1371/journal.pcbi.1002173

Kulkarni, P., Shiraishi, T., Kulkarni, R.V. (2013). Cancer: tilting at windmills? *Mol Cancer*, Sep 24; *12(1)*, 108. doi: https://doi.org/10.1186/1476-4598-12-108 PMID: 24063528; PMCID: PMC3848908

Laukien, F.H. (2021). The evolution of evolutionary processes in organismal and cancer evolution. *Prog Biophys Mol Biol*, Oct; *165*, 43–8. doi: https://doi.org/10.1016/j.pbiomolbio.2021.08.008

Laukien, F.H. (2022). *Active biological evolution: feedback-driven, actively accelerated organismal and cancer evolution*. Evolution Press, North Hampton.

Levan, A. (1967). Some current problems of cancer cytogenetics. *Hereditas*, *57(3)*, 343–55. doi: https://doi.org/10.1111/j.1601-5223.1967.tb02117.x

Levin, M. (2021). Bioelectrical approaches to cancer as a problem of the scaling of the cellular self. *Prog Biophys Mol Biol*, Oct; *165*, 102–13. doi: https://doi.org/10.1016/j.pbiomolbio.2021.04.007

Li, X., Zhong, Y., Zhang, X., Sood, A.K., Liu, J. (2023). Spatiotemporal view of malignant histogenesis and macroevolution via formation of polyploid giant cancer cells. *Oncogene*, Feb; *42(9)*, 665–78. doi: https://doi.org/10.1038/s41388-022-02588-0 Epub 2023 Jan 3. PMID: 36596845; PMCID: PMC9957731.

Lipsick, J. (2019). A history of cancer research: tyrosine kinases. *Cold Spring Harb Perspect Biol*, *11(2)*, a035592. doi: https://doi.org/10.1101/cshperspect.a035592

Liu, G., Stevens, J.B., Horne, S.D., Abdallah, B.Y., Ye, K.J., Bremer, S.W., Ye, C.J., Chen, D.J., Heng, H.H. (2014). Genome chaos: survival strategy during crisis. *Cell Cycle (Georgetown, Tex.)*, *13(4)*, 528–37. doi: https://doi.org/10.4161/cc.27378

Liu, J. (2018). The dualistic origin of human tumors. *Semin Cancer Biol*, Dec; *53*, 1–16. doi: https://doi.org/10.1016/j.semcancer.2018.07.004

Liu, J. (2020). The "life code": a theory that unifies the human life cycle and the origin of human tumors. *Semin Cancer Biol*, Feb; *60*, 380–97. doi: https://doi.org/10.1016/j.semcancer.2019.09.005

Liu, J. (2023). *Giant cells: a new hypothesis for understanding embryogenesis and tumorigenesis across the tree of life*. Polyploidy 2023: Across the Tree of Life, Palm Coast, FL, May 9–12.

Marshall, P. (2021). Biology transcends the limits of computation. *Prog Biophys Mol Biol*, Oct; *165*, 88–101. doi: https://doi.org/10.1016/j.pbiomolbio.2021.04.006

McClintock, B. (1984). Significance of responses of the genome to challenge. *Science*, *226*, 792e801.

Murat, F., Zhang, R., Guizard, S., Gavranović, H., Flores, R., Steinbach, D., Quesneville, H., Tannier, E., Salse, J. (2015). Karyotype and gene order evolution from reconstructed extinct ancestors highlight contrasts in genome plasticity of modern rosid crops. *Genome Biol Evol*, Jan 29; *7(3)*, 735–49. doi: https://doi.org/10.1093/gbe/evv014

Ottaiano, A., Ianniello, M., Santorsola, M., Ruggiero, R., Sirica, R., Sabbatino, F., Perri, F., Cascella, M., Di Marzo, M., Berretta, M., Caraglia, M., Nasti, G., Savarese, G. (2023). From chaos to opportunity: decoding cancer heterogeneity for enhanced treatment strategies. *Biology*, *12(9)*, 1183. https://doi.org/10.3390/biology12091183

Park, S.Y., Gönen, M., Kim, H.J., Michor, F., Polyak, K. (2010). Cellular and genetic diversity in the progression of in situ human breast carcinomas to an invasive phenotype. *J Clin Investig*, *120*, 636–44. doi: https://doi.org/10.1172/JCI40724

Paul, D. (2021). Cancer as a form of life: musings of the cancer and evolution symposium. *Prog Biophys Mol Biol*, Oct; *165*, 120–39. doi: https://doi.org/10.1016/j.pbiomolbio.2021.05.003

Pellestor, F., Gatinois, V. (2020). Chromoanagenesis: a piece of the macroevolution scenario. *Mol Cytogenet*, Jan 28; *13*, 3. doi: https://doi.org/10.1186/s13039-020-0470-0

Russo, G., Tramontano, A., Iodice, I., Chiariotti, L., Pezone, A. (2021). Epigenome chaos: stochastic and deterministic DNA methylation events drive cancer evolution. *Cancers (Basel)*, Apr 9; *13(8)*, 1800. doi: https://doi.org/10.3390/cancers13081800

Schubert, I. (2007). Chromosome evolution. *Curr Opin Plant Biol*, Apr; *10(2)*, 109–15. doi: https://doi.org/10.1016/j.pbi.2007.01.001

Schultz, D.T., Haddock, S.H.D., Bredeson, J.V., Green, R.E., Simakov, O., Rokhsar, D.S. (2023). Ancient gene linkages support ctenophores as sister to other animals. *Nature*, May 17. doi: https://doi.org/10.1038/s41586-023-05936-6 Epub ahead of print. PMID: 37198475.

Shapiro, J.A. (2021). What can evolutionary biology learn from cancer biology? *Prog Biophys Mol Biol*, Oct; *165*, 19–28. doi: https://doi.org/10.1016/j.pbiomolbio.2021.03.005

Shapiro, J.A. (2022). *Evolution: a view from the 21th century, fortified*. Cognition Press, Chicago.

Shapiro, J.A., Noble, D. (2021). The value of treating cancer as an evolutionary disease. *Prog Biophys Mol Biol*, Oct; *165*, 1–2. doi: https://doi.org/10.1016/j.pbiomolbio.2021.08.010

Shibata, D. (1997). Molecular tumour clocks. *Ann Med*, Feb; *29(1)*, 5–7. doi: https://doi.org/10.3109/07853899708998738

Shimizu, K.K. (2022). Robustness and the generalist niche of polyploid species: genome shock or gradual evolution? *Curr Opin Plant Biol*, Oct; *69*, 102292. doi: https://doi.org/10.1016/j.pbi.2022.102292 Epub 2022 Sep 2. PMID: 36063635

Simakov, O., Bredeson, J., Berkoff, K., Marletaz, F., Mitros, T., Schultz, D.T., O'Connell, B.L., Dear, P., Martinez, D.E., Steele, R.E., Green, R.E., David, C.N., Rokhsar, D.S. (2022). Deeply conserved synteny and the evolution of metazoan chromosomes. *Sci Adv*, Feb 4; *8(5)*, eabi5884. doi: https://doi.org/10.1126/sciadv.abi5884

Somarelli, J.A., DeGregori, J., Gerlinger, M., Heng, H.H., Marusyk, A., Welch, D.R., Laukien, F.H. (2022). Questions to guide cancer evolution as a framework for furthering progress in cancer research and sustainable patient outcomes. *Med Oncol (Northwood, London, England)*, *39*(9), 137. https://doi.org/10.1007/s12032-022-01721-z

Somarelli, J.A., Gardner, H., Cannataro, V.L., Gunady, E.F., Boddy, A.M., Johnson, N.A., Fisk, J.N., Gaffney, S.G., Chuang, J.H., Li, S., Ciccarelli, F.D., Panchenko, A.R., Megquier, K., Kumar, S., Dornburg, A., DeGregori, J., Townsend, J.P. (2020). Molecular biology and evolution of cancer: from discovery to action. *Mol Biol Evol*, *37(2)*, 320–6. doi: https://doi.org/10.1093/molbev/msz242

Sottoriva, A., Kang, H., Ma, Z., Graham, T.A., Salomon, M.P., Zhao, J., Marjoram, P., Siegmund, K., Press, M.F., Shibata, D., Curtis, C. (2015). A big bang model of human colorectal tumor growth. *Nat Genet*, *47*, 209–16. doi: https://doi.org/10.1038/ng.3214

Stephens, P.J., Greenman, C.D., Fu, B., Yang, F., Bignell, G.R., Mudie, J., Pleasance, E.D., Lau, K.W., Beare, D., Stebbings, L.A., McLaren, S., Lin, M.L., McBride, D.J., Varela, I., Nik-Zainal, S., Leroy, C., Jia, M., Menzies, A., Butler, A.P., Teague, J.W., et al. (2011). Massive genomic rearrangement acquired in a single catastrophic event during cancer development. *Cell*, *144(1)*, 27–40.

Stevens, J.B., Abdallah, B.Y., Liu, G., Ye, C.J., Horne, S.D., Wang, G., Savasan, S., Shekhar, M., Krawetz, S.A., Hüttemann, M., Tainsky, M.A., Wu, G.S., Xie, Y., Zhang, K., Heng, H.H. (2011). Diverse system stresses: common mechanisms of chromosome fragmentation. *Cell Death Dis*, Jun 30; *2(6)*, e178. doi: https://doi.org/10.1038/cddis.2011.60 PMID: 21716293; PMCID: PMC3169002.

Stevens, J.B., Liu, G., Bremer, S.W., Ye, K.J., Xu, W., Xu, J., Sun, Y., Wu, G.S., Savasan, S., Krawetz, S.A., Ye, C.J., Heng, H.H. (2007). Mitotic cell death by chromosome fragmentation. *Cancer Res*, *67(16)*, 7686–94. doi: https://doi.org/10.1158/0008-5472.CAN-07-0472

Sun, R., Hu, Z., Curtis, C. (2018). Big bang tumor growth and clonal evolution. *Cold Spring Harb Perspect Med*, *8*, a028381. doi: https://doi.org/10.1101/cshperspect.a028381

van Valen, L., Mairorana, V.C. (1991). Hela, a new microbial species. *Evol Theor*, *10*, 71–4.

Vargas-Rondón, N., Villegas, V.E., Rondón-Lagos, M. (2017). The role of chromosomal instability in cancer and therapeutic responses. *Cancers*, *10*, 4. doi: https://doi.org/10.3390/cancers10010004

Vendramin, R., Litchfield, K., Swanton, C. (2021). Cancer evolution: Darwin and beyond. *EMBO J*, Sep 15; *40(18)*, e108389. doi: https://doi.org/10.15252/embj.2021108389 Epub 2021 Aug 30. PMID: 34459009; PMCID: PMC8441388.

Venkatesan, S., Swanton, C. (2016). Tumor evolutionary principles: how intratumor heterogeneity influences cancer treatment and outcome. *Am Soc Clin Oncol Educ Book*, *35*, e141–9. doi: https://doi.org/10.1200/EDBK_158930 PMID: 27249716.

Vincent, M.D. (2010). The animal within: carcinogenesis and the clonal evolution of cancer cells are speciation events sensu stricto. *Evolution*, *64*, 1173–83. doi: https://doi.org/10.1111/j.1558-5646.2009.00942.x

Vincent, T.L., Gatenby, R.A. (2008). An evolutionary model for initiation, promotion, and progression in carcinogenesis. *Int J Oncol*, Apr; *32(4)*, 729–37.

Vineis, P., Berwick, M. (2006). The population dynamics of cancer: a Darwinian perspective. *Int J Epidemiol*, Oct; *35(5)*, 1151–9. doi: https://doi.org/10.1093/ije/dyl185

Waddington, C.H. (1957). *The strategy of the genes*. George Allen & Unwin, London.

Wang, Y., Waters, J., Leung, M. L., Unruh, A., Roh, W., Shi, X., Chen, K., Scheet, P., Vattathil, S., Liang, H., Multani, A., Zhang, H., Zhao, R., Michor, F., Meric-Bernstam, F., Navin, N.E. (2014). Clonal evolution in breast cancer revealed by single nucleus genome sequencing. *Nature*, *512(7513)*, 155–60. https://doi.org/10.1038/nature13600

Weinberg, R.A. (2014). Coming full circle-from endless complexity to simplicity and back again. *Cell*, Mar 27; *157(1)*, 267–71. doi: https://doi.org/10.1016/j.cell.2014.03.004 PMID: 24679541.

Weiss, K.M. (2006). Commentary: evolution of action in cells and organisms. *Int J Epidemiol*, Oct; *35(5)*, 1159–60. doi: https://doi.org/10.1016/10.1093/ije/dyl186 Epub 2006 Sep 20. PMID: 16990286.

Wilkins, A.S., Holliday, R. (2009). The evolution of meiosis from mitosis. *Genetics*, Jan; *181(1)*, 3–12. doi: https://doi.org/10.1534/genetics.108.099762 PMID: 19139151; PMCID: PMC2621177.

Ye, C.J., Horne, S., Zhang, J.Z., Jackson, L., Heng, H.H. (2021). Therapy induced genome chaos: a novel mechanism of rapid cancer drug resistance. *Front Cell Dev Biol*, Jun 10; *9*, 676344. doi: https://doi.org/10.3389/fcell.2021.676344 PMID: 34195196; PMCID: PMC8237085.

Ye, C.J., Liu, G., Bremer, S.W., Heng, H.H. (2007). The dynamics of cancer chromosomes and genomes. *Cytogenet Genome Res*, *118(2–4)*, 237–46. doi: https://doi.org/10.1159/000108306

Ye, C.J., Regan, S., Liu, G., Alemara, S., Heng, H.H. (2018). Understanding aneuploidy in cancer through the lens of system inheritance, fuzzy inheritance and emergence of new genome systems. *Mol Cytogenet*, *11*, 31. doi: https://doi.org/10.1186/s13039-018-0376-2

Ye, C.J., Sharpe, Z., Alemara, S., Mackenzie, S., Liu, G., Abdallah, B., Horne, S., Regan, S., Heng, H.H. (2019). Micronuclei and genome chaos: changing the system inheritance. *Genes (Basel)*, *10(5)*, 366. doi: https://doi.org/10.3390/genes10050366

Ye, C.J., Sharpe, Z., Heng, H.H. (2020). Origins and consequences of chromosomal instability: from cellular adaptation to genome chaos-mediated system survival. *Genes (Basel)*, Sep 30; *11(10)*, 1162. doi: https://doi.org/10.3390/genes11101162 PMID: 33008067; PMCID: PMC7601827.

Zhang, S., Mercado-Uribe, I., Xing, Z., Sun, B., Kuang, J., Liu, J. (2014). Generation of cancer stem-like cells through the formation of polyploid giant cancer cells. *Oncogene*, *33(1)*, 116–28. https://doi.org/10.1038/onc.2013.96; https://en.wikipedia.org/wiki/Chaos_theory

8 Evolutionary and Ecological Perspective on the Multiple States of T Cell Exhaustion

Irina Kareva and Joel S. Brown

8.1 INTRODUCTION

The notion that the immune system can eradicate cancer has, until relatively recently, remained on the fringes of anti-cancer therapy. Indeed, it was believed that cancer cells are too similar to normal cells to be recognized by the immune system. The first major breakthrough in deliberately harnessing the immune system against cancer was made by William Coley, a surgical oncologist who, in the late 1880s, observed that some cancer patients who survived serious bacterial infections underwent spontaneous tumor remissions. After observing a series of such occurrences (Coley), Coley developed a mixture of bacterial strains that he would administer to cancer patients to evoke an anti-tumor immune response, which was effective likely through generation of cross-immunity, leading to development of "Coley's toxins" (Carlson, Flickinger Jr, and Snook). While remarkable for the time, the preparation of the toxins was never standardized, and they eventually ceased being used in 1963.

The next major advance in anti-cancer therapy was pioneered by Steven Rosenberg, who in 1968 encountered a former cancer patient who a decade prior was given weeks to live only to return for an unrelated issue. Rosenberg believed that it was the patient's immune system that had spontaneously eradicated the cancer, and he began his search for ways to deliberately augment a patient's anti-tumor response. The culmination of his efforts was the identification of cytokines that can stimulate anti-cancer immunity and introduction of IL-2 based immunotherapy, approved by the FDA in 1992 for treatment of metastatic renal cell carcinoma and in 1998 for metastatic melanoma. Its use was limited by its toxicity, but it did further advance the notion that the immune system can be unleashed against cancer if given some help.

An alternative approach to achieve the same goal was pursued after discovery of antigen presenting cells, and specifically dendritic cells, which, as the name suggests, serve to prime the adaptive immune system to recognize disease-specific antigens. Improving antigen recognition became the focus of another attempt to energize the immune system against cancer (Steinman and Banchereau). One of the discoverers and advocates of this approach, Ralph Steinman, used the first vaccines developed by his colleagues on himself after receiving a diagnosis of pancreatic cancer in 2007. It is believed that the vaccines helped extend his life well beyond what would have been expected; he died 3 days before being awarded the 2011 Nobel Prize.

The key advance that has finally allowed us to harness the power of the immune system against cancer has been the identification of immune cell checkpoints, such as CTLA-4 in 1991 and PD-1 in 1992. Immune checkpoints serve to provide an additional break on the immune system to prevent autoimmunity. Checkpoints therefore need to be blocked in order to truly enable the immune system to attack cancer cells, which is achieved through a class of drugs known as checkpoint inhibitors (for an excellent review of the history of immunotherapy see (Graeber; Davis)). The first checkpoint inhibitor, ipilimumab, which targets CTLA-4, was approved in 2011 for the treatment of melanoma, followed by anti-PD-1 inhibitor pembrolizumab in 2016, finally ushering in the new generation of immune-based anti-cancer therapy.

DOI: 10.1201/9781003307921-8

Unfortunately, T-cell exhaustion is one of the major barriers to long-term success of immune checkpoint inhibitors, which aim to reverse the state of exhaustion and restore T cell cytotoxic capacity. One of the possible culprits is the growing understanding of the fact that exhaustion is not a single state but instead, a T cell can undergo a series of transitional states, and only some of them are susceptible to checkpoint inhibition.

In (B. C. Miller et al.), Miller et al. indicated that both chronic viral infection and tumors elicit analogous subsets of exhausted CD8+T cells, including "progenitor exhausted" cells that express checkpoints, such as PD1, and are thus more responsive to checkpoint inhibitor therapy (Blackburn et al.; Huang et al.), and "terminally exhausted" T cells, which do not. In subsequent work, Beltra et al. (Beltra et al.) have expanded on this framework and proposed a four-cell-stage developmental framework for exhausted CD8+T cells (Tex): quiescent resident, proliferative circulating, circulating and mildly cytotoxic, and terminally exhausted resident cells, with each subset defined by molecular, transcriptional, and epigenetic mechanisms. As such, it appears that the process of T cell exhaustion can broadly be functionally described as T cells transitioning from an active cytotoxic state to reversible intermediate exhausted state characterized by increased checkpoint expression to an irreversible terminally exhausted state.

This multi-step process of immune cell exhaustion most likely evolved to protect the host against autoimmunity, where a long-term chronic infection that could not be fully eliminated must still be less harmful to an organism's fitness than a state of ongoing active cytotoxicity. This is supported by McKinney et al. (McKinney et al.), who show that transcriptional signatures associated with CD8+T cell exhaustion during chronic infection predict worse clearance of the infection but better prognosis with respect to development of various autoimmune diseases.

Given the critical nature of immune cell exhaustion for treatment of both cancer and chronic infections, it might be informative to explore possible insights from ecological literature, and specifically, different types of predator–prey interactions. As the tumor (prey) starts to grow, it eventually elicits a cytotoxic immune response (predator); the general applicability of the broader predator-prey framework to immune-cancer interactions were discussed in detail in (Kareva et al.; Hamilton, Anholt, and Nelson). Here we expand on this analogy to gain insights into better understanding of evolutionary purpose of such "protracted" exhaustion of the predator. In what follows we 1) contrast the classic predator–prey interactions with that of the immune-cancer cell interaction, 2) discuss the opportunities and constraints placed upon the immune system, 3) examine key elements of cancer cells' antipredator adaptations, and 4) consider immune therapies in the context of biological control agents. As a footnote, throughout we will use the shorthand of "in nature" or "natural systems" to describe ecological interactions between species that do not involve the cancer patient or cancer-immune system interactions while fully recognizing that cancer as an evolving system is natural, and these interactions are not in any way artificial.

8.2 THE IMMUNE SYSTEM AND CANCER AS A PREDATOR–PREY SYSTEM, OR NOT

We view cancer as a speciation event where the cancer cell has jumped levels of selection. Where it was once part of the whole organism's program it has gained agency as an individual organism that evolves in accordance with natural selection. Cancer cells have heritable variation, a struggle for existence, and this variation includes how a cancer cell succeeds ecologically within the confines of its tumor ecosystem. Consequently, they evolve adaptations to exploit opportunities for space and resources and to mitigate hazards from toxins and other cells, including being killed by immune cells. From the point of view of cancer cells, cytotoxic immune cells, such as activated CD8+T cells, are predators. We shall refer to these activated T-cells as killer T cells or often just as T-cells when the context is obvious.

Killer T-cells must make direct contact with cancer cells. They can kill cancer cells by adhering to the cell surface and then puncturing the cell membrane of the cancer cells and injecting cytotoxic agents, such as perforin (Trapani and Smyth), to either induce apoptosis or to create an osmotic imbalance, causing the cell to swell and burst. This is quite different from the way macrophages "prey" upon bacteria or other pathogens by engulfing and ingesting the entire target cell or cell debris (Aminin and Wang). The way T-cells kill cancer cells is similar to natural systems where the predator injects venom or poisons directly into its prey as a means of subduing it. In nature this occurs with numerous species, including spiders and snakes. Single-celled protists such as ciliates experience predation from other ciliate species. Predatory ciliates have three ways of subduing single cell prey: filter feeding through a buccal cavity on smaller prey and particles, suctorial feeding by sessile ciliates that ensnare prey with cilia, and raptorial feeding that involves killing other ciliates via extrusomes (Buonanno and Ortenzi). This last form of predation of one single-celled organism on another comes closest to what killer T-cells do to cancer cells. Like killer T-cells, predator ciliates use the extrusome to release toxic chemicals that paralyze or kill their prey. This release occurs when the predatory ciliate has made direct contact with its prey. In this way cancer cells as prey see a very particular form of predation from killer T-cells.

However, do the killer T-cells conform to the ecology of predators such as raptorial ciliates? From the cancer cells' perspective, cytotoxic immune cells act as predators (Kareva et al.). Cancer cells evolve three categories of immune suppression responses that can be thought of as their antipredator adaptations. Roughly, these include mimicking normal cells (akin to Batesian mimicry), fighting back and temporarily or permanently disabling the killer T-cells (by presenting programmed cell death ligands, PD-L1, and PD-L2 that bind with the PD-1 receptor on T-cells), and signaling (altering cytokines such as interleukin-2, IL-2, or interferon β, INT-β) to other types of immune cells, such as regulatory T-cells, to suppress killer T-cells (akin to plants signaling wasps to prey upon herbivorous insects). These shall be discussed at length in a subsequent section.

While killer T-cell interactions with cancer cells can be abstracted and modelled as a predator-prey system, there are several notable differences. In ecology, a predator is an organism that gains fitness (increased survival and proliferation) by consuming all or part of its prey as its resource to the detriment of the prey's fitness. By this definition, killer T-cells are not really predators. Killer T-cells do not consume the cancer cells. An individual killer T-cell, following a kill gains no resources and does not experience an increase in proliferation or survivorship. Upon killing the cancer cell, the killer T-cell moves on to detect and kill another cancer cell, or it may transition into an exhausted state, or it can die. The "carcass" of the dead cancer cell is degraded or removed by other components of the immune system or even scavenged upon by surrounding cancer cells (pinocytosis [Palm]). Thus, the killer T-cell functions more like a honeybee sacrificing itself by stinging a potential threat to the hive or a soldier ant protecting the ant colony.

The killer T-cells and the cancer cells are also competitors. Both require and consume similar resources for survival and proliferation. Cancer cells may sufficiently deplete glucose so as to starve existing T-cells and slow their proliferation. From an ecological perspective, this has an element of intra-guild predation except that a T-cell does not profit directly from killing a cancer cell. Also, the interaction between killer T-cells and cancer cells have an interference competition component. Clearly, killer T-cells can have a negative effect on a cancer cell's survival, but the reverse is also true whereby cancer cells can directly damage or inhibit T-cells.

8.3   ROLES OF SUBSIDIES AND AUTOIMMUNITY IN IMMUNE INTERACTIONS WITH CANCER CELLS

The killer T-cell exists within a policing bureaucracy of diverse immune cells acting both locally in the tumor and systemically in the lymph nodes, bone marrow, and circulatory system. The different cell types and cell states of the immune system represent an immune-web of interactions that impacts and is impacted by the cancer cells (Kareva et al.). The sophistication and complexities of the immune system likely represent an adaptation for the whole organism. It is a balancing act between autoimmunity and fighting infections, dysregulated normal cells, and cancer. The diversity of pathogens, from viruses, to bacteria, to protists, to multicellular invertebrate parasites, necessitates the diverse set of killer immune cells (macrophages, natural killer cells, and killer T-cells).

There also need to be regulatory feedbacks that prevent immune responses from becoming a runaway process (e.g., the life-threatening consequences of the cytokine storm elicited by the Sars-CoV-2 virus (Fajgenbaum and June; Rana, Chauhan, and Mubayi) and to prevent immune cells from killing too many normal cells as collateral damage (autoimmunity). This second restraint on the immune system involves a variety of regulatory cells (e.g., T-regs) that suppress or curtail both systemic and local proliferation and survival of killer immune cells. Finally, components of the immune system (certain states of macrophages) actually promote the proliferation of other cell types as a component of wound healing. Via inflammatory response signals, cancer cells evolve ways to stimulate such macrophages in a manner that provides additional resources to the cancer cells. In sum, the components of the immune system that either suppress killer immune cells or actively supply growth factors and nutrients to the cancer cells are referred to as a pro-tumor immune response. The action of the killer T-cells in suppressing cancer cells is referred to as an anti-tumor immune response.

Figure 8.1 describes the special features of the interaction of killer T-cells and cancer cells, particularly with regard to different stages of exhaustion. The detection and localization (via antigens) by dendritic cells or cancer cells elicits the recruitment and then activation of T-cells into an active, cytotoxic state. These T-cells now have the capacity to kill cells, regardless of whether they are normal or cancer. Once activated and in the tumor, killer T-cells must make direct contact with the cancer cells. This direct contact serves two purposes, the first is recognition and the second is to use the contact to break the cancer cell's membrane and inject toxins.

The recognition process is analogous to recognition time by foragers in nature (Hughes; Kotler and Mitchell). In nature a forager may facultatively use recognition time to separate desirable from undesirable food or prey items, particularly if the food items are quite similar in terms of visual, tactile, or olfactory cues. Theory suggests that the use of recognition time will increase with the frequency of the undesirable prey relative to the desirable one and with the negative consequences of attacking or consuming the undesirable prey. Thus, when there is a high likelihood of encountering an undesirable prey or the undesirable prey is toxic or life threatening to the predator, the predator

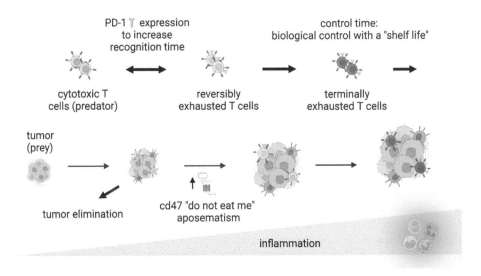

FIGURE 8.1 Different stages of immune cell exhaustion as a function of interactions with cancer cells and inflammation. The figure is adapted from Kareva and Gevertz (2023b).

should make an effort to avoid such prey. In the case of the killer T-cell, the penalty of killing the wrong cell is not directly to itself but rather to the detriment of the whole organism. Hence, safeguards analogous to recognition time have been built into the characteristics of killer T-cells.

Prior to killing an encountered cell, the antigen receptor on the killer T-cell must match and bind to the antigen presenting protein on the surface of the cancer cell. This may take multiple encounters between the T-cell and a particular cancer cell. This may explain the observation that T-cells seem to bounce off of antigen presenting cancer cells several times before effecting a kill (M. J. Miller, Safrina, et al.). Failure to eventually bind with an antigen presenting cell can result in a reversible form of T-cell exhaustion. Such a T-cell may deactivate, but reactivation can occur with signals from other components of the immune system based on antigen release by living or dying cancer cells. Thus, continued killing of cancer cells by other T-cells or the mere turnover of a large cancer cell population can result in re-activating the reversibly exhausted T-cells. Prolonged periods of time in the exhausted state can result in terminally exhausted T-cells that will enter apoptosis (programmed cell death).

Two additional surface proteins may be upregulated by normal cells to discourage accidental killing by T-cells. One of these, CD47, causes the T-cell to disengage from the encountered prey (Takimoto et al.). If cancer cells or those cells targeted by the killer T-cells are relatively rare, this "do not eat me" surface protein minimizes collateral damage. PD-L1 represents a more forceful response by normal cells to prevent unintended killing. By deactivating or even initiating programmed cell death in killer T-cells, the normal cells contribute to suppressing the activity and numbers of activated T-cells. This provides an adaptation for situations when the cells targeted by the killer T-cells are very rare and the number of killer T-cells undesirably high. Cancer cells evolve ways to upregulate one or both of these surface proteins as a way to evade the immune system.

From the perspective of the whole organism, if the ratio of prey (tumor cells) to predator (immune cells) is sufficiently large, and if there are low costs to occasional damage to other species (normal cells), then the optimal strategy is going to involve more active cytotoxicity both directly by the T-cells and indirectly by the local normal cells downregulating CD47 and PD-L1. This is seen in nature, where a high frequency of palatable prey attracts the predator resulting in the increased likelihood of consuming the unpalatable prey (Greenwood, Cotton, and Wilson). The presence of palatable leopard frog tadpoles resulted in dragonfly nymphs killing more of the unpalatable toad tadpoles than they otherwise would have (Wang et al.). This phenomenon is known as "shared doom" (Hay).

However, as the ratio of prey to predator decreases, perhaps as a result of immunoediting (when immune cells eliminate immune-sensitive cancer cells, thereby effectively selecting for immune-resistant cells (Dunn, Old, and Schreiber; Schreiber, Old, and Smyth), such "indiscriminate" predation starts to incur costs to healthy tissues, necessitating a change in strategy. One approach to mitigate the costs of accidentally targeting the wrong prey (normal cells) is to increase recognition time, i.e., the time that is necessary for the predator to access and recognize its prey. The resulting drop in predator efficiency at killing palatable prey because of a high frequency of unpalatable prey is known as an "associational refuge" (Hambäck, Pettersson, and Ericson; Betras et al.). Sparrows foraging for highly palatable millet seeds will harvest fewer when co-occurring with unpalatable ones such as cone flower seeds (Whelan, Brown, and Hank), and fox squirrels harvesting sunflower seeds will similarly harvest fewer palatable ones when they are mixed with sunflower seeds tainted with oxalic acid, a toxin (Emerson et al.).

The process of immune cell exhaustion corresponds to increases in checkpoints as a crucial set of pathways that maintain self-tolerance and serve to prevent indiscriminatory activity of the immune system (Ishida et al.; Ishida). Programmed death protein 1 (PD-1) is one of the most known of the family of checkpoints due to the success of its inhibition for cancer therapy (Eggermont, Maio, and Robert; Esfahani et al.). It is also PD-1 that becomes expressed during the reversible "intermediate" state of immune cell exhaustion, supporting the analogy with its expression corresponding to increased recognition time in other ecological systems. Such checkpoints serve to modulate the T-cells functional response (defined as the rate at which a T-cell kills cancer cells as a function of cancer cell density), as well as its numerical response (defined as the change in the number of T-cells in response to the population size of cancer cells). Both the functional and numerical responses of T-cells seem to change with the ratio of T-cells to cancer cells, as well as the frequency of cancer cells to normal cells. The functional response of the T-cells seems to decline with the ratio of T-cells to cancer cells. This is in line with what in nature is termed ratio dependent predation rates (Akcakaya, Arditi, and Ginzburg; De Troyer et al.).

The recruitment and activation of T-cells, and the transitioning of T-cells to the terminally exhausted state drive the numerical response. Both have some characteristics of standard predator–prey dynamics, but they are driven by quite different processes. Like in predator–prey systems, the increase in T-cell numbers can correlate with the abundance and presence of cancer cells, but this increase in numbers is not directly tied to their functional response on cancer cells. The terminal exhaustion of T-cells has some parallels to predator populations declining from starvation. However, the terminal exhaustion of T-cells may be decoupled from their functional response and dictated by whole system homeostasis and a built-in "shelf-life" that dictates apoptosis. This "control time" has some commonality with "sterile insect techniques" of pest control that include releasing sterilized males to swamp female breeding with non-sterilized males (Kapranas et al.), sterilized pests to boost the number of predators on non-sterilized pests, and sterilized predators to ensure that they themselves do not become a pest (Hendrichs et al.).

It is not entirely clear what triggers the transition from activated T-cells into the reversible or terminal state of exhaustion. The trigger mechanisms are likely density-dependent, with an aforementioned immunoediting-induced reduction in prey density increasing the likelihood of an immune cell targeting a normal cell, requiring a strategy shift. It is also possible that the shift is driven not by absolute density of the prey but the ability of the predator to recognize it, which might be additionally hindered by the expression of "do not eat me" signals, such as CD47, by cancer cells (Jaiswal et al.).

8.4 BATESIAN MIMICRY, NOXIOUSNESS, AND SIGNALING AS CANCER'S ANTIPREDATOR ADAPTATION

In predator–prey systems there are sequences of events that lead to a predator successfully capturing a prey. First comes prey encounter or detection. Predators can increase their encounter rate by way of more rapid movement through an environment, by biasing search efforts toward habitats or areas higher in average prey abundance, and by having the ability to use direct and indirect cues of prey presence usually through sensory cues of vision, hearing, and smell. Such is the case for killer T-cells. In terms of speed, T-cells are not particularly fast. They are not classic pursuit predators such as African hunting dogs hunting wildebeest or adult dragonflies pursuing aerial insects. Nor are they sit-and-wait ambush predators such as many web-building spiders, moray eels striking fish from a concealed lair, or venomous snakes patiently waiting to strike a passing small mammal. Their search strategies seem more akin to an herbivore moving across a landscape with some distribution of palatable vegetation, not unlike a grazing snail feeding on periphyton. T-cells are relatively slow with movement speeds of approximately 11 microns per minute, with peak velocities not exceeding 25 microns/minute (M. J. Miller, Wei, et al.). Sometimes their movement matches that of Brownian motion (random walk) or Levey motion with greater frequencies than by chance of linear motion for extended periods. Much of the motion of T-cells will be influenced by the tissue and density of cells and the stiffness of the extracellular matrix.

In nature, many prey reduce encounters with predators via camouflage whereby a cryptic prey eludes the senses of the predator. Cancer cells use mimicry but not camouflage. For encounters, T-cells literally need to bump into other cells be they normal cells or cancer cells. Because this direct contact seems relatively undirected, cancer cells cannot camouflage their presence. However, this does not mean that cancer cells lack evolved adaptations for thwarting or slowing encounters with killer T-cells. In some cases, cancer cells upregulate the production of collagens and fibronectins that renders the extracellular matrix of the tumor more stiff and less penetrable to immune cells (Rick et al.). Through what is known as the Warburg Effect, many cancer cells exhibit glycolytic metabolism even in the presence of oxygen (Vander Heiden, Cantley, and Thompson), resulting in creation of local acidic microenvironments. Furthermore, to tolerate the acidity many cancer cells upregulate carbonic anhydrase IX (CAIX) that creates a protective boundary layer (Pastorekova and Gillies). Killer T-cells lose functionality under acidic conditions and will avoid such areas. Finally, cancer cells may release signaling proteins, such as cytokines, that promote wound-like environmental cues (Foster et al.; Dvorak). This inflammatory response can 1) promote an environment that killer T-cells avoid, 2) recruit regulatory T-cells (T-regs) that suppress T-cell proliferation, and 3) recruit pro-tumor macrophages that may provide nutrients and resources to the cancer cells.

Cancer cells co-opt normal immune homeostatic processes when they suppress the number and proliferation of killer T-cells by signaling other components of the immune system to place killer T-cells into a permanently exhausted state. This serves to decrease the likelihood of cancer cells encountering an activated killer T-cell. Similar signaling processes can be found in nature. A number of plants will release pheromones to attract wasps that then prey upon the herbivorous arthropods infesting the plant. Mopane trees (*Colophospermum mopane*) in southern Africa can emit volatile signals upon being browsed by mammals such as kudu or elephant. This stimulates neighboring trees to upregulate the production of toxins in their leaves, making them less palatable to the herbivores. While T-regs do not exist as a predator of killer T-cells, they create a cascade of indirect effects that have dynamical similarities to three trophic levels systems in nature where the top trophic level (e.g., a predator = T-regs) benefits the lowest trophic level (e.g., plants = cancer cells) by consuming or suppressing the intermediate trophic level (e.g., herbivores = killer T-cells).

After encountering a potential prey, a second step toward a successful kill is to recognize it as a worthwhile prey. Examples include squirrels investigating an acorn for whether it is fresh (desirable) or infested with weevils (undesirable) or a seed-eating bird testing to determine whether the item is indeed a seed or a small stone or grain of sand with a similar size and appearance. A foraging animal should invest considerable effort into recognition when 1) the frequency of undesirable items is high relative to desirable ones, 2) the subsequent effort in handling a prey item is high, and 3) the cost of consuming an undesirable item is high, perhaps because of toxicities. House sparrows (*Passer domesticus*) alter their harvest of millet seeds (a highly desirable food) when the food patch also contains equal amounts of an undesirable food (Whelan, Brown, and Hank).

Successfully subduing or killing the prey is a third step for the foraging predator. For non-motile, immobile, and sessile prey such as the acorns to a squirrel (though the acorns may be chemically defended so as to be less palatable as described above for many prey) or organic particles to a filter feeder, this final step simply involves handling time. Here, the thicker the shell of a bivalve (e.g., clam) the longer it takes for a moon snail (Naticidae) to use its radula to drill a hole though the shell. Single-celled dinoflagellates possess what to us under the microscope appear as beautiful and varied calcareous shells. Yet these shells serve a key purpose as a handling-time deterrent to would be predators such as copepods (small crustacean zooplankton). For predators that use venom or toxins to immobilize or kill prey, the predator must successfully envenomate the prey, and the prey must be susceptible to the toxin. The California ground squirrel has blood molecules that provide some resistance to the venom of rattlesnakes especially as compared to ground squirrel species that are beyond the range of rattlesnakes (Poran, Coss, and Benjamini; Biardi, Coss, and Smith). Finally, prey may have defense mechanisms that act as an injury deterrent. Injury deterrents by prey can include actively fighting back (e.g., African buffalo defending against lions), emitting noxious chemicals (e.g., the chemical repellent of the bombardier beetle or a chemical release from skunks), or possessing spines (e.g., sea urchins or porcupines). Single-celled Paramecium possess trichocyts (hairlike filaments embedded in the outer cytoplasm that can be discharged into the cell's surroundings) that can repel or even harm their single-celled ciliate predators such as *Dileptus margaritifer*.

In the extreme there is Batesian mimicry where a prey item such as the viceroy butterfly has the same color pattern or appearance to a bird as a toxic prey item such as the monarch butterfly.

In these cases, the toxic prey exhibits aposematic coloration to warn predators. Predators know the toxicity of such prey as the poison dart frogs (*Dendrobatidae*) (Maan and Cummings) and fire salamanders (*Salamandra salamandra*) (Sanchez et al.) through their bright coloration. Such warning colorations are valuable to both predator and prey. The predator does not waste time on an undesirable prey, and the prey avoids being consumed. Normal cells have two forms of "aposematism" to keep killer T-cells from inflicting mortality on the wrong cells. The first, and simplest, is to lack the antigen that allows the T-cells to separate desirable (those that are antigen presenting) from undesirable "prey" (non-antigen presenting). A second, more direct signal involves normal cells upregulating CD47 as a surface protein that communicates "don't eat me" to the killer T-cell that then disengages and moves on. In response to killer T-cells, cancer cells exhibit Batesian mimicry to both of these forms of aposematism. Cancer cells, perhaps at some cost to themselves, will downregulate whatever surface protein is providing the antigenic response and might upregulate CD47 (Jaiswal et al.). Thus, normal cells are the model and cancer cells the mimic.

Another way cancer cells can act as "mimics" is to induce inflammation. In this case, inflammatory cytokines, such as interferons, interleukins, tumor necrosis factors, among others (Lan, Chen, and Wei), which can accumulate as a result of cytotoxic activity, may provide a signal to initiate transitions between different states of CD8+T cell exhaustion (Figure 8.1). Inflammation signals to cytotoxic T cells that an excessive amount of damage has occurred. In response, cytotoxic T cells, to avoid collateral killing of normal cells and slow wound healing, increase their recognition time. This effectively incapacitates the cytotoxic T cells. As such, one might even suggest that cancer cells' stimulation of inflammation acts as a "public good" (Pepper; Stiglitz), stimulating transition of the cytotoxic T cells into an exhausted dysfunctional state by mimicking a chronic infection or tissue wounding.

8.5   IMMUNE THERAPIES AND BIOLOGICAL CONTROL

Immunotherapy and biological control have tantalizing similarities as well as some important differences. Biological control involves introducing, subsidizing, or promoting predators of a pest species (Lacey et al.; Huffaker). Some were ill conceived, such as introducing the Indian mongoose into Hawaii to control rats. Unfortunately, the mongoose found easier pickings preying upon native birds and their nests. Others have now become standard practice, such as introducing predatory mites that can be purchased commercially into greenhouses to control outbreaks of plant-damaging herbivorous mites (Paulitz and Bélanger). For centuries, people have partially-subsidized the feeding of barn cats as a control agent for mice and rats. Fruit flies have been agricultural pests that have spread with human orchards and agriculture. Biological control for fruit flies began with the introduction of parasitoid wasps and other fruit fly predators (Garcia et al.). With some introductions of fruit fly predators came collateral damage as some of these biological control species, like the mongoose, became invasive pests. Newer approaches for fruit flies now include breeding and releasing sterilized males to swamp the breeding pool, predatory nematodes, and sterilized predatory wasps (Shaurub). Currently, biological control agents for invertebrate pests include genetically engineering pathogens (Velivelli et al.). Like immunotherapies aiming to stimulate the patient's immune system, there is a field of conservation biological control with the goal of promoting and facilitating the predators or control species already present in the ecosystem of interest (Begg et al.).

Immunotherapy represents diverse therapies that 1) boost the patient's own immune response to the cancer cells, 2) target the immune-suppressive adaptations of the cancer cells, 3) render the cancer cells more apparent to the immune system, 4) alter the environment of the tumor to make the cancer cells more accessible, and 5) entrain immune cells to the patient's cancer cells and inject them back into the patient. Examples of such "biological control" in cancer immunotherapy include both exogenous and endogenous agents. Exogenous agents include, for instance, chimeric antigen receptor (CAR)-T cell therapies, where immune cells are collected from a patient, modified by adding CARs to program them to recognize a specific antigen and then introduced back into the patient. These therapies have been effective particularly in relapsed and refractory hematologic cancers with complete remission rates as high as 68 to 93% (Zhang et al.; Maude et al.; Park et al.; Lee et al.; Park, Geyer, and Brentjens); however, the risk of autoimmune events, such as cytokine storms, limits their utility.

Some endogenous therapies, in contrast, aim to augment an existing immune response, either by facilitating immune cell access to the tumor through normalizing the vasculature (anti-VEGF drugs), breaking down stromal barriers, treating aberrant chemokines (TGFβ-inhibitors), and reducing hypoxia and acidity (Liu and Sun) or by augmenting the responses of existing immune cells, such as through checkpoint inhibition (i.e., drugs that, within the context of this discussion, aim to revert the initiation and duration of the recognition stage by blocking signals, such as PD-1 or CD47).

An exploration of the literature on biological control and immunotherapy reveals similarities with parallels for each form (Cunningham). Both biological control and immunotherapy attempt to harness, introduce, or stimulate natural processes by which undesirable "pests" are killed by their "predators" or "pathogens". In both, the goal is to complement and reduce the use of biocidal chemicals. Both are routinely studied and implemented in concert with available pesticides or herbicides (in the case of biological control) and with chemotherapies, radiation, and surgery (in the case of cancer treatment).

While biological control and immunotherapy share similarities, there are also some potentially interesting differences. Scale is important. There is a closed system when applying biological control agents in greenhouses or when adding and stimulating the patient's immune response. The infused CAR-T cells do not spread beyond the patient, and the predatory invertebrates released to control herbivorous insects in a greenhouse are supposed to remain in the greenhouse (J. C. Van Lenteren et al.). When added to the greenhouse, the biological control agents may become reproductively self-sustaining or die off either from starvation (having consumed most or all of their prey) or due to engineered infertility (J. Van Lenteren et al.). Of course, there is always the risk that the pest-control species introduced into a greenhouse will escape and spread. Furthermore, unlike the fixed patient-scale of immunotherapies, most uses of biological control species involve unbounded spatial scales. The control species can spread beyond the crop fields or forests of interest into adjacent or even more distant habitats and ecosystems (Pratt and Center). This can be particularly risky when using genetically engineered bacterial or viral pathogens.

The diversity of species available for biological control is vastly greater and more complex than immunotherapies. Immunotherapies always involve using or stimulating the patient's immune cells. These can involve other agents such as monoclonal antibodies or viruses to stimulate an anti-tumor immune response, but the scale of interaction is ultimately between immune cells and cancer cells. This is not so for biological control. First, the target pest species can be plants, fungi, bacteria, as well as all matter of invertebrates and vertebrates (Tscharntke et al.). Second, the pest species can reside in virtually all biomes including terrestrial and aquatic. Third, the diversity of biological control species is also diverse functionally and taxonomically. For example, in the control of water hyacinth, an invasive weed of waterways in Africa, Asia, and Papua New Guinea, two species of weevil and a species of moth provided early biological control (Julien et al.). Agents have expanded to include a plant bug (*Eccritotarsus catarinensis*) (Taylor, Downie, and Paterson) and fungal pathogens.

Nevertheless, despite differences, the field of biological control offers many potential insights into the types of interventions that are more or less likely to succeed in situations that may be similar to immune control systems within the ecosystem of the human body. And perhaps better understanding of the underlying dynamics of immune cell exhaustion, for example, might offer new approaches for immune cell-based therapeutic interventions. Currently, checkpoint inhibitors fight the body's natural adaptation against autoimmunity, where a cancer in a way "mimics" a chronic infection. Checkpoint inhibitors selectively deplete the reversibly exhausted subset of PD-1 expressing T cells, without regard for other ongoing anti-autoimmune adaptations and signals, such as inflammatory cytokines. Consequently, they could be driving the T cell exhaustion process toward the terminally exhausted state, diminishing the treatment's efficacy, since the pool of targetable reversibly exhausted PD-1 expressing cells might not be replenishing.

This would provide an example of development of non-genetic resistance, which can be managed and potentially reversed, in contrast to genetic resistance, in which killing sensitive cells permanently leaves only the resistant subpopulation, and where the only path forward is to seek alternative therapy. In the case of non-genetic resistance, one can potentially enrich the population of susceptible cells either potentially through microenvironmental modification (Kareva and Brown) or through changing the dose and schedule of administration (Pasquier, Kavallaris, and André; Kerbel et al.; Kareva and Gevertz). Understanding the evolutionary forces that have shaped a biological behavior of cancer-immune interactions, even and perhaps especially in another context, and drawing upon learnings from other attempts at biological control, might allow us to circumvent the therapeutic hurdle and expand effectiveness of one of the most promising and effective anti-cancer therapies of our generation.

CONFLICTS OF INTEREST

Irina Kareva is an employee of EMD Serono, business of Merck KGaA. Views and opinions expressed in this manuscript do not necessarily represent the views of EMD Serono.

REFERENCES

Akcakaya, H Resit, Roger Arditi, and Lev R Ginzburg. "Ratio-dependent predation abstraction that works." *Ecology* 76.3 (1995): 995–1004. Print.

Aminin, Dmitry, and Yun-Ming Wang. "Macrophages 'weapon' Anticancer cellular Immunotherapy." *The Kaohsiung Journal of Medical Sciences* 37.9 (2021): 749–758. Print.

Begg, Graham S et al. "A functional overview conservation biological control." *Crop Protection* 97 (2017): 145–158. Print.

Beltra, Jean-Christophe et al. "Developmental relationships four exhausted CD8+ T cell subsets reveals underlying transcriptional epigenetic landscape control mechanisms." *Immunity* 52.5 (2020): 825–841. Print.

Betras, Tiffany L et al. "Do invasive species provide refuge browsers: A test associational resistance peri urban habitat plagued deer." *Forest Ecology and Management* 510 (2022): 120086. Print.

Biardi, James E, Richard G Coss, and David G Smith. "California ground squirrel (spermophilus beecheyi) blood sera inhibits crotalid venom proteolytic activity." *Toxicon* 38.5 (2000): 713–721. Print.

Blackburn, Shawn D et al. "Selective expansion subset exhausted CD8 T cells αPD-L1 blockade." *Proceedings of the National Academy of Sciences* 105.39 (2008): 15016–15021. Print.

Buonanno, Federico, and Claudio Ortenzi. "Predatorprey interactions ciliated protists." *Extremophilic Microbes and Metabolites-Diversity, Bioprospecting and Biotechnological Applications* (2018): 13–40. Print.

Carlson, Robert D, John C Flickinger Jr, and Adam E Snook. "Talkin'toxins Coley's modern cancer immunotherapy." *Toxins* 12.4 (2020): 241. Print.

Coley, William B. "Thetreatment malignant tumors repeated inoculations erysipelas: With report ten original cases 1." *The American Journal of the Medical Sciences (1827–1924)* 105.6 (1893): 487. Print.

Cunningham, Jessica J. "Acall integrated metastatic management." *Nature Ecology & Evolution* 3.7 (2019): 996–998. Print.

Davis, Daniel M. *The beautiful cure: The revolution immunology what it means your health.* University of Chicago Press, 2018. Print.

De Troyer, Niels et al. "Ratio dependent functional response two common Cladocera present farmland ponds Batrachochytrium dendrobatidis." *Fungal Ecology* 53 (2021): 101089. Print.

Dunn, Gavin P, Lloyd J Old, and Robert D Schreiber. "The three Es cancer immunoediting." *Annual Review of Immunology* 22 (2004): 329–360. Print.

Dvorak, Harold F. "Tumors wounds that do not heal—redux." *Cancer Immunology Research* 3.1 (2015): 1–11. Print.

Eggermont, Alexander MM, Michele Maio, and Caroline Robert. "Immune check point inhibitors melanoma provide cornerstones curative therapies." *Seminars Oncology* 42 (2015): 429–435. Elsevier. Print.

Emerson, Sara E et al. "Scale dependent neighborhood effects shared doom associational refuge." *Oecologia* 168 (2012): 659–670. Print.

Esfahani, K et al. "A review cancer immunotherapy: Past, present, future." *Current Oncology* 27.s2 (2020): 87–97. Print.

Fajgenbaum, David C, and Carl H June. "Cytokine storm." *New England Journal of Medicine* 383.23 (2020): 2255–2273. Print.

Foster, Deshka S et al. "The evolving relationship wound healing tumorstroma." *JCI Insight* 3.18 (2018): n. pag. Print.

Garcia, Flávio RM et al. "Biological control tephritid fruit flies Americas Hawaii: A review use parasitoids predators." *Insects* 11.10 (2020): 662. Print.

Graeber, Charles. *The break through immunotherapy race cure cancer.* Twelve, 2018. Print.

Greenwood, Jeremy JD, Peter A Cotton, and Duncan M Wilson. "Frequency dependent selection aposematic prey: Some experiments." *Biological Journal of the Linnean Society* 36.1–2 (1989): 213–226. Print.

Hambäck, PA, J Pettersson, and L Ericson. "Are associational refuges species specific?" *Functional Ecology* (2003): 87–93. Print.

Hamilton, Phineas T, Bradley R Anholt, and Brad H Nelson. "Tumour immunotherapy lessons predator–prey theory." *Nature Reviews Immunology* (2022): 1–11. Print.

Hay, Mark E. "Marine chemical ecology what's Known what's next?" *Journal of Experimental Marine Biology and Ecology* 200.1–2 (1996): 103–134. Print.

Hendrichs, Jorge et al. "Improving cost-effectiveness, trade safety biological control agricultural insect pests using nuclear techniques." *Biocontrol Science and Technology* 19.sup1 (2009): 3–22. Print.

Huang, Alexander C et al. "T cell invigoration tumour burden ratio associated Anti PD1 response." *Nature* 545.7652 (2017): 60–65. Print.

Huffaker, Carl B. *Theory practice biological control.* Elsevier, 2012. Print.

Hughes, Roger N. "Optimal diets energy maximization premise effects recognition time learning." *The American Naturalist* 113.2 (1979): 209–221. Print.

Ishida, Yasumasa et al. "Induce dexpression PD-1, novel member immunoglobulin gene superfamily, programmed cell death." *The EMBO Journal* 11.11 (1992): 3887–3895. Print.

———. "PD1 its discovery, involvement cancer immunotherapy." *Cells* 9.6 (2020): 1376. Print.

Jaiswal, Siddhartha et al. "CD47 is upregulated circulating hematopoietic stem cells leukemia cells avoid phagocytosis." *Cell* 138.2 (2009): 271–285. Print.

Julien, MH et al. "International cooperation linkages management water hyacinth emphasis biological control." *Proceedings IX International Symposium Biological Control Weeds.* Vol. 9. University of Cape Town Stellenbosch, 1996. 273–282. Print.

Kapranas, Apostolos et al. "Review role sterile insect technique biologically-based pest control–an appraisal existing regulatory frameworks." *Entomologia Experimentalis et Applicata* 170.5 (2022): 385–393. Print.

Kareva, Irina, and Joel S Brown. "Estrogen essential resource coexistence ER+ER cancer cells." *Frontiers in Ecology and Evolution* (2021): 534. Print.

Kareva, Irina, and J Gevertz. "Cytokine storm mitigation exogenous immune agonists." *Mathematics of Control, Signal and Systems* (2023a): 1–22. Print.

———. "Mitigating non-genetic resistance checkpoint inhibition based multiplestates immune exhaustion." *Research Square Preprint* (2023b). https://doi.org/10.21203/rs.3.rs-3358908/v1. Print

Kareva, Irina et al. "Predator prey tumor-immune interactions: a wrong model just incomplete one?" *Frontiers in Immunology* (2021): 3391. Print.

Kerbel, RS et al. "Continuous low doseanti angiogenic/metronomicchemotherapy research laboratory oncology clinic." *Annals of Oncology* 13.1 (2002): 12–15. Print.

Kotler, Burt P, and William A Mitchell. "The effect costly information diet choice." *Evolutionary Ecology* 9 (1995): 18–29. Print.

Lacey, LA et al. "Insect pathogens biological control agents future." *Journal of Invertebrate Pathology* 132 (2015): 1–41. Print.

Lan, Tianxia, Li Chen, and Xiawei Wei. "Inflammatory cytokines cancer: Comprehensive understanding clinical progress gene therapy." *Cells* 10.1 (2021): 100. Print.

Lee, Daniel W et al. "T cells expressing CD19 chimericantigen receptors acute lymphoblastic leukaemia children young adults phase1 dose escalation trial." *The Lancet* 385.9967 (2015): 517–528. Print.

Liu, Yuan-Tong, and Zhi-Jun Sun. "Turning cold tumors hot tumors improving T-cell infiltration." *Theranostics* 11.11 (2021): 5365. Print.

Maan, Martine E, and Molly E Cummings. "Sexual dimorphism directional sexual selection Aposematic signals poison frog." *Proceedings of the National Academy of Sciences* 106.45 (2009): 19072–19077. Print.

Maude, Shannon L et al. "Tisagenlecleucel children young adults B-cell lymphoblastic leukemia." *New England Journal of Medicine* 378.5 (2018): 439–448. Print.

McKinney, Eoin F et al. "T cell exhaustion, co-stimulation clinical outcome autoimmunity infection." *Nature* 523.7562 (2015): 612–616. Print.

Miller, Brian C et al. "Subsets exhausted CD8+ T cells differentially mediate tumor control respond checkpoint blockade." *Nature Immunology* 20.3 (2019): 326–336. Print.

Miller, Mark J, et al. "Autonomous T cell trafficking examined vivo intravital two photon microscopy." *Proceedings of the National Academy of Sciences* 100.5 (2003): 2604–2609. Print.

———. "Imaging single cell dynamics CD4+T cell activation dendritic cells lymphnodes." *The Journal of Experimental Medicine* 200.7 (2004): 847–856. Print.

Palm, Wilhelm. "Metabolic functions macropinocytosis." *Philosophical Transactions of the Royal Society B* 374.1765 (2019): 20180285. Print.

Park, Jae H, Mark B Geyer, and Renier J Brentjens. "CD19 targeted CART cell therapeutics hematologic malignancies: Interpreting clinical outcomes date." *Blood, the Journal of the American Society of Hematology* 127.26 (2016): 3312–3320. Print.

Park, Jae H et al. "Long term follow CD19 CAR therapy acutelymphoblastic leukemia." *New England Journal of Medicine* 378.5 (2018): 449–459. Print.

Pasquier, Eddy, Maria Kavallaris, and Nicolas André. "Metronomic chemotherapy: New rationale new directions." *Nature Reviews Clinical Oncology* 7.8 (2010): 455–465. Print.

Pastorekova, Silvia, and Robert J Gillies. "Therole carbonic anhydrase IX cancer development links hypoxia, acidosis." *Cancer and Metastasis Reviews* 38.1–2 (2019): 65–77. Print.

Paulitz, Timothy C, and Richard R Bélanger. "Biological control greenhouse systems." *Annual Review of Phytopathology* 39.1 (2001): 103–133. Print.

Pepper, John W. "Drugs that target pathogen public goods are robust evolved drug resistance." *Evolutionary Applications* 5.7 (2012): 757–761. Print.

Poran, Naomie S, Richard G Coss, and ELI Benjamini. "Resistance California ground squirrels (Spermophilus beecheyi) venom Northern Pacific rattlesnake (Crotalus viridis oreganus) study adaptive variation." *Toxicon* 25.7 (1987): 767–777. Print.

Pratt, PD, and TD Center. "Biocontrol borders unintended spread introduced weed biological control agents." *BioControl* 57.2 (2012): 319–329. Print.

Rana, Payal, Sudipa Chauhan, and Anuj Mubayi. "Burden cytokines storm prognosis SARS-CoV-2 infection immune response dynamic analysis optimal control immuno modulatory therapy." *The European Physical Journal Special Topics* (2022): 1–19. Print.

Rick, Jonathan W et al. "Fibronectin malignancy: Cancer-specific alterations, protumoral effects, therapeutic implications." *Seminars Oncology* 46 (2019): 284–290. Elsevier. Print.

Sanchez, Eugenia et al. "The conspicuous post metamorphic coloration fire salamanders, not their toxicity, is affected larval background albedo." *Journal of Experimental Zoology Part B: Molecular and Developmental Evolution* 332.1–2 (2019): 26–35. Print.

Schreiber, Robert D, Lloyd J Old, and Mark J Smyth. "Cancer immunoediting integrating immunity's roles cancer suppression promotion." *Science* 331.6024 (2011): 1565–1570. Print.

Shaurub, El-Sayed H. "Review entomopathogenic fungi nematodes biological control agents tephritid fruit flies current status future vision." *Entomologia Experimentalis et Applicata* 171.1 (2023): 17–34. Print.

Steinman, Ralph M, and Jacques Banchereau. "Taking dendritic cells medicine." *Nature* 449.7161 (2007): 419–426. Print.

Stiglitz, Joseph E. "The theory local public goods." *The Economics Public Services* (1977): 274–333. Springer. Print.

Takimoto, CH et al. "The macrophage 'Do not eat me' signal, CD47, is clinically validated cancer immunotherapy target." *Annals of Oncology* 30.3 (2019): 486–489. Print.

Taylor, SJ, DA Downie, and ID Paterson. "Genetic diversity introduced populations water hyacinth biological control agent eccritotarsus catarinensis (Hemiptera Miridae)." *Biological Control* 58.3 (2011): 330–336. Print.

Trapani, Joseph A, and Mark J Smyth. "Functional significance perforin/granzyme cell death pathway." *Nature Reviews Immunology* 2.10 (2002): 735–747. Print.

Tscharntke, Teja et al. "Reprint 'Conservation biological control enemy diversity landscape scale' [Biological Control 43 (2007) 294–309]." *Biological Control* 45.2 (2008): 238–253. Print.

Vander Heiden, Matthew G, Lewis C Cantley, and Craig B Thompson. "Understanding Warburg effect metabolic requirements cell proliferation." *Science* 324.5930 (2009): 1029–1033. Print.

Van Lenteren, Joop C et al. "Aphelinid parasitoids sustainable biological control agents greenhouses." *Journal of Applied Entomology* 121.1–5 (1997): 473–485. Print.

———. "Biological control agents control pests greenhouses." *Integrated Pest and Disease Management in Greenhouse Crops* (2020): 409–439. Print.

Velivelli, Siva LS et al. "Biological control agents field market, problems, challenges." *Trends in Biotechnology* 32.10 (2014): 493–496. Print.

Wang, Cuiyan et al. "A chain mediation model COVID-19 symptoms mental health outcomes Americans, Asians Europeans." *Scientific Reports* 11.1 (2021): 6481. Print.

Whelan, Christopher J, Joel S Brown, and Amy E Hank. "Diet preference house Sparrow Passer domesticus: Hooked Millet?" *Bird Study* 62.4 (2015): 569–573. Print.

Zhang, Xian et al. "Efficacy safety anti CD19 CART cell therapy 110 patients B cell Acutelymphoblastic leukemia high-risk features." *Blood Advances* 4.10 (2020): 2325–2338. Print.

9 Landscape Genetics for Cancer Biology

Erin L. Landguth and Norman A. Johnson

9.1 CANCER AND THE LANDSCAPE

Cancer is unlike most other diseases in that it arises from a caricature of tissue renewal and other normal cellular processes (Pierce and Speers 1988). It is a disease where our own cells rebel (Aktipis 2020; Arney 2020). Cancer's cellular rebellion is not only an ecological and evolutionary process (Somarelli 2021), but also is a spatial process (Bressan et al. 2023). The unfolding of cancer from the first mutated cell to the metastatic tumor sending propagules of tumors to other tissues occurs within a landscape—the human body. Thus, improved understanding of cancer progression and possibly the development of treatments should arise from a better understanding of how tumor cells divide, move, and interact with their environment in an explicit spatial context.

We contend that the field of landscape genetics can be applied to provide this spatial context for cancer progression. Landscape genetics, which emerged in the early 2000s, examines how spatial genetic patterns emerge from the movement of individuals and their interactions with the environments they experience. Here, we discuss how the concepts and information from landscape genetics can be applied to cancer progression. We discuss the similarities and the differences between the ecological processes that cancer cells experience within the landscape of the human body with those processes that organisms experience. We begin with a short history of landscape genetics. Then, we define terminology between the two fields and where these parallel connections can be made. We follow with a presentation of ongoing research examples in landscape genetics with potential applications in cancer biology that includes (1) understanding how landscape structure and composition impact genetic processes and (2) key methodologies unique to landscape genetics. We hope this chapter will inspire future work and collaborations at the intersection of landscape genetics and cancer biology, spurring innovative research to translational applications.

9.2 A BRIEF HISTORY OF LANDSCAPE GENETICS THROUGH THE LENS OF GENETIC CONNECTIVITY

Individuals of species (animals, microbes, plants, or plant propagules) move across landscapes, which are often heterogenous. In addition to persisting in established landscapes, individuals can also colonize new habitats and foster new populations. A fundamental component of these population movements is connectivity. Genetic connectivity is defined by Lowe and Allendorf (2010) as the "degree to which gene flow affects evolutionary processes within subpopulations". The nature of genetic connectivity of populations impacts the maintenance of genetic variation between and within populations. Such variation affects the current and future viability of populations as well as their ability to adapt to environmental change (Lewontin 1974; Frankham et al. 2010; Kardos et al. 2021).

The inception of these ideas began through the theoretical lenses of the fields of population genetics and evolutionary biology in the early 20th century (e.g., Hardy 1908; Wright 1917; Fisher 1918; reviewed in Provine 1971). During the second half of the 20th century various technologies emerged that permitted the quantification of genetic differences within and among natural populations. Notably, geneticists used electrophoretic protein variants (allozymes) (Lewontin and Hubby 1966; Lewontin 1974) and then DNA sequencing (Kreitman 1983; Aquadro and Greenberg 1983)

DOI: 10.1201/9781003307921-9

coupled with the polymerase chain reaction (PCR; Mullis et al. 1986) to study the genetic basis of evolution. Application of technologies generate metrics for patterns of genetic variation that shed light on four main evolutionary processes: mutation, selection, gene flow, and genetic drift. These metrics can also be used to study recombination, demography, admixture, and connectivity in populations. Metrics for allelic diversity, inbreeding, genetic differentiation, and effective population sizes have provided conservation genetics with important tools for assessing species conservation status and guiding the management of populations/species. Conservation genetics was then recognized as an important component for the budding field of conservation biology (e.g., Dasmann 1968) roughly at the same time that human-caused loss and fragmentation of habitats were determined to be major drivers of the long-term conservation for populations ((e.g., Levins 1969). See also Soulé and Wilcox (1980) for an early history of conservation biology).

The ecological discipline of landscape ecology emerged shortly after conservation biology in the late 20th century to address species spatial distributions, patterns, and processes (e.g., Levin 1992; Turner 1989; Turner et al. 2001). The key focus of landscape ecology is understanding the effects of landscape heterogeneity (structure, composition, scale) on the interactions between individuals (Wiens 1992). As with many fields, technological advances propelled landscape ecology forward through remote sensing and geographical informational systems (GIS) and specifically advanced areas in ecology that consider spatial and temporal variability across scales (Kareiva and Wennergren 1995). But as broadscale environmental challenges emerged (biodiversity loss, habitat fragmentation, climate change, non-native species invasions), so too did much of the developments in landscape ecology.

During the 1980s and 1990s, conservation biologists began to fuse ideas from population and conservation genetics with those of conservation biology and landscape ecology (e.g., Pamilo 1988; Gaines et al. 1997). However, landscape genetics did not fully form as a field until the seminal paper from Manel et al. (2003). These authors argued that landscape genetics differs from other genetic approaches, such as phylogeography (the study of historical processes affecting the geographic distribution of genetic relationships and lineages), in that it tends to focus on processes at finer spatial scales and at the individual level. Landscape genetics models are also different than habitat suitability or species distribution in that landscape genetics measures successful dispersal via reproduction, and habit suitability models measure home range habitat dispersal. Much excitement grew around this field, and many newly applied methods emerged (see Section 9.4.3) from network theory (Murphy et al. 2008) to individual-based landscape simulation modeling (Landguth et al. 2010a, 2010b) to spatial statistical methods (van Strien et al. 2012) with the field stimulating research in ecology, evolution, and conservation (see Manel and Holderegger 2013; Dauphin et al. 2023). Balkenhol et al. (2015a, p. 3) nicely consolidated early definitions (Manel et al. 2003; Holderegger et al. 2006; Storfer et al. 2007) into the following: landscape genetics involves research that combines population genetics, landscape ecology, and spatial analytical techniques to explicitly quantify the effects of landscape composition, configuration, and matrix quality on micro-evolutionary processes, such as gene flow, drift, and selection, using neutral and adaptive genetic data."

9.3   THE LEVELS OF ORGANIZATION PERTINENT TO LANDSCAPE GENETICS AND CANCER BIOLOGY

It is important to first define field-specific terminology that will allow us to draw connections between landscape genetics and cancer biology. Figure 9.1 identifies the levels of organization within each field and the corresponding connections through hierarchical levels. The organization levels are familiar concepts within each respective field: left-hand side of triangle corresponds to units within ecology for landscape genetics concepts and right-hand side of triangle refers to units within biology for cancer biology concepts. However, we can connect each field's level and study units starting with a single entity—the *individual* level for ecology and the *cell* level for biology. For example, in cancer biology cancer cells (level 1) are grouped together to form tumors (level 2).

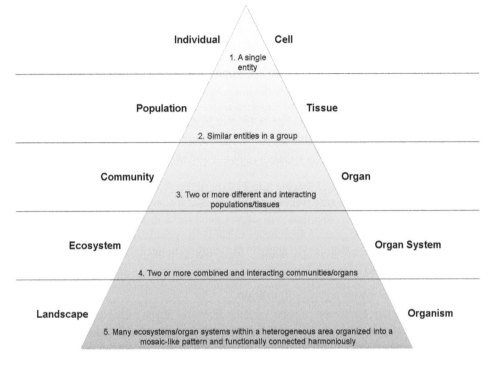

FIGURE 9.1 Organizational levels of ecology and biology. Landscape genetics follows the levels of ecology (on left), and cancer biology follows the organizational levels of biology (on right). (1) A single entity begins each level with cascading groupings up to the final study units of *landscape* (for ecology) and *organism* (for cancer).

A tumor microenvironment aggregates to form biological communities of coordinating units within their biotic and abiotic environments (levels 3–4). Eventually, microenvironments connect through heterogenous areas organized in mosaic-like patterns that are functionally connected harmoniously (level 5). We note that the organization chart could be further refined to start at even a smaller unit of study (e.g., molecular and to study how DNA-level changes influence all cascading levels) or expanded to consider biomes and global organismal genetics. But for the purposes of this chapter, we will begin here with the connections made from the individual to the landscape and from the cell to the organism for both landscape genetics and cancer biology, respectively.

9.4 KEY RESEARCH AREAS WITHIN LANDSCAPE GENETICS WITH PARALLELS TO CANCER BIOLOGY

9.4.1 How Landscape Structure and Composition Affect Population Genetic Processes

The habitats in which individuals and tumors live vary spatially and across scales. The structure and composition of landscapes are key drivers of both demographics (population dynamics and life history strategies) and genetic processes (gene flow, genetic variation, adaptive evolution). Any disruption in the landscape or organism (Figure 9.1) structure or composition in which individuals or tumors reside, respectively, may interfere with the ecological processes necessary for population persistence (Fahrig 2002). For example, habitat loss and fragmentation create "patchy" landscapes that in turn may disturb critical resources or create discontinuities that prevent movement and dispersal, reducing genetic connectivity. Note again, that genetic connectivity is measured as the degree to which the single entity unit (individual or cell) successfully disperses to a new location, reproduces there, and subsequently passes on its genetic material. What constitutes healthy/unhealthy or successful/unsuccessful genetic connectivity depends on the single entity's genetic makeup and its ability to navigate key habitat and infiltrate landscape barriers. Disruption to key habitat can lead to reduced fitness, population declines, and/or adaptive evolution under directional selection. The degree to which the landscape structure and composition affects population genetic processes is still an ongoing research direction in landscape genetics but arguably dominated by (1) habitat composition and configuration and (2) scale effects (in both time and space) and ultimately how these characteristics of landscape structure and composition influence population processes and persistence.

The microhabitats that tumors and nascent tumors experience also affect their ability to survive, grow, and disperse. Regions will vary with respect to oxygen concentrations, nutrient availability, and several other characteristics. We thus will expect patchiness in suitability of tissues for the tumor to thrive. The landscape genetic approaches to study the population processes of organisms in patchy environments can be modified to apply to the study of tumor cells across the patchy environments of the human body.

9.4.1.1 Habitat Composition and Configuration

One of the most important landscape impacts on gene flow and fitness is habitat composition and configuration (or the distribution and amount of natural environment for a species). Habitat composition and configuration determines the availability of resources. Disruption of this key habitat through reduced area, introduced structural discontinuities (barriers), or fragmentation/loss can affect individual foraging behavior possibly changing the ability to acquire the resources needed to survive and reproduce (Mangel and Clark 1986; Wiens et al. 1993). If food resources become more patchily distributed on the landscape, it may be more costly to acquire them (Mahan and Yahner 2000). Most species require at least a minimum area of habitat in order to meet all life history requirements (e.g., Robbins et al. 1989) with the general rule of the larger the species the more habitat requirement. Theoretical studies predict a threshold habitat level below which the population cannot sustain itself (Fahrig 2002; Flather and Bevers 2002; Hill and Caswell 1999). Within cancer biology, parallels exist for understanding this corresponding threshold habitat level for how tumor cells utilize available food resources (nutrients in blood or surrounding proteins) within different microenvironments.

9.4.1.2 Temporal Scale Effects

Temporal scale and dynamics of landscape genetic processes—or how rates and patterns of landscape change affect the emergence, change, and loss of genetic structure—is a primary area of research in landscape genetics (Ezard and Travis 2006; Murphy et al. 2008; Cushman and Landguth 2010). These temporal dynamics interact with aspects of the biology of a species. For example, consider dispersal capabilities of a species and the ability of genetic analyses to detect barriers (anthropogenic or natural). Landscape genetics studies have explored the interaction between species-specific dispersal characteristics and the emergence and loss of genetic structure following establishment and removal of dispersal barriers (Landguth et al. 2010b). When organisms have naturally high dispersal ability relative to their natural range, signals of barriers will quickly appear. In contrast, in studies of organisms with reduced dispersal, the appearance of a barrier signal will take much longer to detect. Hence, in genetic studies of species with low dispersal, effects of landscape fragmentation may not be detected for many generations even if the disturbance has resulted in complete isolation of subgroups (e.g., Epperson 2007; Spear and Storfer 2008; Murphy et al. 2010). Furthermore, for the groups that have very limited dispersal abilities, legacy effects from past landscape conditions could be detected for dozens to hundreds of generations. These landscape genetics barrier detection studies highlight the importance of considering species vagility and life history within the context of a species range extent as well as temporally changing landscapes (Figure 9.2). One largely open field of inquiry is how variation in the plasticity of populations (in this case, ability to switch life history strategies) will impact interaction with the landscape.

In the case of cancer, some cancer cell populations are able to switch from more stationary, proliferative phenotypes to more migratory, invasive, proliferative-low phenotypes (Jolly et al. 2017). For cancer biology, understanding potential effects of microenvironments may require understanding the historical life history strategy of cells (e.g., long-distance dispersal, migratory) and the evolutionary history derived from past signals of genetic structure. Moreover, profound changes in tumor microenvironments occur during normal aging. These include changes in the numbers of types of fibroblasts, the cells that produce collagen and other substances; alterations in the populations of immune cells; changes to the extracellular matrix; and an increase in chronic inflammation (inflammaging) (reviewed in Fane and Weeraratna 2020). These aging-related, tumor microenvironment changes likely help explain why cancer incidences rapidly rise with age (DeGregori 2018). Landscape genetic tools can be modified to study these changes. See also Section 9.4.2.5.

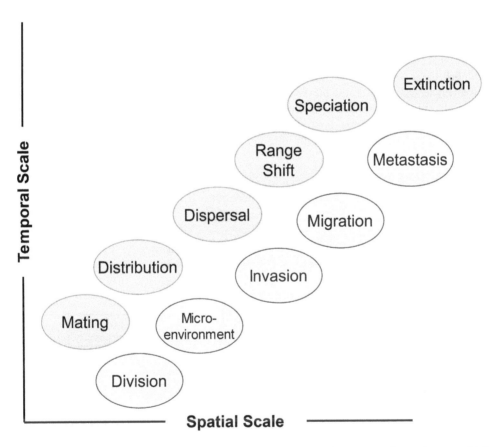

FIGURE 9.2 Spatiotemporal scales and relevant processes in landscape genetics and cancer biology. Different landscape genetics (green) and cancer biology (grey) processes occur at different spatial and temporal scales. The process of mating among individuals of a population as well as cellular division (mitosis) occurs at short time scales and fine spatial scales relative to dispersal and invasion distances, which occur at larger spatial scales. Invasion of tumor cells is the process of spreading of the cells into neighborhood normal microenvironments. The species distribution that corresponds to the tumor cells' initial microenvironment typically encompasses the individual/cell extent for both mating/division and dispersal/invasion. The range of a species distribution may shift at longer time scales. Through migration eventually metastasis occurs, and speciation and extinction of the species typically occur at the longest time scales.

9.4.1.3   Spatial Scale Effects

Landscape genetics focuses on the interaction of patterns and processes that are inherently scale-dependent and hierarchical in concept (Wiens 1989; Levin 1992). Spatial scale has been shown to influence relationships of genetic variation and landscape structure (van Strien et al. 2012). Despite the acknowledgment that scale must be considered in landscape genetics studies (Balkenhol et al. 2015b), the vast majority of studies do not address these issues (but see Galpern and Manseau 2013; Balkenhol et al. 2020). This importance cannot be stressed enough as landscape structure and composition affect processes at characteristic scales (Figure 9.2). This means that processes act at characteristic scales that are in turn affected by landscape structure and composition. For example, processes of mating or cellular division occur at the finest scale. Eventually, landscape structure and composition may affect species distribution resulting in range shift. Equivalently for cancer biology this could mean that tumor cells migrate out of current microenvironments and metastasize in new locations. Ultimately, once a species has exhausted all possible scales, extinction will occur. Similar ideas can be gleaned for cancer biology as tumor cells exhaust space or overwhelm an organism. How landscapes interact with these extinction vortexes is discussed next.

9.4.1.4   The Extinction Vortex and Evolutionary Potential

The extinction vortex is a helpful concept from conservation biology (first coined by Gilpin and Soulé 1986) that could parallel ideas in cancer biology research—that is, creating mutual reinforcements among biotic and abiotic processes to drive tumor "species" population sizes downward to extinction (Figure 9.3). The dominant landscape characteristics of habitat composition, configuration, and scale (discussed previously) influence landscape genetic processes in several ways that, in turn, create this extinction vortex. First, through habitat loss (or ecological disturbances decreasing habitat extent), populations initially decline through reduced vital rates and population growth rates,

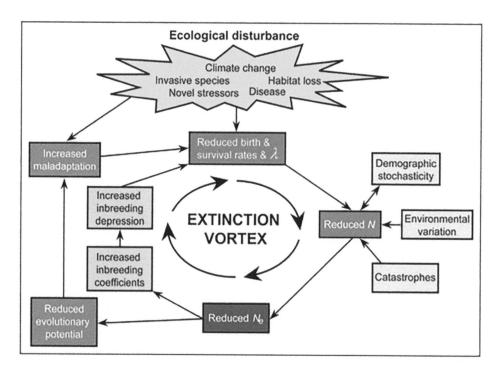

FIGURE 9.3   The extinction vortex. Habitat loss and ecological disturbances (yellow) reduce population dynamic processes (red) and population sizes (blue) and effective sizes (purple) and ultimately lead to more inbreeding (light green) and reduced evolutionary potential (dark green). Figure from Forester et al. (2022).

which results in reduction in effective population sizes, acceleration of genetic drift, lower equilibrium heterozygosity and allelic richness, and ultimately, increasing inbreeding and causing inbreeding depressions. Landscape composition and structure can comprise of fragmented habitat patches, creating a mosaic that determines the permeability of the landscape to movement (known as landscape "resistance", discussed more in Section 9.4.2.1). Increasing the landscape resistance can result in local populations becoming functionally isolated due to this habitat loss or edge effects acting as a barrier that impedes movement. As habitat loss approaches the extinction threshold in some areas of the landscape, these local populations may rapidly decline to extinction, resulting in gaps in distribution, which lead to attenuated gene flow, spatial genetic structure, and further reductions in heterozygosity and allelic richness. All of this further degrades populations and reduces evolutionary potential (the capacity to evolve genetically-based changes in traits that increase population-level fitness in response to changing environmental conditions; Forester et al. 2022) resulting in maladaptive phenotypes and reduced fitness. As the habitat area is further reduced beyond the extinction threshold the entire regional population will become extinct. While difficult to quantify the extinction vortex, proxies and efforts to measure evolutionary potential are currently being considered in species extinction risk assessments (Forester et al. 2022).

A recent example by Forester et al. (2023) used individual-based, spatially explicit dynamic eco-evolutionary simulations to evaluate the extinction risk of an endangered desert songbird, the southwestern willow flycatcher (*Empidonax traillii extimus*) in response to climate change. They illustrate how evolutionary potential (or lack of) together with compounding impacts of drought and warming temperatures influence extinction risk (Figure 9.4). Analogies to extinction risk assessments such as those in Forester et al. (2023) include devising adaptive therapies of cancer. Adaptive therapy (Gatenby et al. 2009) seeks to manage cancer, in opposition to the standard cancer

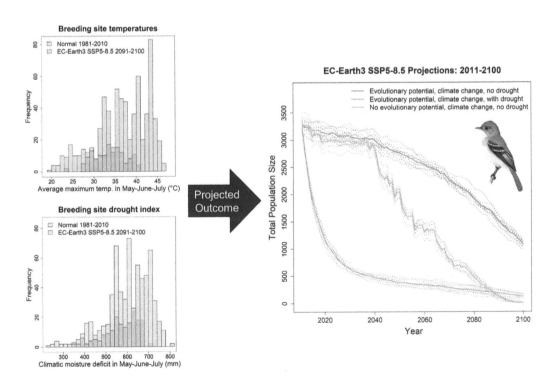

FIGURE 9.4 Evolutionary potential using adaptive landscape genetics simulation modeling. Population trajectories for southwestern willow flycatchers under variable selection due to climate change with and without drought effects of nest success. Figure from Forester et al. 2023.

treatments wherein high doses of a killing agent are used. This adaptive treatment "evolves in response to the temporal and spatial variability of the tumor microenvironment and cellular phenotype as well as therapy-induced perturbations" (Gatenby et al. 2009, p. 4894). The fine-tuning of the therapy is designed to keep the cancer in check while minimizing harm to the patient from the therapy and limiting evolution of resistance of the tumor to the treatments. As knowledge of the evolutionary potential is useful in assessing extinction risk as in Forester et al. (2022), knowing the evolutionary potential of the tumor (among other parameters and variables) will be useful in guiding adaptive therapy regimes.

9.4.2 Landscape Genetics Methods and Applications Relevant for Cancer Biology

Landscape genetics has become an important analytical framework for understanding and measuring genetic connectivity (Manel et al. 2003). Ultimately, understanding how the landscape mediates genetic connectivity can inform strategies for restoring or improving functional connectivity and evolutionary potential among populations. Next, we highlight key methodologies that landscape genetics uses in determining a spatial landscape genetics relationship through creating landscape resistance hypotheses, identifying the optimal resistance hypothesis, validating or simulating pattern-process relationships, developing connectivity models, isolating selection-driven patterns from contemporary neutral genetic signatures, and using environmental DNA (eDNA)—and linking to cancer biology where possible.

9.4.2.1 Creating Resistance to Gene Movement Landscapes

Hypotheses about how the landscape affects genetic connectivity can be as simple as isolation by distance (IBD), where geographic distance is the sole predictor of genetic distance (metrics of pairwise similarities in gene frequencies). In contrast, isolation by resistance (IBR; McRae 2006) or isolation-by-landscape resistance (IBLR) hypotheses involve assigning resistance values to landscape variables according to their propensity to impede genetic movement. Creating resistance surfaces is an important component of landscape genetics and can be further defined as spatial layers that represent the extent to which the conditions at each grid cell constrain movement or gene flow. The approach is based on a gridded map of cells (i.e., raster layer/map) that represent the cell level resistance to movement or gene flow related to different habitat or vegetation types, elevation, slope, or other landscape features (see Cushman et al. 2006; Spear and Storfer 2008). Cells or pixels are given weights or "resistance values" reflecting the presumed influence of each variable on dispersal of the species in question.

For example, in riverscape genetics much attention is paid to IBB (isolation by barriers) from dams or culverts (Figure 9.5) and how gene flow and movement passibility changes directionally (one-way vs two-way direction barriers) (Mims et al. 2019). As with fish passage, tumor cells could have similar asymmetrical resistance to movement ideas applied throughout their movement matrix mediums within an organism (e.g., through endothelium and vasculature or extracellular matrix). As such, riverine landscapes and riverscape genetics could provide more direct applications to cancer biology. Because streams are linear in nature with direction flow, they provide similar analogies to the circulatory system within an organism. Barriers and IBB in this context can play significant roles in limiting movement of individuals and cells, thus facilitating gene flow. Barriers such as culverts and dams have historically fragmented populations and blocked the passage of migratory fish returning to natal grounds for spawning. Construction of these barriers has resulted in declines of migratory fish species (Lawrence et al. 2016), loss of genetic diversity (Zarri et al. 2022), changes to community structure (Freedman et al. 2014), altered geomorphology and flow regimes (Ligon et al. 1995), and loss of Indigenous and commercial fishing practices (Montgomery 2003). The removal of barriers such as dams and culverts has garnered much attention and support in recent decades. Countless studies have quantified the effects of discontinuities due to the presence of culverts and

FIGURE 9.5   Riverine landscape genetics. In riverscape genetics, identifying asymmetrical barriers within a stream network and the passibility probabilities associated with fish movement is an important component for creating resistant to movement surfaces.

dams on aquatic communities (Muhlfeld et al. 2012), the benefits of barrier removal for improving gene flow and connectivity among populations (Landguth et al. 2010b; Prunier et al. 2023), riverine substrate, habitat, and climate on population connectivity (Escalante et al. 2018), and many methods for identifying and prioritizing barrier removals have been developed. How these riverscape genetics studies might translate to circulatory substrate, habitat, and climatic conditions for moving tumor cells, along with maximizing impediments to tumor connectivity, could provide convergent research avenues.

9.4.2.2   Statistical Methods for Identifying Optimal Landscape Resistance

As with most fields, the best mathematical or statistical methods are continually improved over time. Likewise for the field of landscape genetics, much attention has been given recently to the development of statistical methods for identifying the optimal IBR hypothesis for a given population (Shirk et al. 2010, 2012; Peterman and Pope 2021). The basic framework for using resistance to movement surfaces in landscape genetics involves the steps of variable selection, parameterizing values for resistance surfaces, correlating resistance with genetic data, and, in many cases, employing the best supported resistance surfaces in downstream analyses (see Section 9.4.2.4 on connectivity modeling). Important considerations for variable selection include whether the variables are hypothesized to influence genetic connectivity, the appropriate spatial and thematic scale, and the accuracy of spatial data used to create resistance surfaces. Parameterizing, correlating, and finding the best supported resistance surface model with genetic data can be conducted through empirical methods, such as maximum-likelihood population-effects models (MLPE; Clarke et al. 2002) with quasi-optimization (Row et al. 2017) or machine learning optimization (ResistanceGA; Peterman 2018 or radish; Pope and Peterman 2020) or least-cost transect analysis methods (van Strien et al. 2012) with machine learning optimization (Pless et al. 2021; Vanhove and Launey 2023).

9.4.2.3　Landscape Genetic Simulators and Their Applications for Cancer Biology

Simulation modeling has high value for science and society. It can be used to predict and explain, guide data collection, illuminate core dynamics of a system, discover new questions, bound outcomes to plausible ranges, quantify uncertainties, offer crisis options in near-real time, demonstrate trade-offs and suggest efficiencies, challenge the robustness of prevailing theory through perturbations, expose prevailing wisdom as incompatible with available data, train students and practitioners, educate the general public, and reveal the apparently simple to be complex as well as vice versa (Epstein 2007). Consequently, simulations are increasingly used in all areas of scientific research, have provided many important findings in various disciplines (e.g., Grimm et al. 2005), and are increasingly accepted by empiricists (Jeltsch et al. 2013). A detailed treatment of general simulation modeling is beyond the scope of this chapter, but an excellent and comprehensive introduction can be found in Grimm and Railsback (2005, 2012). In the remainder of the section, we will specifically focus on simulation modeling in landscape genetics, how it is different from past population genetic simulators, and how simulation modeling in landscape genetics can be used with particular emphasis on its role for validating landscape resistance to movement hypotheses. Finally, we will discuss possible ways simulation modeling can be used for studying the movement of tumors through the body.

Simulations have been used in genetic research for many decades (e.g., Kimura and Ohta 1974), and the availability of software for simulating genetic data is increasing steadily (https://surveillance.cancer.gov/genetic-simulation-resources/home/). However, by comparing landscape genetic simulations to "classic" population genetic simulations we can begin to tease apart differences. First, many population genetic simulation approaches generate genetic data only at the population level, meaning that the output of these simulations is typically in the form of summary statistics for each simulated population, for example, in terms of population-specific allele frequencies or inbreeding coefficients. In contrast, many landscape genetic simulations actually produce genetic data for every individual, even if these individuals are grouped into populations. Thus, landscape genetic simulations generally rely on individual-based models, classes of computational models used to simulate the actions and interactions of autonomous individuals, with a key aspect being the variability among these individuals. These interindividual differences increase the realism of individual-based models but also increase their complexity.

A second major characteristic of landscape genetic simulations is the fact that they are always based on spatially explicit models. These models are defined by placing individuals or groups of individuals (i.e., populations) on one-or two-dimensional regular lattices or in irregular (x-, y-) coordinate space. Specific rules in the simulation model then define how individuals move and interact across space, for example, by defining the distances that individuals can move away from their birth location or the distance within which they can find a mating partner. Clearly, population genetic simulations are often also spatially explicit (e.g., Balloux 2001), but while simulating population genetic data without space is possible, this is not the case for landscape genetic simulations.

In addition to space, another vital characteristic of landscape genetic simulations is the direct incorporation of environmental heterogeneity into the underlying model (i.e., putting the "landscape" into landscape genetics). This usually requires that in addition to the locations of individuals or populations in space, there also needs to be a spatial representation of the environment that individuals are placed in. Importantly, this environment is variable (i.e., non-homogeneous) in space, and potentially also in time, and it directly affects some or all of the essential processes included in the model. Thus, the rules that govern the actions and reactions of simulated individuals not only depend on pure space but also on the user-defined environmental heterogeneity included in the model. The fact that essential processes are directly affected by environmental heterogeneity is the key distinguishing feature between population genetic and landscape genetic simulation modeling. In sum, the landscape genetic simulation modeling framework is generally individual-based, always spatially explicit, and always incorporates environmental heterogeneity as a major influence on the actions and reactions of simulated organisms (Landguth et al. 2015).

Drawing linkages, simulation modeling that is cell-based, spatially explicit within the organism, and incorporates microenvironment heterogeneity as a major influence on the actions and reactions of simulated cells could provide interesting opportunities for cancer biology. It's possible to use these hypothetical models to test existing theories and help identify underlying mechanisms as the true driving processes are rarely ever known, that is, pattern-process modeling for cancer biology (e.g., Stichel et al. 2017). Accurate simulation of cancer cells acting in microenvironments and moving through organ systems is a significant computational undertaking that balances costs with realism of sufficient complexity to capture relevant behavior (Balogh et al. 2021). However, we believe that we are close to achieving the computational strength needed to create these models that answer questions of mechanism and pattern of complex behavior.

9.4.2.4 Connectivity Modeling for Identifying Optimal Gene Flow Corridors

Once resistance surfaces have been created/parameterized, identified as the best supporting hypothesis, and validated for a landscape genetics application (possibly with simulations; pattern-process modeling), a next step in most analyses is to translate to a connectivity model. Connectivity models are important tools for conservation action and can be used to identify key habitat corridors between populations. Preserving these key corridors has been shown to increase the movement of individuals and counteract population isolation (Gilbert-Norton et al. 2010). Likewise, identifying these key habitats for cancer biology could be used to decrease movement of tumor cells and preserve population isolation.

Creating connectivity models requires two steps: a resistance to movement landscape surface (described in previous section) and a connectivity algorithm. Common connectivity algorithms include Dijkstra's algorithm (Dijskstra 1959) for calculating factorial least-cost paths and randomized shortest paths (Saerens et al. 2009) implemented through programs like UNICOR (Landguth et al. 2012) or the package "gdistance" in R (van Etten 2017), resistant kernels (Compton et al. 2007), or CircuitScape (McRae and Beier 2007) (see Figure 9.6). Factorial least-cost paths calculate the cumulative cost from all source to destination pairs of points across a resistance surface. Factorial least-cost paths can then create multiple paths, and troughs can be identified where the cumulative cost and distance between the two points are minimized. However, there is little reason to assume that an animal knows the route or destination of the least-cost path. To remedy this, the resistant kernels method was developed. It estimates connectivity as a function of source locations, landscape resistance, and dispersal thresholds, without requiring knowledge of destination points. In comparison, CircuitScape, based on electrical circuit theory, runs current across resistance surfaces where each pixel is a node and the resistance values between pixels are the resistors. CircuitScape results in an estimated current density for each pixel on the landscape, with higher values of current representing higher probabilities that a random walker will pass through a pixel. As with resistance kernels, CircuitScape also does not assume an individual has complete knowledge of the landscape through which it is travelling.

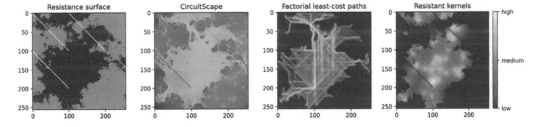

FIGURE 9.6 Connectivity models for resistance to movement surfaces. An example of a resistance surface (left), and the resulting connectivity predictions of CircuitScape, factorial least-cost paths, and resistant kernels applied to this resistance surface. Kumar and Cushman compared all three methods and showed that resistant kernels are the most appropriate model for conservation applications, except for when the movement is strongly directed toward a known location. Figure from Unnithan Kumar and Cushman (2022).

In conclusion for connectivity modeling, improving structural landscape connectivity will allow for increased movement of genes at a rate sufficient to reduce population vulnerability to extirpation. Hence, increased connectivity can be beneficial and understanding these linkages of important landscape factors is a major objective in landscape genetics. Positive outcomes are generally expected when connectivity is increased for conservation and landscape genetics research, such as improved demographic and genetic health and spread of important adaptive alleles facing climate change and disturbance (Creech et al. 2017; Razgour et al. 2018). But connectivity can also lead to negative consequences; examples in conservation biology and landscape genetics include invasions by non-native species (With 2002), spread of agricultural pests (Margosian et al. 2009), spread of maladaptive genotypes (Lowe and Allendorf 2010; Fitzpatrick and Reid 2019), and loss of overall genetic variation and/or evolutionary lineages. Increasing the connectivity through tumor environments is likened to when tumor cells with malignant phenotypes have detached from primary tumor masses, invaded through basement membranes and into the circulation, and reached a new location where tumor cells have successfully reproduced/multiplied. Increased connectivity in a tumor system should correlate with increased rates of metastasis.

9.4.2.5 Adaptive Landscape Genetics

Landscape genomics attempts to understand how landscape factors influence spatial patterns of genome-wide genetic variation. For conservation biology, many management actions revolve around delineating conservation units for monitoring programs and predicting how climate or land use change will potentially affect loss of adaptive genetic variation. And, over the past few decades, improvements in next-generation sequencing technologies (Goodwin et al. 2016) have provided an opportunity to study local adaptation and natural selection in conservation research. Thus, adaptive landscape genetics emerged, which uses these data to study how landscapes and spatial processes influence the distribution and amount of selection-driven genetic variation and can be used to identify the factors underlying local adaptation. This has, in turn, informed our understanding of evolutionary potential and how best to manage and conserve the adaptive capacity of populations in the face of climate and land use changes (e.g., Funk et al. 2018; Forester et al. 2023). Creech et al. (2017) used a simulation approach to investigate the spread of adaptive genotypes in desert bighorn sheep (*Ovis canadensis nelsoni*), a habitat specialist expected to be threatened by habitat loss due to climate change. In this novel approach, landscape resistance models were developed for desert bighorn sheep in three different regions that varied in habitat quantity and configuration, using data from noninvasively collected neutral genetic markers. Simulations based on these resistance models were used to investigate how the spread of an adaptive allele varied based on selection strength and whether the adaptive variant was derived from standing genetic variation or a new mutation. Adaptation from standing genetic variation had a much higher incidence of spread and likelihood of persistence than a novel mutation, especially when landscapes were more highly connected (Figure 9.7). These results are in line with empirical and conceptual work (reviewed in Hendry 2013) and highlight the importance of maintaining standing genetic variation in desert bighorn sheep populations, as well as the potential need for assisted gene flow targeting multiple locations in isolated populations.

Adaptive landscape genetics includes a range of techniques for identifying spatially structured selection-driven genetic variation. Here, we describe one of the most widely used methods for analysis of adaptive variants that incorporates environmental variation across selection-driven genes: genotype-environmental associations (GEA). GEA analyses refer to techniques that partition neutral from potentially adaptive genetic variation. Candidate adaptive loci are first identified based on associations between allele distributions and landscape variables hypothesized to drive selection and based on the physiology and ecology of the focal species. A pattern of selection-driven alleles will occur in higher frequency in certain environments over neutral alleles (Rellstab et al. 2015). GEA methods are very good at detecting divergent selection but can also detect weaker selective pressures, such as selection on standing genetic variation (Forester et al. 2018). Many methods and

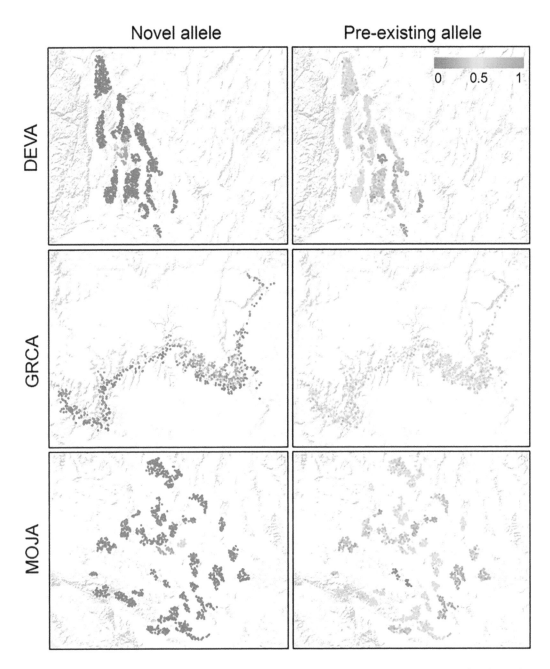

FIGURE 9.7 Novel mutations vs standing genetic variation using adaptive landscape genetics simulation modeling. Simulated spread of an adaptive allele in populations of desert bighorn sheep (*Ovis canadensis nelsoni*) in different regions of the United States over 100 years. Regions are Death Valley in the northern Mojave Desert (DEVA), the Grand Canyon in northern Arizona (GRCA), and the southern Mojave Desert (MOJA). Colored dots are individual locations with color gradient reflecting the proportion of simulations in which the adaptive allele is present, assuming strong selection and a medium dispersal threshold. Left and right columns show presence of the adaptive allele after novel mutation and selection on standing genetic variation, respectively. Adaptation from standing genetic variation had a much higher incidence of spread and likelihood of persistence than a novel mutation, especially when landscapes were more highly connected. Figure from Creech et al. (2017).

statistical approaches have been used for GEAs, from generalized linear models (e.g., Joost et al. 2007) to redundancy analysis (Forester et al. 2018), but it is important that the method is able to analyze the high dimensionality of genomic data sets. That is, a method should be able to analyze all loci and predictor variables simultaneously for many thousands of individuals and thousands of loci in order to detect multilocus selection, and to consider groups of markers and how they covary in response to environmental predictors (Forester et al. 2018).

Adaptive landscape genetics could be of service to the study of cancer during changes in tumor microenvironments. DeGregori (2018) and others have presented a convincing argument that some of the aging-related microenvironment changes alter the selection regimes in proto-tumors such that cancer is more favored in older individuals. We contend that concepts and tools from landscape genetics can be useful in the study of the cancer-promoting tumor microenvironment changes. Although some biomarkers of aging-related inflammaging (including some microRNAs) have been found (Fane and Weeraratna 2020), more of such would be useful in understanding the inflammaging process as well as possibly diagnosis. Methods like GEA used in landscape genetics could assist in finding and assessing biomarkers.

9.4.2.6   eDNA

Environmental DNA (eDNA) for landscape genetics has a parallel, cell-free DNA (cfDNA), in cancer biology. eDNA is an emerging technique that can be used to create multi-species datasets for landscape genetics studies. eDNA detects species via trace amounts of their DNA expelled into the environment (Rees et al. 2014). The incorporation of eDNA into landscape genetics applications can provide new insights for population dynamics and multi-species interactions with predators, competitors, or invasive species (Parsley et al. 2020). Lin et al. (2021) used eDNA analyses with environmental data to predict biodiversity indicators (alpha, beta, zeta diversity) showing habitat classification corelating with community composition. Similarly for cancer biology, different bodily fluids (mostly blood) can be sampled and multiple stages to detect alterations (mutations, chromosomal rearrangements, epigenetic stages of methylation, etc.) in cancer cells through circulating cfDNA (e.g., Gao et al. 2022). Understanding how different landscapes and microenvironments might associate with eDNA or cfDNA could be an area of fruitful discovery.

9.5   CONCLUDING THOUGHTS

In this chapter, we aimed to summarize the current state of knowledge of landscape genetics and draw parallels with cancer biology. Since its inception in the early 2000s, landscape genetics involves research that combines population genetics, landscape ecology, and spatial analytical techniques to explicitly quantify the effects of landscape structure, composition, configuration, and matrix quality on microevolutionary processes, such as gene flow, drift, and selection, using neutral and adaptive genetic data. Overall, landscape genetics has led to several exciting findings for conservation and evolutionary biology. We also anticipate that the rapid development of new and improved methods will likely ensure future progress. Several of these techniques have direct comparisons or immediate applications relevant for cancer biology questions—from creating resistance to movement landscapes and isolating selection-driven patterns associated with environments to developing connectivity models for identifying key corridors of gene flow. We also note that rapid advances in spatial 'Omics and other technologies (Bressan et al. 2023) will be instrumental in observing and measuring cancer cells as they move through the landscapes of human bodies. There are many similarities between tumor connectivity and landscape genetic connectivity—or how tumors spread and metastasize to distant sites with how individuals migrate and disperse across landscapes. Understanding the relative role of organism structural features (e.g., basement membranes and extracellular matrix) and tumor dispersal can provide insights for researchers managing critical components of metastasis. Convergence research aims to solve complex problems employing transdisciplinarity (or the deep integration of multiple disciplines) (NRC 2014). We hope this chapter might provide scientific opportunities enabled by convergence of landscape genetics and cancer biology, spurring collaboration from innovative research to translational applications, and making fundamental contributions to some of the most difficult and pressing research challenges facing us as a society.

ACKNOWLEDGMENTS

Erin L. Landguth was supported by National Institute of General Medical Sciences of the National Institutes of Health, United States (Award numbers P20GM130418). We thank K. Zeller and J. Somarelli for comments.

REFERENCES

Aktipis A (2020) *The cheating cell: how evolution helps us understand and treat cancer*. Princeton University Press.

Aquadro CF, Greenberg BD (1983) Human mitochondrial DNA variation and evolution: analysis of nucleotide sequences from seven individuals. Genetics, 103: 287–312.

Arney K (2020) *Rebel cell: cancer, evolution, and the new science of life's oldest betrayal*. BenBella Books, Inc.

Balkenhol N, Cushman SA, Storfer A, Waits LP (2015a) Landscape genetics—concepts, methods, applications. In: Balkenhol N, Cushman SA, Storfer AT, Waits LP (eds) *Landscape genetics: concepts, methods, applications*. Wiley Blackwell.

Balkenhol N, Cushman SA, Waits LP, Storfer A (2015b) Current status, future opportunities, and remaining challenges in landscape genetic. In: Balkenhol N, Cushman SA, Storfer AT, Waits LP (eds) *Landscape genetics: concepts, methods, applications*. Wiley Blackwell.

Balkenhol N, Schwarta MK, Inman RM, Copeland JP, Squires JS, Anderson NJ, Waits LP (2020) Landscape genetics of wolverines (*Gulo gulo*): scale-dependent effects of bioclimatic, topographic, and anthropogenic variables. Journal of Mammalogy, 101(3), 790–803, 22 May. https://doi.org/10.1093/jmammal/gyaa037

Balloux, F (2001) EASYPOP (Version 1.7): a computer program for population genetic simulations. Journal of Heredity, 92, 301–2.

Balogh P, Gounley J, Roychowdhury S, et al. (2021) A data-driven approach to modeling cancer cell mechanics during microcirculatory transport. Scientific Reports, 11, 15232. https://doi.org/10.1038/s41598-021-94445-5

Bressan D, Battistoni G, Hannon GJ (2023) The dawn of spatial omics. Science, 381, eabq4964.

Creech TG, Epps CW, Landguth EL, Wehausen JD, Crowhurst RS, Holton B, Monello RJ (2017) Simulating the spread of selection-driven genotypes using landscape resistance models for desert bighorn sheep. PloS One, 12, e0176960.

Clarke RT, Rothery P, Raybould AF (2002) Confidence limits for regression relationships between distance matrices: estimating gene flow with distance. Journal of Agricultural Biological and Environmental Statistics, 7, 361–72.

Compton BW, McGarigal K, Cushman SA, Gamble LR (2007) A resistant kernel model of connectivity for amphibians that breed in vernal pools. Conservation Biology, 21(3), 788–99.

Cushman SA, Landguth EL (2010) Spurious correlations and inferences in landscape genetics. Molecular Ecology, 19. https://doi.org/10.1111/j.1395-294X.2010.04656.x.

Cushman SA, McKelvey KS, Hayden J, Schwartz MK (2006) Gene flow in complex landscapes: testing multiple hypotheses with causal modeling. The American Naturalist, 168(4), 486–99.

Dasmann RFA (1968) *Different kind of country*. MacMillan Company.

Dauphin B, Rellstab C, Wüest RO, Karger DN, Holderegger R, Gugerli F, Manel S. (2023) Re-thinking the environment in landscape genomics. Trends in Ecology & Evolution, 38(3), 261–74, March. https://doi.org/10.1016/j.tree.2022.10.010.

DeGregori J (2018) *Adaptive oncogenesis: a new understanding of how cancer evolves inside us*. Harvard University Press.

Dijskstra EW (1959) A note on two problems in connexion with graphs. Numerische Mathematik, 1, 269–71.

Epperson BK (2007) Plant dispersal, neighborhood size and isolation by distance. Molecular Ecology, 16, 3854–65.

Epstein JM (2007) Remarks on the role of modeling in infectious disease mitigation and containment. In: Lemon SM, Hamburg MA, Sparling PF, Choffnes ER, Mack A (eds) *Ethical and legal considerations in mitigating pandemic disease: workshop summary: forum on microbial threats*. National Academies Press.

Escalante M, Leon G, Arturo RL, Landguth EL, Manel S (2018) The interplay of riverscape features and exotic introgression on the genetic structure of the Mexican golden trout (*Oncorhynchus chysogaster*). Journal of Biogeography, 45, 1500–14. https://doi.org/10.1111/jbi.13246

Ezard THP, Travis JMJ (2006) The impact of habitat loss and fragmentation on genetic drift and fixation time. Oikos, 114, 367–75.

Fahrig L (2002) Effect of habitat fragmentation on the extinction threshold: a synthesis. Ecological Applications, 12, 346–53.

Fane M, Weeraratna AT (2020) How the ageing microenvironment influences tumour progression. Nature Reviews Cancer, 20, 89–106.

Fisher RA (1918) The correlation between relatives on the supposition of mendelian inheritance. Transactions of the Royal Society of Edinburgh, 53, 399–433.

Fitzpatrick SW, Reid BN (2019) Does gene flow aggravate or alleviate maladaptation to 627 environmental stress in small populations? Evolutionary Applications, 12, 1402–16.

Flather CH, Bevers M (2002) Patchy reaction-diffusion and population abundance: the relative importance of habitat amount and arrangement. The American Naturalist, 159, 40–56.

Forester BR, Beever EA, Darst C, Szymanski J, Funk WC (2022) Linking evolutionary potential to extinction risk: applications and future directions. Frontiers in Ecology and the Environment, 20(9), 507–15. https://doi.org/10.1002/fee.2552.

Forester BR, Day CC, Ruegg K, Landguth EL (2023) Evolutionary potential mitigates extinction risk under climate change in the endangered southwestern willow flycatcher. Journal of Heredity, esac067. https://doi.org/10.1093/jhered/esac067.

Forester BR, Landguth EL, Hand BK, Balkenhol N (2018) Landscape genomics for wildlife research. In: Hohenlohe PA, Rajora OP (eds) *Population genomics: wildlife*. Springer. https://doi.org/10.1007/13836_2018_56

Frankham R, Ballou JD, Briscoe DA (2010) *Introduction to conservation genetics*. Cambridge University Press.

Freedman JA, Lorson BD, Taylor RB, Carline RF, Stauffer JR (2014) River of the dammed: longitudinal changes in fish assemblages in response to dams. Hydrobiologia, 727, 19–33.

Funk WC, Forester BR, Converse SJ, Darst C, Morey S (2018) Improving conservation policy with genomics: a guide to integrating adaptive potential into U.S. endangered species act decisions for conservation practitioners and geneticists. Conservation Genetics, 20. https://doi.org/10.1007/s10592-018-1096-1.

Gaines MS, Diffendorfer JE, Tamarin RH, Whittam TS (1997) The effects of habitat fragmentation on the genetic structure of small mammal population. Journal of Heredity, 88, 294–304.

Galpern P, Manseau M (2013) Finding the functional grain: comparing methods for scaling resistance surfaces. Landscape Ecology, 28, 1269–81.

Gao Q, Zeng Q, Wang Z, Li C, Xu Y, Cui P, Zhu X, Lu H, Wang G, Cai S, Wang J, Fan J (2022) Circulating cell-free DNA for cancer early detection. Innovation (Cambridge), 3(4), 100259, 6 May. https://doi.org/10.1016/j.xinn.2022.100259. PMID: 35647572; PMCID: PMC9133648.

Gatenby RA, Silva AS, Gillies RJ, Frieden BR (2009). Adaptive therapy. *Cancer Research*, 69: 4894–904.

Gilbert-Norton L, Wilson R, Stevens JR, Beard KH (2010) A meta-analytic review of corridor effectiveness. Conservation Biology, 24, 660–8.

Gilpin ME, Soulé ME (1986) Minimum viable populations: processes of species extinction. In: Soulé ME (ed) *Conservation biology: the science of scarcity and diversity*. Sinauer.

Goodwin S, McPherson JD, McCombie WR (2016) Coming of age: ten years of next-generation sequencing technologies. Nature Reviews Genetics, 17, 333–51.

Grimm V, Railsback SF (2005) *Individual-based modeling in ecology*. Princeton University Press.

Grimm V, Railsback SF (2012) Designing, formulating and communicating agent-based models. In: Heppenstall AJ, Crooks AT, See LM, Batty M (eds) *Agent-based models of geographical systems*. Springer.

Grimm V, et al. (2005) Pattern-oriented modeling of agent-based complex systems: lessons from ecology. Science, 310, 987–91.

Hardy GH (1908) Mendelian proportions in a mixed population. Science, 28, 49–50.

Hendry AP (2013) Key questions in the genetics and genomics of eco-evolutionary dynamics. Heredity, 111, 456–66.

Hill MF, Caswell H (1999) Habitat fragmentation and extinction thresholds on fractal landscapes. Ecology Letters, 2, 121–7.

Holderegger R, Kamm U, Gugerli F (2006) Adaptive vs. neutral genetic diversity: implications for landscape genetics. Landscape Ecology, 21, 797–807

Jeltsch F, et al. (2013) Integrating movement ecology with biodiversity research—exploring new avenues to address spatiotemporal biodiversity dynamics. Movement Ecology, 1, 1–13.

Jolly MT, Ware KE, Gilja S, Somarelli JA, Levine H (2017) EMT and MET: necessary or permissive for metastasis? Molecular Oncology, 11, 755–69.

Joost S, Bonin A, Bruford MW, Després L, Conord C, Erhardt G, Taberlet P (2007) A spatial analysis method (SAM) to detect candidate loci for selection: towards a landscape genomics approach to adaptation. Molecular Ecology, 16, 3955–69.

Kardos M, Armstrong EE, Fitpatrick SW, Hauser S, Hedrick PW, Miller JM, Tallmon DA, Funk C (2021) The crucial role of genome-wide genetic variation in conservation. PNAS, 118(48), e2104642118.

Kareiva P, Wennergren U (1995) Connecting landscape patterns to ecosystem and population processes. Nature, 373, 299–302.

Kimura M, Ohta T (1974) On some principles governing molecular evolution. Proceedings of the National Academy of Sciences of the United States of America, 71, 2848–52.

Kreitman M (1983) *The neutral theory of molecular evolution.* Cambridge University Press.

Landguth EL, Cushman SA, Balkenhol N (2015) Chapter 6: simulation modeling in landscape genetics. In: Balkenhol N, Waits L, Cushman S (eds) *Landscape genetics.* Wiley.

Landguth EL, Cushman SA, Murphy M, Luikart G (2010a) Relationships between migration rates and landscape resistance assessed using individual-based simulations. Molecular Ecology, 10, 854–62.

Landguth EL, Cushman SA, Schwartz MK, McKelvey KS, Murphy M, Luikart G (2010b) Quantifying the lag time to detect barriers in landscape genetics. Molecular Ecology, 19, 4179–91.

Landguth EL, Hand BK, Glassy JM, Cushman SA, Sawaya M (2012) UNICOR: a species corridor and connectivity network simulator. Ecography, 12, 9–14.

Lawrence ER, Kuparinen A, Hutchings JA (2016) Influence of dams on population persistence in Atlantic salmon (Salmo Salar). Canadian Journal of Zoology, 94, 329–38.

Levin SA (1992) The problem of pattern and scale in ecology: the Robert H. MacArthur award lecture. Ecology, 73, 1943–67.

Levins R (1969) Some demographic and genetic consequences of environmental heterogeneity for biological control. Bulletin of the Entomological Society of America, 15, 237–40.

Lewontin RC (1974) *The genetic basis of evolutionary change.* Columbia University Press.

Lewontin RC, Hubby JL (1966) A molecular approach to the study of genetic heterozygosity in natural populations. II: amount of variation and degree of heterozygosity in natural populations of *Drosophila pseudoobscura.* Genetics, 54, 595–609.

Ligon FK, Dietrich WE, Trush WJ (1995) Downstream ecological effects of dams. BioScience, 45, 183–92.

Lin M, Simons AL, Harrigan RJ, Curd EE, Schneider FD, Ruiz-Ramos DV, Gold Z, Osborne MG, Shirazi S, Schweizer TM, Moore TN, Fox EA, Turba R, Garcia-Vedrenne AE, Helman SK, Rutledge K, Mejia MP, Marwayana O, Munguia Ramos MN, Wetzer R, Pentcheff ND, McTavish EJ, Dawson MN, Shapiro B, Wayne RK, Meyer RS. (2021) Landscape analyses using eDNA metabarcoding and Earth observation predict community biodiversity in California. Ecological Applications, 31(6), e02379, Sep. https://doi.org/10.1002/eap.2379.

Lowe WH, Allendorf FW (2010) What can genetics tell us about population connectivity? Molecular Ecology, 19, 3038–51. https://doi.org/10.1111/j.1365-294X.2010.04688.x.

Mahan CG, Yahner RH (2000) Effects of forest fragmentation on behaviour patterns in the eastern chipmunk (Tamias striatus). Canadian Journal of Zoology, 77, 1991–7.

Manel S, Holderegger R (2013) Ten years of landscape genetics. Trends in Ecology & Evolution, 28(10), 614–21, Oct. https://doi.org/10.1016/j.tree.2013.05.012.

Manel S, Schwartz MK, Luikart G, Taberlet P (2003) Landscape genetics: combining landscape ecology and population genetics. Trends in Ecology and Evolution, 18, 189–97.

Mangel M, Clark CW (1986) Towards a unified foraging theory. Ecology, 67, 1127–38.

Margosian ML, Garrett KA, Hutchinson JS, With KA (2009) Connectivity of the American agricultural landscape: assessing the national risk of crop pest and disease spread. BioScience, 59, 141–51.

McRae BH (2006) Isolation by resistance. Evolution, 60, 1551–61.

McRae BH, Beier P (2007) Circuit theory predicts gene flow in plant and animal populations. Proceedings of the National Academy of Sciences of the United States of America, 104, 19885–90. https://doi.org/10.1073/pnas.0706568104.

Mims M, Day C, Hinkle J, Fuller M, Burkhart J, Bearlin A, DeHann P, Holden Z, Landguth EL (2019) Simulating demography, genetics, and spatially-explicit processes to inform reintroduction of a threatened char. Ecosphere, 10. https://doi.org/10.1002/ecs2.2589.

Montgomery DR (2003) *King of fish: the thousand-year run of salmon.* Westview Press.

Muhlfeld CC, D'Angelo V, Kalinowski ST, Landguth EL, Downs CC, Tohtz J, Kershner JL (2012) A fine-scale assessment of using barriers to conserve native stream salmonids: a case study in Akokala Creek, Glacier National Park, USA. The Open Fish Science Journal, 5, 9–20.

Mullis K, Faloona F, Schaffer S, Horn G, Erlich H (1986) Specific enzymatic amplification of DNA in vitro: the polymerase chain reaction. Cold Spring Harbor Symposia on Quantitative Biology, 51, 263–73.

Murphy M, Evans J, Cushman S, Storfer A (2008) Representing genetic variation as continuous surfaces: an approach for identifying spatial dependency in landscape genetic studies. Ecography, 31, 685–97.

Murphy M, Evans J, Stofer A (2010b) Quantifying Bufo boreas connectivity in Yellowstone National Park with landscape genetics. Ecology, 91, 252–61.

NRC (National Research Council) (2014) *Convergence: facilitating transdisciplinary integration of life sciences, physical sciences, engineering, and beyond.* The National Academies Press, Engineering, and Medicine. https://doi.org/10.17226/18722.

Pamilo P (1988) Genetic variation in heterogeneous environments. Annales Zoologici Fennici, 25, 99–106.

Parsley MB, Torres ML, Banerjee SM, et al. (2020) Multiple lines of genetic inquiry reveal effects of local and landscape factors on an amphibian metapopulation. Landscape Ecology, 35, 319–35. https://doi.org/10.1007/s10980-019-00948-y.

Peterman WE (2018) ResistanceGA: an R package for the optimization of resistance surfaces using genetic algorithms. Methods in Ecology and Evolution, 9, 1638–47. https://doi.org/10.1111/2041-210X.12984.

Peterman WE, Pope NS (2021) The use and misuse of regression models in landscape 742 genetic analyses. Molecular Ecology, 30, 37–47.

Pierce GB, Speers WC (1988) Tumors as caricatures of the process of tissue renewal: prospects for therapy by directing differentiation. Cancer Research, 48, 1996–2004.

Pless E, Saarman NP, Powell JR, Caccone A, Amatulli G (2021) A machine-learning approach to map landscape connectivity in Aedes aegypti with genetic and environmental data. Proceedings of the National Academy of Sciences of the United States of America, 118, 1–8.

Pope NS, Peterman W (2020) *Radish R package.* https://github.com/nspope/radish.

Provine, WB (1971) *The origins of theoretical population genetics.* University of Chicago Press.

Prunier GP, Loot G, Veyssiere C, Poulet N, Blanchet S (2023) Novel operation index reveals rapid recovery of genetic connectivity in freshwater fish species after riverine restoration. Conservation Letters, e12939.

Razgour O, Taggart JB, Manel S, Juste J, Ibanez C, Rebelo H, Alberdi A, Jones G, Park K (2018) An integrated framework to identify wildlife populations under threat from climate change. Molecular Ecology Resources, 18, 18–31.

Rees HC, Maddison BC, Middleditch DJ, Patmore JRM, Gough KC (2014) The detection of aquatic animal species using environmental DNA: a review of eDNA as a survey tool in ecology. Journal of Applied Ecology, 51, 1450–9.

Rellstab C, Gugerli F, Eckert AJ, Hancock AM, Holderegger R (2015) A practical guide to environmental association analysis in landscape genomics. Molecular Ecology, 24, 4348–70.

Robbins CS, Dawson DK, Dowell BA (1989) Habitat area requirements of breeding forest birds of the middle Atlantic states. Wildlife Monographs, 103, 1–34.

Row JR, Knick ST, Oyler-McCance SJ, Lougheed SC, Fedy BC (2017) Developing approaches for linear mixed modeling in landscape genetics through landscape-directed dispersal simulations. Ecology and Evolution, 7(11), 3751–3761. https://doi.org/10.1002/ece3.2825.

Saerens M, Achbany Y, Fouss F, Yen L (2009) Randomized shortest-path problems: two related models. Neural Computation, 21, 2363–404.

Shirk AJ, Cushman SA, Landguth EL (2012) Simulating pattern–process relationships to validate landscape genetic models. International Journal of Ecology, 539109.

Shirk AJ, Wallin DO, Cushman SA, Rice CG, Warheit KI (2010) Inferring landscape effects on gene flow: a new model selection framework. Molecular Ecology, 17, 3603–19. https://doi.org/10.1111/j.1365-294X.2010.04745.x.

Somarelli J (2021) The hallmarks of cancer as ecologically-driven phenotypes. Frontiers in Ecology and Evolution, 9, 226.

Soulé ME, Wilcox B (1980) *Conservation biology: an evolutionary-ecological perspective.* Sinauer Associates, Inc.

Spear SF, Storfer A (2008) Landscape genetic structure of coastal tailed frogs (*Ascaphus truei*) in protected vs. managed forests. Molecular Ecology, 17, 4642–56.

Stichel D, Middleton AM, Müller BF, et al. (2017) An individual-based model for collective cancer cell migration explains speed dynamics and phenotype variability in response to growth factors. NPJ Systems Biology and Applications, 3(5). https://doi.org/10.1038/s41540-017-0006-3

Storfer A, Murphy MA, Evans JS, Goldberg CS, Robinson S, Spear SF, Dezzani R, Delmelle E, Vierling L, Waits LP (2007) Putting the "landscape" in landscape genetics. Heredity, 98, 128–42.

Turner MG (1989) Landscape ecology: the effect of pattern on process. Annual Review of Ecology and Systematics, 20, 171–97.

Turner MG, Gardner R, O'Neill RV (2001) *Landscape ecology in theory and practice: pattern and process.* Springer.

Unnithan Kumar S, Cushman SA (2022) Connectivity modelling in conservation science: a comparative evaluation. Scientific Reports, 12, 16680. https://doi.org/10.1038/s41598-022-20370-w

van Etten J (2017) R package G distance: distances and routes on geographical grids. Journal of Statistical Software, 76(13), 1–21. https://doi.org/10.18637/jss.v076.i13.

van Strien MJ, Keller D, Holderegger R (2012) A new analytical approach to landscape genetic modelling: least-cost transect analysis and linear mixed models. Molecular Ecology, 21, 4010–23. https://doi.org/10.1111/j.1365-294X.2012.05687.x.

Vanhove M, Launey S (2023) Estimating resistance surfaces using gradient forest and allelic frequencies. Molecular Ecology Resources, 1–15. https://doi.org/10.1111/1755-0998.13778.

Wiens JA (1989) Spatial scaling in ecology. Functional Ecology, 3, 385–97.

Wiens, JA (1992) What is landscape ecology, really? Landscape Ecology, 7, 149–50.

Wiens JA, Stenseth NC, van Horne B, Ims RA (1993) Ecological mechanisms and landscape ecology. Oikos, 66, 369–80.

With KA (2002) The landscape ecology of invasive spread. Conservation Biology, 16, 1192–203.

Wright S (1917) Color inheritance in mammals. Journal of Heredity, 8, 224–35.

Zarri LJ, Palkovacs EP, Post DM, Therkildsen NO, Flecker AS (2022) The evolutionary consequences of dams and other barriers for riverine fishes. BioScience, 72, 431–48.

10 Tumor Island Biogeography
Theory and Clinical Applications

Antonia Chroni

10.1 INTRODUCTION

Integration of ecological and evolutionary features enables investigation of the interplay of tumor heterogeneity, microenvironment, and metastatic potential. Developing a theoretical framework that considers the complex architecture of tumor heterogeneity is intrinsic to deciphering tumors' tremendous spatial and longitudinal variation patterns in patients. In this pursuit, tumors can be considered evolutionary island-like ecosystems, i.e., isolated systems undergoing evolutionary and spatiotemporal dynamic processes that 1) shape tumor heterogeneity and its associated microenvironment, and 2) drive migration, invasion, and metastasis of cancer cells. This conceptual framework can leverage the extensive changes observed in tumor heterogeneity and its associated microenvironment by using various cellular modalities, such as genomes and transcriptomes of cancer cells. Furthermore, the tumor island framework enables one to 1) investigate the observed (high or low) tumor spatial and longitudinal variation patterns in patients as a critical function of important island features that shape overall diversity, i.e., tumor size and carrying capacity as well as physical distance (hereinafter distance) and connectivity between anatomical sites; 2) explore the relationship between different cell populations in tumors and their microenvironments illuminating the metastatic potential of cancer cells; and 3) facilitate the translation of evolutionary and ecological elements for clinicians to develop clinical decision-making tools.

10.2 TUMORS ARE EVOLUTIONARY ISLANDS

Cancer is an evolutionary disease. Cancer cells accumulate genetic aberrations that can be traced back to one cancer cell formed through mutations of a normal cell (Nowell, 1976). "The formation of tumors is nothing like walking along a straight line but rather like dwelling at the center of the Minotaur's labyrinth" (Chroni & Kumar, 2021). Cancer cells are equipped with a complex architecture of unique features that allow them to survive, engineer their microenvironment, and form their own ecosystems. These include immortality, genomic instability, sustained proliferation, induction of angiogenesis, resisting apoptosis, altered metabolism, and immune response, increased inflammation, and the ability for invasion/metastasis (the hallmarks of cancer) (Hanahan & Weinberg, 2011; Welch & Hurst, 2019). This is because cancer cells undergo dynamic evolutionary and spatiotemporal processes as well as "ecological" adaptations caused by genetic, transcriptomic, and epigenetic changes over time and across space (Somarelli, 2021). Moreover, these dynamic processes cause extensive intra- and inter-tumoral genetic heterogeneity within and among patients and across different cancers (Gerlinger et al., 2012; Mitchell et al., 2022; Mroz & Rocco, 2017). This high heterogeneity is of clinical relevance for patient stratification, drug design, and treatment decision-making.

Moreover, some cancer cells can disseminate from their initial site of growth; enter the bloodstream or lymphatic system; survive, invade, and colonize other anatomical sites; and essentially, develop secondary tumors in a process known as metastasis (Turajlic & Swanton, 2016; Williams et al., 2019). Manifestations of metastasis are responsible for 90% of cancer-associated morbidity

DOI: 10.1201/9781003307921-10

and mortality (Steeg, 2006; Welch & Hurst, 2019). Cancer cells with metastatic potential have additional features that drive their metastatic behavior. These include motility and invasion, engineering of their microenvironment, plasticity, and the ability for colonization (Welch & Hurst, 2019). Metastasis includes not only the mechanisms and processes of a cancer cell for disseminating and circulating outside of its tumor of origin, but also the by-product of these processes (Welch & Hurst, 2019). Understanding and properly modeling these metastatic phenomena is critical for understanding disease evolution and progression, treatment resistance, and outcomes to develop better clinical management and more appropriate treatment strategies.

To date, two main theories have been proposed for exploring and understanding the mechanisms and aetiologies of the metastatic behavior of cancer cells. The first theory assumes that metastasis occurs randomly as cancer cells detach from the original tumor and circulate within the body following blood flow distribution (Virchow, 1856) and are mechanically arrested in the first tissue they encounter (Ewing, 1928). The second theory supports that metastasis can be formed in anatomical sites that provide the ideal (pre)niche for cancer cells, and it is known as the "seed and soil" theory (Paget, 1989). In this context, cancer cells are the seeds, and the tissue where they land is the soil. The latter has been proposed in various types of cancer for explaining metastatic propensities for specific tissues/organs (de Groot et al., 2017).

There is emerging evidence supporting a rather mutual metastatic theory where chemical signals produced specifically in anatomical sites create ideal environmental conditions and together with mechanical obstruction of cancer cells coordinate the invasion, survival, and colonization of the tissue (Chu & Allan, 2012). The presence of circulating tumor cells (CTCs) in the blood of patients afflicted with metastatic disease indicates that CTCs have the ability to colonize new anatomical sites and form metastases (Galizia et al., 2013; Ulz et al., 2017). Additionally, studies show that organ tropism often occurs, meaning that certain cancer types will preferentially metastasize to other tissue (Gao et al., 2019), e.g., prostate cancer has a high affinity for metastasis to bones (de Groot et al., 2017).

In pursuit of understanding metastatic progression, there have been efforts to decipher the tissue microenvironment. Characterization of malignant cell populations and inference of tumor clones (i.e., a set of cancer cells with identical genotypes) as well as their interactions with the healthy microenvironment (e.g., immune cells) are intrinsic for illuminating the mechanisms and processes of tumor evolution and metastasis. These allow an overall description of the tumor ecosystem and, hence, elucidate the 'hows' and 'whys' of cancer cells' fitness and survival in the tissue, as well as dormancy and extinction. These investigations can allow us to further understand how tumors experience waves of clonal migrations that lead to metastatic malformations at local and distant anatomical sites. The latter is supported by a growing body of evidence showing that new tumors arise as a result of single (Chroni et al., 2022; Turajlic et al., 2018a), multiple (Chroni et al., 2022; Macintyre et al., 2017; Noorani et al., 2020), and reseeding (Chroni et al., 2022; Kim et al., 2009; Norton & Massagué, 2006; Savas et al., 2016; Yates et al., 2017) clones from either primary and/or metastatic tumors (Chroni et al., 2022; Macintyre et al., 2017) in various cancer types.

Recent studies suggest that processes and mechanisms driving and shaping tumor evolution and its associated microenvironment are similar to the ones observed in natural ecosystems (Maley et al., 2006; Somarelli et al., 2017). That has led to the popularization of theoretical concepts and computational approaches from the organismal evolution, ecology, and biogeography fields (Alves et al., 2019; Chroni et al., 2019, 2021; Maley et al., 2006). For example, phylogenetic frameworks allow the deconvolution of tumor clones and deciphering clonal exchanges between tumors in breast cancer (Chroni et al., 2022; El-Kebir et al., 2018; Kumar et al., 2020), colorectal cancer (Alves et al., 2019), and other types of cancer (Chroni et al., 2022). These efforts showcase the potential to bring these outcomes into the clinic to refine treatment strategies, enhance treatment responsiveness, and prevent disease relapse that will ultimately lead to increased patient survival.

The integration of evolutionary and ecological perspectives into the investigations of cancer evolution and progression has undoubtedly stimulated and promoted exciting advances in the cancer field (Maley et al., 2017; Thomas et al., 2020). This integration has also promoted the investigation of methods and frameworks from the field of biogeography to infer clonal exchanges and metastatic migration histories (Alves et al., 2019; Chroni et al., 2019, 2021). Several popular biogeographic approaches leverage phylogenetic, longitudinal, and spatial signals for inferring migrations and exploring dispersals due to physical distance (Landis et al., 2013; Ree & Smith, 2008; Ronquist & Huelsenbeck, 2003; Ronquist & Sanmartín, 2011; Yu et al., 2010) (Table 10.1). Interestingly, these frameworks consider evolutionary processes important for inferring the origin and trajectory of migrations, and many are relevant to cancer cells, such as duplication (genetic divergence within an area), extinction, founder-events, and migration (dispersal, expansion, and vicariance) (Chroni et al., 2021) (Table 10.1). These studies showcase attributes that could capture the fine architecture of tumors' evolution and microenvironment and might be particularly insightful for modeling and estimating metastatic migrations. These attributes are also related to clonal exchange and "migration" thinking; both are still to be appreciated during conceptualization and within mathematical and computational frameworks for investigating tumor evolution and metastasis.

To bridge the gap between biogeographic approaches and cancer biology, the analogy of tumors as evolutionary insular ecosystems was recently introduced as a novel conceptual approach for exploring intra-tumor heterogeneity and assessing clonal exchanges between tumors over disease progression and metastatic malformations (Chroni & Kumar, 2021). In Box 10.1, the analogy between island and tumor island biogeography is further explained through relevant terminology of ecosystems and processes.

Islands are ecosystems experiencing isolation to different degrees. This insularity attribute drives migration and extinction rates and ultimately reflects on the overall species abundance and diversity of the island habitats. Islands have unique features that can dictate the evolutionary and ecosystem balances therein: 1) limitations in size and habitat diversity affecting the total carrying capacity and driving competition for resources between members of the ecosystems; and 2) variations in physical distance and connectivity to adjacent areas allowing for new colonizations and subsequently increasing migration and diversity rates. Overall, "islands are reminders of arrivals and departures" (Ehrlich, 2017); and hence, the diversity of their ecosystems will be shaped by the newcomers (migration rate) and their opportunities for migration and colonization of additional sites (physical distance and connectivity between anatomical sites).

Interestingly, other scientific medical fields have drawn similar parallels from the field of biogeography for exploring the composition of communities in different tissues and organs within a patient's body, e.g., delineation of lung microbiome of patients with respiratory infections (tuberculosis (Adami & Cervantes, 2015)) or other disorders (cystic fibrosis (Whiteson et al., 2014)). Similarly, the clonal exchange between tumors is analogous to migrations of species and populations between islands and island-like ecosystems. Molecular, histological, and morphological intra- and inter-tumor heterogeneity are anticipated to be affected by the clonal exchange between tumors and any other attributes shaping tumor insularity that may or may not facilitate the mechanisms and processes of metastatic migrations. In this context, conceptualization and modeling of tumors as evolutionary islands that exchange cancer cells at any given time affecting the cancer's heterogeneity and microenvironment and subsequently, response to treatment, are relevant for understanding and predicting the metastatic potential and behavior of cancer cells.

TABLE 10.1

Matching Tumor Evolutionary Processes to Biogeographic Events and Existent Biogeographic Methods Would Provide More Insights into Cancer Cells' Evolution and Migration

Tumor evolutionary processes	Biogeographic events	Properties that define islands	Biogeographic methods			
			BBM	BayArea	DEC	DIVA
Mutation (genetic divergence)	Duplication	n/a	yes	no	yes	yes
Extinction	Extinction	n/a	yes	yes	yes	yes
Founder-event	Founder-event effect	n/a	no	customize	customize	customize
Migration	Expansion	n/a	yes	yes	yes	no
	Distant dispersal	n/a	yes	yes & distance-dependent effect on the dispersal probability	yes	yes
n/a	Vicariance	n/a	no	no	yes	yes
Sampling time or mutation rate	Time	n/a	customize	customize	customize	no
n/a	n/a	Area size	customize	no	no	no
n/a	n/a	Isolation (distance between areas)	customize	no	no	no

BOX 10.1　TERMS AND DEFINITIONS

Field	Term	Definition	Term	Definition	Field
		Island		**Tumor evolutionary island**	
Island biogeography	Islet	A very small piece of landmass surrounded by water	Tumor islet	Only primary tumors that are contained within the organ they started in; no signs of metastatic disease	Tumor island biogeography
	Island	A landmass surrounded by water	Tumor island	Primary tumor and at least one metastasis; limited connectivity observed between tumors	
	Archipelago	A cluster or chain formation of islands	Tumor Archipelago	Primary tumor and at least two metastases with a degree of connectivity observed between tumors	
	Ecosystem	A biological community of interacting living organisms and their physical environment	Tumor microenvironment	The environment around the site where a tumor grows, including blood vessels, immune cells, fibroblasts, signaling molecules, and the extracellular matrix	
	Landscape or habitat	Spatially heterogeneous patches	NA	NA	
	Ecological niche	The role of an organism in its ecosystem	Tumor niche	Pre-stage of the formation of the tumor microenvironment	
	Species diversity	Number of different species found in an ecosystem	Tumor heterogeneity	Genotypic, morphological and phenotypic changes in the clones observed in a tumor	
	Genetic divergence	Accumulation of genetic changes (mutations) over time leads to substantial differences between individuals or populations	Tumor genetic heterogeneity	Accumulation of genetic changes (mutations) over time leads to substantial differentiation between cancer cells or clones (cancer cells with identical genotypes)	
	Species migration	Movements of populations or species as a response to changes in the ecosystem	Clonal seeding	Movements of cancer cells as a response to changes in the tumor microenvironment	
	Colonization	The formation of a new ecosystem because of migration	Metastasis	Development of secondary malignant growths at a distance from a primary site of cancer because of clonal seeding	
	Source-area	Mainland	Source-area	Primary and/or metastatic tumors	
	Recipient-area	Islands	Recipient-area	Metastatic tumors, and in the case of reseeding events, primary tumors as well	

10.3   TUMOR MICROENVIRONMENT

Tumors form their own "tumor microenvironments" (TME). In the TME, cancer cells form ecosystems along with the surrounding cellular (tumor stromal fibroblasts, adipocytes, endothelial cells, immune cells, etc.) and non-cellular components of the extracellular matrix (e.g., collagen) (Baghban et al., 2020). Interactions between cancer cells and the healthy counterparts of their ecosystem can restructure and engineer the TME over the evolution and progression of the disease (Qian & Akçay, 2018; Ungefroren et al., 2011). Cancer cells' ability to engineer their microenvironment subsequently influences their heterogeneity and the tumor's overall fitness and survival (Yang et al., 2014). Moreover, these cellular interactions in the TME may induce inflammatory responses, treatment resistance, and disease relapse (Lucas, 2021; van Galen et al., 2019). Essentially, tumors are part of complex, dynamic ecosystems, making the characterization of their associated microenvironment essential for understanding tumor growth, tissue invasion, and metastatic migrations of clones.

In addition, the characterization of the TME will provide insights into the tumor's heterogeneity and evolution (Tissot et al., 2019). A proper description of the cellular composition and configuration of the TME will pinpoint landscapes defined by spatially heterogeneous patches. These are known as landscape mosaics (Daoust et al., 2013) and are habitats with different capacities for supplying and sustaining diversity (Maley et al., 2017). Tumor landscape mosaics will further indicate the relationship of the cellular and non-cellular populations of the TME, their potential competition for limited resources, tumor clonal turnover (the rate at which clones go extinct, and new clones are formed), and, ultimately, the "pressure" to metastasize (MacArthur & Wilson, 1963).

During metastasis, cancer cells invade, disseminate, and colonize other adjacent and distant anatomical sites. Studies provide emerging evidence of the existence of a pre-metastatic niche that allows for the disseminating cancer cells to survive and colonize the tissue, indicating that metastasis is the product of niche construction or ecological engineering (Qian & Akçay, 2018; Solary & Laplane, 2020). Unraveling the components and mechanisms that are part of the tumor's niche construction is relevant to understanding the metastatic processes. This is because, in the context of the tumor niche, cancer cells will share but also compete for resources, promoting the tumor's growth in the tissue. The latter could further disturb TME and jeopardize the tumor's survival ability by invoking hypoxia, dysfunctional vascularization, and epithelial-to-mesenchymal transition, potentially resulting in cell mobility and metastasis (Sleeman, 2012).

Studies showcase the clinical relevance of merging evolutionary and ecological perspectives when clinicians attempt to address therapy resistance and treatment failures (Maley et al., 2017; Thomas et al., 2020). Considering tumor insularity and migration components in the context of the TME may lead to integrative conceptual and computational frameworks for properly exploring tumor clonality and migrations of cancer cells between anatomical areas that are sources and recipients of metastases. In the following section, I will discuss the tumor island biogeography theory and suggest hypothesis testing that will allow one to investigate tumor evolution and insularity in the context of metastatic disease and clinical settings.

10.4   TUMOR ISLAND BIOGEOGRAPHY THEORY (TIB)

Historically, islands come in many forms and sizes and are not necessarily strictly defined by the traditional concept of "a land surrounded by water" (Figure 10.1a). Any system that experiences isolation because of its limited connection to adjacent areas can be considered an island or "island-like" system or "biological" island (Itescu, 2019), such as reef coasts (Figure 10.1b). Overall, island-like systems are spatially fragmented areas and/or spatially and temporally isolated ecosystems, attributes that decrease the available niches and habitat diversity, as well as their opportunity to function as a supply for colonization (Itescu, 2019). The diversity within an island-like system can be affected by the size of the island and its connectivity to adjacent areas as well as the species

(a) Island

(b) Island-like system

FIGURE 10.1   (a) Example of a true island (islets close to Lesbos Island, Greece). (b) An island-like system (reef coast in Ano Koufonisi Island, Greece). Photo credit: A. Chroni et al. (2019).

composition (which is affected by seasonality) and genetic divergence and to the evolution of new populations and species radiations (besides the presence of endemic species) (Borregaard et al., 2017; Itescu, 2019; Matthews et al., 2019; Rosindell & Phillimore, 2011).

In the context of cancer metastasis, tumor islands are isolated systems undergoing evolutionary and spatiotemporal dynamic processes that shape intra- and inter-tumor microenvironments and drive the migration of cancer cells. Tumor heterogeneity and its associated microenvironment can be conceptualized by tumor insularity and its attributes. Therefore, tumor insularity will be shaped by 1) the size and carrying capacity of the tumor site(s); 2) the physical distance and connectivity between tumors; and 3) stochastic processes. Essentially, tumor insularity can be used to infer tumor clonality and model exchange of cancer cells between primary and metastatic tumors, allowing exploration of those as a consequence of migration and extinction phenomena within and between tumors. Investigating attributes of tumor insularity might provide more information about the idiosyncratic patterns of tumor heterogeneity under different hypotheses among patients and cancer types regarding stochastic or rather predetermined processes based on tissue-cancer cell dynamics at genetic, transcriptomic, and/or epigenetic levels.

In this regard, the presence of tumors in different anatomical sites of the body is anticipated to increase the complexity of clonal exchanges and sources for metastatic migrations, as shown in previous studies (Chroni et al., 2022). This complexity is depicted via the toy example in Figure 10.2. For simplicity, I use the example of a patient with lung cancer (primary tumor) and two metastases in the liver and kidney to explore possibilities of sources of metastasis (Figure 10.2a). The number of anatomical sites where tumors have been detected and potential interconnections (proximity and connectivity) between tumors will create a network of insular areas (Figure 10.2b). Tumor insularity, and subsequently, heterogeneity and microenvironment, can be further explored as a function of the size of the anatomical area and carrying capacity (i.e., the availability of resources) (Chen et al., 2011; van Zijl et al., 2011; Wai et al., 2013). This is because the size of the insular areas will differ in their carrying capacities and resources available in ways similar to islands (Sanmartín et al., 2008), and as consequence, this will also affect the migration of cancer cells in the body. In this example network of tumor insular ecosystems, primary and metastatic tumors can act both as source- and recipient-areas depending on whether it is the starting or ending point of colonization of clones (i.e., a set of cancer cells with identical genotypes; Figure 10.2b).

In addition, investigation of the degree of connection between source- and recipient-areas will identify the source of metastasis, i.e., primary-to-metastasis, metastasis-to-metastasis, or metastasis-to-primary migrations and spread. Such investigations will reveal anatomical sites that are more often subject to metastasis and thus, will facilitate predictions of clonal exchanges between tumors over the course of the disease. Physical distance and connectivity between source- and recipient-areas are expected to drive migration and dispersal of cancer cells (Acevedo et al., 2015; Pein & Oskarsson, 2015). One may reasonably assume that organs better connected or close to the primary tumor site are more likely to be invaded and colonized by CTCs, as indicated by the contribution of the blood vascularization for cancer cell migration and colonization to distant organs (Liu et al., 2017; Palazzolo et al., 2020).

Therefore, considering the presence and abundance of metastatic tumors and the degree of heterogeneity and isolation at the time of diagnosis, tumor ecosystems can be categorized into islets, islands, and archipelagos (Box 10.1). Tumor ecosystems will undergo dynamic changes and adaptations over time as a response to treatment or disease stage (i.e., diagnosis, remission, or relapse).

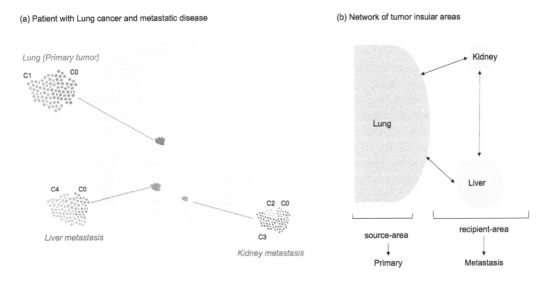

FIGURE 10.2 Colonization of insular areas is the result of migrations of tumor clones between source- and recipient-areas. Here, clones are defined as a set of cancer cells with identical genotypes. (a) Patient with lung cancer afflicted with liver and kidney metastatic disease. Each tumor site is characterized by different heterogeneity, shown by the presence of different clonal populations (C1-C5), with some of them being unique to the tumor site (e.g., C1 in the lung) and some others shared between tumors (e.g., C2 in the lung, liver, and kidney). (b) Example of a possible network between source- (lung, grey) and recipient-areas (liver and kidney metastases, blue). Size of recipient-areas and distance and connectivity between the site of origin of the migration to recipient-area, i.e., from the source- to recipient-area and between recipient-areas matter for successful colonization. In this network of insular systems, we observe areas of different sizes and their potential intraconnections. The size of the circle is proportional to the size of the areas. Arrows show metastatic migrations can happen between any tumor sites, but also, back and forth between areas.

This will have an impact on the overall heterogeneity and clonal exchange, and thus, tumors might transform from an islet to an island, but also to an Archipelago of tumors.

In the following section, I will explain the classification of tumor insularity based on evidence from the literature review about the observed heterogeneity as a function of spatial and temporal isolation of tumors. This classification can be extended for investigating tumor insularity under different hypothesis testing and across various cancer types. Tumor insularity can be assessed via exploration of its attributes, i.e., evaluation of the size and carrying capacity of the tumor site(s) and physical distance and modes of connectivity between tumors. Outcomes from such investigations will hold promise on improving patient stratification in clinical settings, e.g., for clinical trials, or contributing to metastasis prediction models, because they will facilitate assessment and assignment of prognostic scores that account for tumor heterogeneity and its associated microenvironment in a more integrative way.

10.4.1   Tumor Islets

Tumor islets are formed when solitary (primary) tumors are contained within the organ in which they are initiated. That is to say that tumor islets are characterized by the absence of metastatic disease. Indeed, often patients present with only a primary tumor at the time of diagnosis, e.g., in breast cancer (Caswell-Jin et al., 2019), esophageal cancer (Martinez et al., 2016; Yan et al., 2019), and in hepatocellular carcinoma (Ding et al., 2019). In this case, intra-tumor heterogeneity will be the result of spatial and temporal isolation because of interactions between the malignant cell populations and healthy counterparts of the tissue ecosystem, e.g., the immune microenvironment.

10.4.2   Tumor Islands

Patients with a primary tumor and at least one metastatic tumor observed at the time of diagnosis will fall into the category of tumor islands. In this case, primary tumors are the main source of clonal exchange, and metastatic tumors are the recipient-areas. Tumor islands will also be characterized by limited connectivity between anatomical sites that will hinder clonal exchanges and metastatic migrations. The presence of metastatic tumors is anticipated to add to the overall complexity of the tumor heterogeneity and its associated microenvironment but also to the overall dynamic interplay of the cell populations in the tissue ecosystem, e.g., tumor and immune cells, favoring or inhibiting metastatic migrations. That is to say that clonal exchange between anatomical sites occurred at some point, but because of limited connectivity, spatial (and/or temporal) isolation, and evolutionary pressures, there will be extensive variation in inter-tumor heterogeneity. Recent studies support the presence of tumor islands in lung cancer (Govindan et al., 2012; Jamal-Hanjani et al., 2014, 2017; Joshi et al., 2019), prostate cancer (Lindberg et al., 2015), and renal cancer (Turajlic et al., 2018a) showing that clonal heterogeneity is the result of spatial and/or temporal isolation between tumors. Studies in pancreatic cancer further support evidence of tumor islands as they attribute morphological and histological heterogeneity mainly to spatial and/or temporal separation of tumors because of a lack of connectivity between tumors (Chan-Seng-Yue et al., 2020; Hayashi et al., 2020). Similarly, cases with single metastasis (after removal of the primary tumor by surgery and/or radiation) should be classified as tumor islands. This is because tumor islands will require a preexisting main source for clonal exchanges. However, further analyses will indicate cell dynamics and clonal exchanges in the metastasis over time and how these might impact its classification. For example, if the tumor heterogeneity remains steady or decreases over time that might lead to metastasis to be transformed to an islet. On the other hand, a rapid clonal expansion or presence of multiple clones or recurrence of primary tumor will classify as a tumor island.

10.4.3   Tumor Archipelagos

In the event of primary and multiple (at least two) metastatic tumors present at the time of initial diagnosis with evidence of connectivity between the anatomical sites, e.g., regional disease in the lymph nodes, tumor ecosystems are anticipated to be featured by greater complexity and will be forming tumor archipelagos. This is because of multiple anatomical sites acting as a source of metastatic migration along with a degree of connectivity that will facilitate and increase the likelihood of clonal exchanges. The numbers and types of seedings are also expected to increase because of the increased probability of single and multiple clones and reseedings to occur, but also because of multiple sources contributing to *de novo* metastatic malformations (Chroni et al., 2022).

Interestingly, metastasis occurs in local and distant anatomical sites along with regional disease, as an immediate and proximate mean to the primary tumor. Studies show evidence of the contribution of lymph nodes in lung cancer (Abbosh et al., 2017) but also of tumor thrombus, lymph nodes, and adrenal glands in renal cancer (Turajlic et al., 2018a, 2018b). Studies on pancreatic cancer (Hayashi et al., 2020), prostate cancer (Gundem et al., 2015; Hieronymus et al., 2014; Hong et al., 2015; Robinson et al., 2015), and colorectal cancer (Reiter et al., 2020) also support the implication of connectivity on the metastatic migrations through lymph nodes and regional disease and, thus, of tumor archipelagos presence.

10.4.3.1 TIB under the Equilibrium Model and Hypothesis Testing

The conceptualization and classification of tumors as evolutionary island-like ecosystems provide an exciting and promising framework for assessing tumor heterogeneity and predicting clonal exchanges between tumors and metastases. The proposed TIB framework takes into account tumors' size and carrying capacity, physical distance, and the degree of connectivity between tumors, as well as the effect of stochastic processes when estimating tumor heterogeneity (Figure 10.3). Tumor heterogeneity can be estimated in terms of numbers of cells or clones, number of driver (and passenger) mutations, molecular signatures, average heterozygosity across sites, and standard genetic diversity in tumors. These can be used in evolutionary and phylogenetic frameworks to investigate evolutionary dynamics over time and space (Somarelli et al., 2017). This framework can be further expanded by considering the available resources in the tumor microenvironment and the anatomical sites that can be preferentially colonized by CTCs.

The aforementioned parameters will be key elements for the tumor's fitness in the tissue ecosystem, as cancer cells will strive for growth and survival. This will drive an equilibrium between the cell populations in the tumor's microenvironment that will allow cancer cells to survive in the tissue, and/or promote metastatic behavior at the same time. In that pursuit, a variety of hypotheses can be formulated and tested that will allow cancer researchers and clinicians to investigate the underlying mechanisms and predictions of tumor heterogeneity and metastatic disease over time, across space, and in response to treatment.

One hypothesis can explore the effect of tumor size and hence, of the carrying capacity on tumor heterogeneity and its associated microenvironment (Figure 10.3). Under this hypothesis, large-sized tumors are expected to be positively correlated to tumor grade and heterogeneity, and hence, to harbor greater molecular, histological, and morphological heterogeneity comparing to smaller sized tumors (Hatt et al., 2011; Kuo et al., 2017; Li et al., 2018; Payan et al., 2020; Sugiyama et al., 2019). This is also in accordance with the assumption that a larger tumor size will have access to greater resources for its inhabitants. That is the so-called equilibrium in island biogeography theory in which species abundance and diversity observed on an island are associated with the size of the island and fragmentation of habitats therein. This hypothesis is driven by evidence of large tumors showing higher heterogeneity in lung, pancreatic, and renal cancers (Hatt et al., 2011; Kuo et al., 2017; Li et al., 2018; Payan et al., 2020; Sugiyama et al., 2019). Moreover, these are anticipated to act as a force on the dynamics of the migration and extinction rates of cancer cells. Indeed, studies are reporting positive correlations between tumor size and lymph node metastasis incidence, tumor progression, overall survival, and poor prognosis. The latter indicates exciting avenues for exploring the metastatic capabilities and behavior of cancer cells in relevance to the size and carrying capacity of a tumor site.

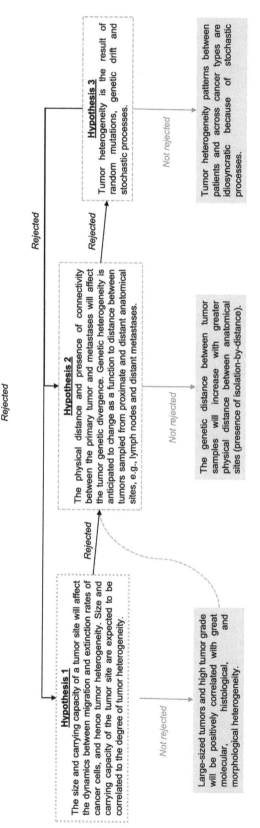

FIGURE 10.3 Flow chart of the hypothesis testing of tumors as evolutionary islands. Rejection of hypotheses #1 and #2 might lead to the validation of hypothesis #3. Confirmation of hypothesis #1 does not necessarily reject hypothesis #2 because tumor heterogeneity patterns could still result from peculiar relationships between cancer cells and tumor sites and the degree of physical distance between anatomical sites. Tumor heterogeneity can be measured in the numbers of cells or clone types, distinct molecular signatures, average heterozygosity across sites, and standard genetic diversity in tumors. Ultimately, all three hypotheses may be true, depending on the cancer type and the number of tumor sites observed in a patient.

The second hypothesis would explore tumor heterogeneity and clonal exchanges as a function of physical distance and connectivity among anatomical sites that are likely or have already developed metastases (Figure 10.3). In theory, genetic distance (and diversity) is expected to be positively correlated to the physical distance, i.e., the genetic distance between two populations will increase with the greater physical distance between sampled areas. This is simply because the physical distance will create a barrier for migrations, population exchanges, and gene flow. In this case, heterogeneity between areas will reflect the physical distance and spatial isolation. This phenomenon has been described before as isolation by distance (IBD) (Wright, 1943).

In the context of tumor heterogeneity, this is anticipated to be correlated to the physical distance between the source- and recipient-areas for metastasis. Under this second hypothesis, tumors in adjacent anatomical sites are expected to have similar (and low) heterogeneity. This is because new clones are more likely to be introduced in adjacent locations, which would increase the metastatic migration rate and clonal seedings between tumors and, as a consequence, will lead to a decrease and homogenization of the overall tumor heterogeneity. If this hypothesis is not rejected, that will indicate that anatomical sites closer to the primary tumor have a higher likelihood of metastatic malformations because migrations and clonal seedings are contributing to the heterogeneity as opposed to tumors detected in distant locals. Moreover, attention is needed in cases of physical barriers that might affect and impede clonal seedings in adjacent tissues, e.g., blood-brain barrier.

Furthermore, low tumor heterogeneity between distant anatomical sites will be evidence of the absence of IBD, i.e., the physical distance of anatomical sites does not affect metastatic migrations, and so clone exchanges still happen at high frequency, thereby shaping and decreasing tumor heterogeneity. In this scenario, the complexity of tumor heterogeneity will also appear to be the result of the number of clones (single or multiple) and the presence of reseedings involved in the formation of tumors, making this of critical consideration when interpreting inferences related to the effect of physical distance on the overall tumor heterogeneity.

On the other hand, in the case of high heterogeneity between adjacent areas not associated with presence of IBD, that will lead to the rejection of the second hypothesis. This will be evidence that tumor heterogeneity is not affected by spatial isolation (physical distance and connectivity) but rather shaped by other forces acting on it, e.g., response to treatment. Notably, high inter-tumor heterogeneity might be primarily the result of past evolutionary events. Once tumor clones migrate and form a metastasis, they will continue to evolve and accumulate mutations over time, leading to high differences in inter-tumor heterogeneity. This process is analogous to how past evolutionary events shape species diversity over time (Martiny et al., 2006). That is to say that caution is needed in the context of metastasis, with consideration of stages of the disease (early and late) and/or treatment administration expected to impact inferences between tumor heterogeneity and its associated microenvironment. Overall, investigations of the second hypothesis will showcase the role of isolation by distance between anatomical sites and the relevance of tissue-specific molecules and mechanisms (e.g., organ tropism) in shaping tumor heterogeneity and driving metastasis.

The third hypothesis is driven by the fundamental concept of randomness and genetic drift that leads to evolution (Figure 10.3). The accumulation of genetic differences across the genomes of cancer cells in the tumor microenvironment will cause genomic divergence and tumor heterogeneity. Under this hypothesis, variability across patients and cancer types is driven by stochastic processes operating within tumors on existing and newly generated genetic variation. Thus, clonal exchanges and the contribution of anatomical sites as tumor-sources in metastasis randomly occur leading to idiosyncratic patterns of tumor heterogeneity and metastatic migrations (Birkbak & McGranahan, 2020). That is, the tumor area relationship, distance, and connectivity will have less impact on shaping tumor heterogeneity than random genetic drift and selection within tumors.

Finally, validation of the first hypothesis will not exclude the possibility of the second being true as well, i.e., tumor heterogeneity might be associated with both tumor size and carrying capacity as well as physical distance and connectivity between tumors. However, the rejection of both hypotheses will provide evidence of the effect of stochastic processes on tumor heterogeneity. Ultimately, all three hypotheses may be true, depending on the cancer type and the number of tumor sites observed in a patient. Hypothesis testing in the context of tumor insularity and its attributes can lead to the establishment of shared and unique features among tumors and their heterogeneity and accommodation of the spatial and environmental dimensions of tumors that are not well understood.

10.5 TUMOR ISLAND BIOGEOGRAPHY: CLINICAL APPLICATIONS

Undeniably, the cancer field has seen substantial progress over the past decade, leveraging breakthroughs in many areas such as sequencing, imaging, immunology, and targeted therapy (Bonaguro et al., 2022). We have entered the era of high-throughput sequencing across a multidimensional scale of spatial and longitudinal tumor multi-omics. We can now sequence whole genomes of cancer cells and infer their somatic evolution. We can use "multi-omics" data from tumor samples to identify clusters of cancer cells with respect to their molecular, morphological, and histological heterogeneity in the tissue ecosystem. These advances should facilitate the investigation of the relationships between tumor heterogeneity and insularity. Nowadays, spatial (multiple regions and tumors) and longitudinal (over the course of the disease) sequencing are becoming a commonality in studying cancer cohorts. These along with information about the size and distances between tumor sites will be required for hypothesis testing of the tumor island framework.

Despite the promise of multi-omic sequencing innovations, limitations should also be considered in data analysis and interpretation. These are related to a low number of sampled somatic variants in bulk whole exome sequencing and errors due to sequencing but also to data sparsity, particularly for single-cell sequencing approaches (Miura et al., 2018). Incomplete sampling of clones per tumor region and anatomical location might also add to overall challenges. These issues are of importance because they might lead to the generation of artifacts and hence, erroneous inferences of heterogeneity on tumor islands, leading to the inference of false and inaccurate heterogeneity patterns (Kumar et al., 2020). Finally, integration of multimodal sequencing data often proves to be a challenge, with computational advances still needed to streamline these integrative studies. Ongoing efforts in sequencing and analyzing tumor data will promote continued development and improvement of sophisticated methods and computational tools to infer tumor heterogeneity and model insularity to illuminate the key processes involved in disease initiation and progression, tumor relapse, and drug resistance.

In Figure 10.4, I use the previous toy example of the patient with lung cancer and metastatic disease (liver and kidney; for more details see Figure 10.2a) to illustrate the capabilities of a multi-omic toolkit in investigating tumor heterogeneity and insularity in metastatic disease. For example, bulk whole exome sequencing could allow the detection of mutations and copy number variation that can be used for tumor clonal deconvolution and phylogenetic inference (Chroni et al., 2022). Knowledge of tumor clones, their genetic differences, and tumor locations will further enable inference of clonal exchanges and metastatic migration histories, e.g., clones C0 and C1 seeding kidney metastasis. This can further indicate the source of metastasis, which in the previous example is from lung to kidney (primary-to-metastasis spread). Such inferences are useful in understanding patient-variability patterns across cancer types (Chroni et al., 2022). Moreover, these inferences will be particularly relevant in cases with treatment involvement because they will allow inferring tumor clonal exchanges and sources of metastasis under different conditions. For example, studies report an increase of metastases and multiple seedings after treatment (Brady et al., 2019).

In addition, the use of single-cell transcriptomic sequencing data can shed light on the ecosystem aspect and indicate the malignant cell populations in the tissue microenvironment across anatomical sites (Figure 10.4b). Such inferences could allow comparisons of malignant cell populations across tissues and provide evidence for interactions between, e.g., cancer and immune cells, and how these cell populations and their interactions impact the TME in response to inflammation and treatment.

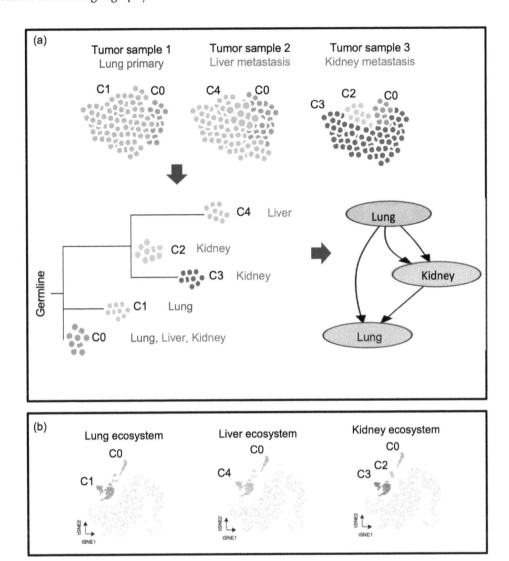

FIGURE 10.4 Tumor heterogeneity, insularity, and migration. This is a toy example of a patient with lung cancer afflicted with liver and kidney metastatic disease (for more details see Figure 10.2). (a) Genomic approaches, such as bulk exome sequencing, allow detection of mutations and copy number alterations that are useful for tumor clone deconvolution and inferences. Knowledge of tumor clones, their genetic differences, and tumor locations enable inferring clone phylogenies, clonal exchanges, and metastatic migration histories, e.g., clones C0 and C1 form lung seeding kidney metastasis (primary-to-metastasis spread). (b) On the other hand, single cell RNA sequencing data allow characterization of the tumor and its surrounded microenvironment. Integration of multimodal sequencing data might shed more light on patient variability, and these should be further explored under the lens of tumor insularity (tumor size and carrying capacity or physical distance and connectivity between anatomical sites).

Essentially, using multimodal data from genomics and transcriptomics across multiple tumor sites, one could infer tumor heterogeneity and migration/dissemination. These inferences might still indicate patient-specific variability that could be further investigated in the setting of insularity and its attributes. These inferences should be further explored under the lens of tumor insularity as tumor size and carrying capacity or physical distance and connectivity between anatomical sites (and stochastic processes) might be affecting or contributing to the observed patterns. Overall, the tumor island biogeography framework offers a sophisticated approach to integrate tumor heterogeneity and micro-environment in association with tumor insularity across disease stages and in response to treatment.

ACKNOWLEDGMENTS

I thank Drs Sudhir Kumar, Sarah R. Amend, Iannis Aifantis, and Sanam Loghavi for their insightful comments.

REFERENCES

Abbosh, C., Birkbak, N. J., Wilson, G. A., Jamal-Hanjani, M., Constantin, T., Salari, R., Le Quesne, J., Moore, D. A., Veeriah, S., Rosenthal, R., Marafioti, T., Kirkizlar, E., Watkins, T. B. K., McGranahan, N., Ward, S., Martinson, L., Riley, J., Fraioli, F., Al Bakir, M., . . . Swanton, C. (2017). Phylogenetic ctDNA analysis depicts early-stage lung cancer evolution. *Nature*, *545*(7655), 446–451.

Acevedo, M. A., Fletcher, R. J., Jr., Tremblay, R. L., & Meléndez-Ackerman, E. J. (2015). Spatial asymmetries in connectivity influence colonization-extinction dynamics. *Oecologia*, *179*(2), 415–424.

Adami, A. J., & Cervantes, J. L. (2015). The microbiome at the pulmonary alveolar niche and its role in Mycobacterium tuberculosis infection. *Tuberculosis*, *95*(6), 651–658.

Alves, J. M., Prado-López, S., Cameselle-Teijeiro, J. M., & Posada, D. (2019). Rapid evolution and biogeographic spread in a colorectal cancer. *Nature Communications*, *10*(1), 5139.

Baghban, R., Roshangar, L., Jahanban-Esfahlan, R., Seidi, K., Ebrahimi-Kalan, A., Jaymand, M., Kolahian, S., Javaheri, T., & Zare, P. (2020). Tumor microenvironment complexity and therapeutic implications at a glance. *Cell Communication and Signaling: CCS*, *18*(1), 59.

Birkbak, N. J., & McGranahan, N. (2020). Cancer genome evolutionary trajectories in metastasis. *Cancer Cell*, *37*(1), 8–19.

Bonaguro, L., Schulte-Schrepping, J., Ulas, T., Aschenbrenner, A. C., Beyer, M., & Schultze, J. L. (2022). A guide to systems-level immunomics. *Nature Immunology*, *23*(10), 1412–1423.

Borregaard, M. K., Amorim, I. R., Borges, P. A. V., Cabral, J. S., Fernández-Palacios, J. M., Field, R., Heaney, L. R., Kreft, H., Matthews, T. J., Olesen, J. M., Price, J., Rigal, F., Steinbauer, M. J., Triantis, K. A., Valente, L., Weigelt, P., & Whittaker, R. J. (2017). Oceanic island biogeography through the lens of the general dynamic model: Assessment and prospect. *Biological Reviews of the Cambridge Philosophical Society*, *92*(2), 830–853.

Brady, S. W., Ma, X., Bahrami, A., Satas, G., Wu, G., Newman, S., Rusch, M., Putnam, D. K., Mulder, H. L., Yergeau, D. A., Edmonson, M. N., Easton, J., Alexandrov, L. B., Chen, X., Mardis, E. R., Wilson, R. K., Downing, J. R., Pappo, A. S., Raphael, B. J., . . . Zhang, J. (2019). The clonal evolution of metastatic osteosarcoma as shaped by cisplatin treatment. *Molecular Cancer Research: MCR*, *17*(4), 895–906.

Caswell-Jin, J. L., McNamara, K., Reiter, J. G., Sun, R., Hu, Z., Ma, Z., Ding, J., Suarez, C. J., Tilk, S., Raghavendra, A., Forte, V., Chin, S.-F., Bardwell, H., Provenzano, E., Caldas, C., Lang, J., West, R., Tripathy, D., Press, M. F., & Curtis, C. (2019). Clonal replacement and heterogeneity in breast tumors treated with neoadjuvant HER2-targeted therapy. *Nature Communications*, *10*(1), 657.

Chan-Seng-Yue, M., Kim, J. C., Wilson, G. W., Ng, K., Figueroa, E. F., O'Kane, G. M., Connor, A. A., Denroche, R. E., Grant, R. C., McLeod, J., Wilson, J. M., Jang, G. H., Zhang, A., Dodd, A., Liang, S.-B., Borgida, A., Chadwick, D., Kalimuthu, S., Lungu, I., . . . Notta, F. (2020). Transcription phenotypes of pancreatic cancer are driven by genomic events during tumor evolution. *Nature Genetics*, *52*(2), 231–240.

Chen, J., Sprouffske, K., Huang, Q., & Maley, C. C. (2011). Solving the puzzle of metastasis: The evolution of cell migration in neoplasms. *PloS One*, *6*(4), e17933.

Chroni, A., & Kumar, S. (2021). Tumors are evolutionary island-like ecosystems. *Genome Biology and Evolution*, *13*(12). https://doi.org/10.1093/gbe/evab276

Chroni, A., Miura, S., Hamilton, L., Vu, T., Gaffney, S. G., Aly, V., Karim, S., Sanderford, M., Townsend, J. P., & Kumar, S. (2022). Clone phylogenetics reveals metastatic tumor migrations, maps, and models. *Cancers*, *14*(17). https://doi.org/10.3390/cancers14174326

Chroni, A., Miura, S., Oladeinde, O., Aly, V., & Kumar, S. (2021). Migrations of cancer cells through the lens of phylogenetic biogeography. *Scientific Reports*, *11*(1), 17184.

Chroni, A., Vu, T., Miura, S., & Kumar, S. (2019). Delineation of tumor migration paths by using a Bayesian biogeographic approach. *Cancers*, *11*(12). https://doi.org/10.3390/cancers11121880

Chu, J. E., & Allan, A. L. (2012). The role of cancer stem cells in the organ tropism of breast cancer metastasis: A mechanistic balance between the "seed" and the "soil"? *International Journal of Breast Cancer*, *2012*, 209748.

Daoust, S. P., Fahrig, L., Martin, A. E., & Thomas, F. (2013). From forest and agro-ecosystems to the micro-ecosystems of the human body: What can landscape ecology tell us about tumor growth, metastasis, and treatment options? *Evolutionary Applications*, 6(1), 82–91.

de Groot, A. E., Roy, S., Brown, J. S., Pienta, K. J., & Amend, S. R. (2017). Revisiting seed and soil: examining the primary tumor and cancer cell foraging in metastasis. *Molecular Cancer Research: MCR*, 15(4), 361–370.

Ding, X., He, M., Chan, A. W. H., Song, Q. X., Sze, S. C., Chen, H., Man, M. K. H., Man, K., Chan, S. L., Lai, P. B. S., Wang, X., & Wong, N. (2019). Genomic and epigenomic features of primary and recurrent hepatocellular carcinomas. *Gastroenterology*, 157(6), 1630–1645.e6.

Ehrlich, G. (2017). *Islands, the universe, home: Essays*. London: Penguin Group.

El-Kebir, M., Satas, G., & Raphael, B. J. (2018). Inferring parsimonious migration histories for metastatic cancers. *Nature Genetics*, 50(5), 718–726.

Ewing, J. (1928). *Neoplastic diseases*, 3rd ed. Philadelphia: W. B. Saunders Company.

Galizia, G., Gemei, M., Orditura, M., Romano, C., Zamboli, A., Castellano, P., Mabilia, A., Auricchio, A., De Vita, F., Del Vecchio, L., & Lieto, E. (2013). Postoperative detection of circulating tumor cells predicts tumor recurrence in colorectal cancer patients. *Journal of Gastrointestinal Surgery: Official Journal of the Society for Surgery of the Alimentary Tract*, 17(10), 1809–1818.

Gao, Y., Bado, I., Wang, H., Zhang, W., Rosen, J. M., & Zhang, X. H.-F. (2019). Metastasis organotropism: Redefining the congenial soil. *Developmental Cell*, 49(3), 375–391.

Gerlinger, M., Rowan, A. J., Horswell, S., Math, M., Larkin, J., Endesfelder, D., Gronroos, E., Martinez, P., Matthews, N., Stewart, A., Tarpey, P., Varela, I., Phillimore, B., Begum, S., McDonald, N. Q., Butler, A., Jones, D., Raine, K., Latimer, C., . . . Swanton, C. (2012). Intratumor heterogeneity and branched evolution revealed by multiregion sequencing. *The New England Journal of Medicine*, 366(10), 883–892.

Govindan, R., Ding, L., Griffith, M., Subramanian, J., Dees, N. D., Kanchi, K. L., Maher, C. A., Fulton, R., Fulton, L., Wallis, J., Chen, K., Walker, J., McDonald, S., Bose, R., Ornitz, D., Xiong, D., You, M., Dooling, D. J., Watson, M., . . . Wilson, R. K. (2012). Genomic landscape of non-small cell lung cancer in smokers and never-smokers. *Cell*, 150(6), 1121–1134.

Gundem, G., Van Loo, P., Kremeyer, B., Alexandrov, L. B., Tubio, J. M. C., Papaemmanuil, E., Brewer, D. S., Kallio, H. M. L., Högnäs, G., Annala, M., Kivinummi, K., Goody, V., Latimer, C., O'Meara, S., Dawson, K. J., Isaacs, W., Emmert-Buck, M. R., Nykter, M., Foster, C., . . . Bova, G. S. (2015). The evolutionary history of lethal metastatic prostate cancer. *Nature*, 520(7547), 353–357.

Hanahan, D., & Weinberg, R. A. (2011). Hallmarks of cancer: The next generation. *Cell*, 144(5), 646–674.

Hatt, M., Cheze-le Rest, C., van Baardwijk, A., Lambin, P., Pradier, O., & Visvikis, D. (2011). Impact of tumor size and tracer uptake heterogeneity in (18)F-FDG PET and CT non-small cell lung cancer tumor delineation. *Journal of Nuclear Medicine: Official Publication, Society of Nuclear Medicine*, 52(11), 1690–1697.

Hayashi, A., Fan, J., Chen, R., Ho, Y.-J., Makohon-Moore, A. P., Lecomte, N., Zhong, Y., Hong, J., Huang, J., Sakamoto, H., Attiyeh, M. A., Kohutek, Z. A., Zhang, L., Boumiza, A., Kappagantula, R., Baez, P., Bai, J., Lisi, M., Chadalavada, K., . . . Iacobuzio-Donahue, C. A. (2020). A unifying paradigm for transcriptional heterogeneity and squamous features in pancreatic ductal adenocarcinoma. *Nature Cancer*, 1(1), 59–74.

Hieronymus, H., Schultz, N., Gopalan, A., Carver, B. S., Chang, M. T., Xiao, Y., Heguy, A., Huberman, K., Bernstein, M., Assel, M., Murali, R., Vickers, A., Scardino, P. T., Sander, C., Reuter, V., Taylor, B. S., & Sawyers, C. L. (2014). Copy number alteration burden predicts prostate cancer relapse. *Proceedings of the National Academy of Sciences of the United States of America*, 111(30), 11139–11144.

Hong, M. K. H., Macintyre, G., Wedge, D. C., Van Loo, P., Patel, K., Lunke, S., Alexandrov, L. B., Sloggett, C., Cmero, M., Marass, F., Tsui, D., Mangiola, S., Lonie, A., Naeem, H., Sapre, N., Phal, P. M., Kurganovs, N., Chin, X., Kerger, M., . . . Hovens, C. M. (2015). Tracking the origins and drivers of subclonal metastatic expansion in prostate cancer. *Nature Communications*, 6, 6605.

Itescu, Y. (2019). Are island-like systems biologically similar to islands? A review of the evidence. *Ecography*, 42(7), 1298–1314.

Jamal-Hanjani, M., Hackshaw, A., Ngai, Y., Shaw, J., Dive, C., Quezada, S., Middleton, G., de Bruin, E., Le Quesne, J., Shafi, S., Falzon, M., Horswell, S., Blackhall, F., Khan, I., Janes, S., Nicolson, M., Lawrence, D., Forster, M., Fennell, D., . . . Swanton, C. (2014). Tracking genomic cancer evolution for precision medicine: The lung TRACERx study. *PLoS Biology*, 12(7), e1001906.

Jamal-Hanjani, M., Wilson, G. A., McGranahan, N., Birkbak, N. J., Watkins, T. B. K., Veeriah, S., Shafi, S., Johnson, D. H., Mitter, R., Rosenthal, R., Salm, M., Horswell, S., Escudero, M., Matthews, N., Rowan, A., Chambers, T., Moore, D. A., Turajlic, S., Xu, H., . . . TRACERx Consortium. (2017). Tracking the evolution of non-small-cell lung cancer. *The New England Journal of Medicine*, 376(22), 2109–2121.

Joshi, K., de Massy, M. R., Ismail, M., Reading, J. L., Uddin, I., Woolston, A., Hatipoglu, E., Oakes, T., Rosenthal, R., Peacock, T., Ronel, T., Noursadeghi, M., Turati, V., Furness, A. J. S., Georgiou, A., Wong, Y. N. S., Ben Aissa, A., Sunderland, M. W., Jamal-Hanjani, M., . . . Chain, B. (2019). Spatial heterogeneity of the T cell receptor repertoire reflects the mutational landscape in lung cancer. *Nature Medicine*, *25*(10), 1549–1559.

Kim, M.-Y., Oskarsson, T., Acharyya, S., Nguyen, D. X., Zhang, X. H.-F., Norton, L., & Massagué, J. (2009). Tumor self-seeding by circulating cancer cells. *Cell*, *139*(7), 1315–1326.

Kumar, S., Chroni, A., Tamura, K., Sanderford, M., Oladeinde, O., Aly, V., Vu, T., & Miura, S. (2020). PathFinder: Bayesian inference of clone migration histories in cancer. *Bioinformatics*, *36*(Suppl_2), i675–i683.

Kuo, H.-T., Que, J., Lin, L.-C., Yang, C.-C., Koay, L.-B., & Lin, C.-H. (2017). Impact of tumor size on outcome after stereotactic body radiation therapy for inoperable hepatocellular carcinoma. *Medicine*, *96*(50), e9249.

Landis, M. J., Matzke, N. J., Moore, B. R., & Huelsenbeck, J. P. (2013). Bayesian analysis of biogeography when the number of areas is large. *Systematic Biology*, *62*(6), 789–804.

Li, D., Hu, B., Zhou, Y., Wan, T., & Si, X. (2018). Impact of tumor size on survival of patients with resected pancreatic ductal adenocarcinoma: A systematic review and meta-analysis. *BMC Cancer*, *18*(1), 985.

Lindberg, J., Kristiansen, A., Wiklund, P., Grönberg, H., & Egevad, L. (2015). Tracking the origin of metastatic prostate cancer. *European Urology*, *67*(5), 819–822.

Liu, Q., Zhang, H., Jiang, X., Qian, C., Liu, Z., & Luo, D. (2017). Factors involved in cancer metastasis: A better understanding to "seed and soil" hypothesis. *Molecular Cancer*, *16*(1), 176.

Lucas, D. (2021). Structural organization of the bone marrow and its role in hematopoiesis. *Current Opinion in Hematology*, *28*(1), 36–42.

MacArthur, R. H., & Wilson, E. O. (1963). An equilibrium theory of insular zoogeography. *Evolution; International Journal of Organic Evolution*, 373–387.

Macintyre, G., Van Loo, P., Corcoran, N. M., Wedge, D. C., Markowetz, F., & Hovens, C. M. (2017). How subclonal modeling is changing the metastatic paradigm. *Clinical Cancer Research: An Official Journal of the American Association for Cancer Research*, *23*(3), 630–635.

Maley, C. C., Aktipis, A., Graham, T. A., Sottoriva, A., Boddy, A. M., Janiszewska, M., Silva, A. S., Gerlinger, M., Yuan, Y., Pienta, K. J., Anderson, K. S., Gatenby, R., Swanton, C., Posada, D., Wu, C.-I., Schiffman, J. D., Hwang, E. S., Polyak, K., Anderson, A. R. A., . . . Shibata, D. (2017). Classifying the evolutionary and ecological features of neoplasms. *Nature Reviews, Cancer*, *17*(10), 605–619.

Maley, C. C., Galipeau, P. C., Finley, J. C., Wongsurawat, V. J., Li, X., Sanchez, C. A., Paulson, T. G., Blount, P. L., Risques, R.-A., Rabinovitch, P. S., & Reid, B. J. (2006). Genetic clonal diversity predicts progression to esophageal adenocarcinoma. *Nature Genetics*, *38*(4), 468–473.

Martinez, P., Timmer, M. R., Lau, C. T., Calpe, S., Sancho-Serra, M. D. C., Straub, D., Baker, A.-M., Meijer, S. L., Kate, F. J. W. T., Mallant-Hent, R. C., Naber, A. H. J., van Oijen, A. H. A. M., Baak, L. C., Scholten, P., Böhmer, C. J. M., Fockens, P., Bergman, J. J. G. H. M., Maley, C. C., Graham, T. A., & Krishnadath, K. K. (2016). Dynamic clonal equilibrium and predetermined cancer risk in Barrett's oesophagus. *Nature Communications*, *7*, 12158.

Martiny, J. B. H., Bohannan, B. J. M., Brown, J. H., Colwell, R. K., Fuhrman, J. A., Green, J. L., Horner-Devine, M. C., Kane, M., Krumins, J. A., Kuske, C. R., Morin, P. J., Naeem, S., Ovreås, L., Reysenbach, A.-L., Smith, V. H., & Staley, J. T. (2006). Microbial biogeography: Putting microorganisms on the map. *Nature Reviews, Microbiology*, *4*(2), 102–112.

Matthews, T. J., Rigal, F., Triantis, K. A., & Whittaker, R. J. (2019). A global model of island species-area relationships. *Proceedings of the National Academy of Sciences of the United States of America*, *116*(25), 12337–12342.

Mitchell, E., Spencer Chapman, M., Williams, N., Dawson, K. J., Mende, N., Calderbank, E. F., Jung, H., Mitchell, T., Coorens, T. H. H., Spencer, D. H., Machado, H., Lee-Six, H., Davies, M., Hayler, D., Fabre, M. A., Mahbubani, K., Abascal, F., Cagan, A., Vassiliou, G. S., . . . Campbell, P. J. (2022). Clonal dynamics of haematopoiesis across the human lifespan. *Nature*, *606*(7913), 343–350.

Miura, S., Huuki, L. A., Buturla, T., Vu, T., Gomez, K., & Kumar, S. (2018). Computational enhancement of single-cell sequences for inferring tumor evolution. *Bioinformatics*, *34*(17), i917–i926.

Mroz, E. A., & Rocco, J. W. (2017). The challenges of tumor genetic diversity. *Cancer*, *123*(6), 917–927.

Noorani, A., Li, X., Goddard, M., Crawte, J., Alexandrov, L. B., Secrier, M., Eldridge, M. D., Bower, L., Weaver, J., Lao-Sirieix, P., Martincorena, I., Debiram-Beecham, I., Grehan, N., MacRae, S., Malhotra, S., Miremadi, A., Thomas, T., Galbraith, S., Petersen, L., . . . Fitzgerald, R. C. (2020). Genomic evidence supports a clonal diaspora model for metastases of esophageal adenocarcinoma. *Nature Genetics*, *52*(1), 74–83.

Norton, L., & Massagué, J. (2006). Is cancer a disease of self-seeding? *Nature Medicine*, *12*(8), 875–878.

Nowell, P. C. (1976). The clonal evolution of tumor cell populations. *Science*, *194*(4260), 23–28.

Paget, S. (1989). The distribution of secondary growths in cancer of the breast. 1889. *Cancer Metastasis Reviews*, *8*(2), 98–101.

Palazzolo, G., Mollica, H., Lusi, V., Rutigliani, M., Di Francesco, M., Pereira, R. C., Filauro, M., Paleari, L., DeCensi, A., & Decuzzi, P. (2020). Modulating the distant spreading of patient-derived colorectal cancer cells via aspirin and metformin. *Translational Oncology, 13*(4), 100760.

Payan, N., Presles, B., Brunotte, F., Coutant, C., Desmoulins, I., Vrigneaud, J.-M., & Cochet, A. (2020). Biological correlates of tumor perfusion and its heterogeneity in newly diagnosed breast cancer using dynamic first-pass 18F-FDG PET/CT. *European Journal of Nuclear Medicine and Molecular Imaging, 47*(5), 1103–1115.

Pein, M., & Oskarsson, T. (2015). Microenvironment in metastasis: roadblocks and supportive niches. *American Journal of Physiology, Cell Physiology, 309*(10), C627–C638.

Qian, J. J., & Akçay, E. (2018). Competition and niche construction in a model of cancer metastasis. *PloS One, 13*(5), e0198163.

Ree, R. H., & Smith, S. A. (2008). Maximum likelihood inference of geographic range evolution by dispersal, local extinction, and cladogenesis. *Systematic Biology, 57*(1), 4–14.

Reiter, J. G., Hung, W.-T., Lee, I.-H., Nagpal, S., Giunta, P., Degner, S., Liu, G., Wassenaar, E. C. E., Jeck, W. R., Taylor, M. S., Farahani, A. A., Marble, H. D., Knott, S., Kranenburg, O., Lennerz, J. K., & Naxerova, K. (2020). Lymph node metastases develop through a wider evolutionary bottleneck than distant metastases. *Nature Genetics, 52*(7), 692–700.

Robinson, D., Van Allen, E. M., Wu, Y.-M., Schultz, N., Lonigro, R. J., Mosquera, J.-M., Montgomery, B., Taplin, M.-E., Pritchard, C. C., Attard, G., Beltran, H., Abida, W., Bradley, R. K., Vinson, J., Cao, X., Vats, P., Kunju, L. P., Hussain, M., Feng, F. Y., . . . Chinnaiyan, A. M. (2015). Integrative clinical genomics of advanced prostate cancer. *Cell, 161*(5), 1215–1228.

Ronquist, F., & Huelsenbeck, J. P. (2003). Mr. Bayes 3: Bayesian phylogenetic inference under mixed models. *Bioinformatics, 19*(12), 1572–1574.

Ronquist, F., & Sanmartín, I. (2011). Phylogenetic methods in biogeography. *Annual Review of Ecology, Evolution, and Systematics, 42*(1), 441–464.

Rosindell, J., & Phillimore, A. B. (2011). A unified model of island biogeography sheds light on the zone of radiation. *Ecology Letters, 14*(6), 552–560.

Sanmartín, I., Van Der Mark, P., & Ronquist, F. (2008). Inferring dispersal: A Bayesian approach to phylogeny-based island biogeography, with special reference to the Canary Islands. *Journal of Biogeography, 35*(3), 428–449.

Savas, P., Teo, Z. L., Lefevre, C., Flensburg, C., Caramia, F., Alsop, K., Mansour, M., Francis, P. A., Thorne, H. A., Silva, M. J., Kanu, N., Dietzen, M., Rowan, A., Kschischo, M., Fox, S., Bowtell, D. D., Dawson, S.-J., Speed, T. P., Swanton, C., & Loi, S. (2016). The subclonal architecture of metastatic breast cancer: Results from a prospective community-based rapid autopsy program "CASCADE". *PLoS Medicine, 13*(12), e1002204.

Sleeman, J. P. (2012). The metastatic niche and stromal progression. *Cancer Metastasis Reviews, 31*(3–4), 429–440.

Solary, E., & Laplane, L. (2020). The role of host environment in cancer evolution. *Evolutionary Applications, 13*(7), 1756–1770.

Somarelli, J. A. (2021). The hallmarks of cancer as ecologically driven phenotypes. *Frontiers in Ecology and Evolution, 9*. https://doi.org/10.3389/fevo.2021.661583

Somarelli, J. A., Ware, K. E., Kostadinov, R., Robinson, J. M., Amri, H., Abu-Asab, M., Fourie, N., Diogo, R., Swofford, D., & Townsend, J. P. (2017). PhyloOncology: Understanding cancer through phylogenetic analysis. *Biochimica et Biophysica Acta, Reviews on Cancer, 1867*(2), 101–108.

Steeg, P. S. (2006). Tumor metastasis: Mechanistic insights and clinical challenges. *Nature Medicine, 12*(8), 895–904.

Sugiyama, Y., Yatsuda, J., Murakami, Y., Ito, N., Yamasaki, T., Mikami, Y., Ogawa, O., & Kamba, T. (2019). Impact of tumor size on patient survival after radical nephrectomy for pathological T3a renal cell carcinoma. *Japanese Journal of Clinical Oncology, 49*(5), 465–472.

Thomas, F., Roche, B., Giraudeau, M., Hamede, R., & Ujvari, B. (2020). The interface between ecology, evolution, and cancer: More than ever a relevant research direction for both oncologists and ecologists. *Evolutionary Applications, 13*(7), 1545–1549.

Tissot, T., Massol, F., Ujvari, B., Alix-Panabieres, C., Loeuille, N., & Thomas, F. (2019). Metastasis and the evolution of dispersal. *Proceedings. Biological Sciences / The Royal Society, 286*(1916), 20192186.

Turajlic, S., & Swanton, C. (2016). Metastasis as an evolutionary process. *Science, 352*(6282), 169–175.

Turajlic, S., Xu, H., Litchfield, K., Rowan, A., Chambers, T., Lopez, J. I., Nicol, D., O'Brien, T., Larkin, J., Horswell, S., Stares, M., Au, L., Jamal-Hanjani, M., Challacombe, B., Chandra, A., Hazell, S., Eichler-Jonsson, C., Soultati, A., Chowdhury, S., . . . TRACERx Renal Consortium. (2018a). Tracking cancer evolution reveals constrained routes to metastases: TRACERx renal. *Cell, 173*(3), 581–594.e12.

Turajlic, S., Xu, H., Litchfield, K., Rowan, A., Horswell, S., Chambers, T., O'Brien, T., Lopez, J. I., Watkins, T. B. K., Nicol, D., Stares, M., Challacombe, B., Hazell, S., Chandra, A., Mitchell, T. J., Au, L., Eichler-Jonsson, C., Jabbar, F., Soultati, A., . . . TRACERx Renal Consortium. (2018b). Deterministic evolutionary trajectories influence primary tumor growth: TRACERx renal. *Cell, 173*(3), 595–610.e11.

Ulz, P., Heitzer, E., Geigl, J. B., & Speicher, M. R. (2017). Patient monitoring through liquid biopsies using circulating tumor DNA. *International Journal of Cancer, Journal International Du Cancer, 141*(5), 887–896.

Ungefroren, H., Sebens, S., Seidl, D., Lehnert, H., & Hass, R. (2011). Interaction of tumor cells with the microenvironment. *Cell Communication and Signaling: CCS, 9*, 18.

van Galen, P., Hovestadt, V., Wadsworth, M. H., II, Hughes, T. K., Griffin, G. K., Battaglia, S., Verga, J. A., Stephansky, J., Pastika, T. J., Lombardi Story, J., Pinkus, G. S., Pozdnyakova, O., Galinsky, I., Stone, R. M., Graubert, T. A., Shalek, A. K., Aster, J. C., Lane, A. A., & Bernstein, B. E. (2019). Single-cell RNA-seq reveals AML hierarchies relevant to disease progression and immunity. *Cell, 176*(6), 1265–1281.e24.

van Zijl, F., Krupitza, G., & Mikulits, W. (2011). Initial steps of metastasis: Cell invasion and endothelial transmigration. *Mutation Research, 728*(1–2), 23–34.

Virchow, R. (1856). *Gesammelte Abhandlungen zur wissenschaftlichen Medicin.* G. Grote.

Wai, L.-E., Narang, V., Gouaillard, A., Ng, L. G., & Abastado, J.-P. (2013). In silico modeling of cancer cell dissemination and metastasis. *Annals of the New York Academy of Sciences, 1284*, 71–74.

Welch, D. R., & Hurst, D. R. (2019). Defining the hallmarks of metastasis. *Cancer Research, 79*(12), 3011–3027.

Whiteson, K. L., Bailey, B., Bergkessel, M., Conrad, D., Delhaes, L., Felts, B., Harris, J. K., Hunter, R., Lim, Y. W., Maughan, H., Quinn, R., Salamon, P., Sullivan, J., Wagner, B. D., & Rainey, P. B. (2014). The upper respiratory tract as a microbial source for pulmonary infections in cystic fibrosis: Parallels from island biogeography. *American Journal of Respiratory and Critical Care Medicine, 189*(11), 1309–1315.

Williams, M. J., Sottoriva, A., & Graham, T. A. (2019). Measuring clonal evolution in cancer with genomics. *Annual Review of Genomics and Human Genetics, 20*, 309–329.

Wright, S. (1943). Isolation by distance. *Genetics, 28.*

Yan, T., Cui, H., Zhou, Y., Yang, B., Kong, P., Zhang, Y., Liu, Y., Wang, B., Cheng, Y., Li, J., Guo, S., Xu, E., Liu, H., Cheng, C., Zhang, L., Chen, L., Zhuang, X., Qian, Y., Yang, J., . . . Cui, Y. (2019). Multi-region sequencing unveils novel actionable targets and spatial heterogeneity in esophageal squamous cell carcinoma. *Nature Communications, 10*(1), 1670.

Yang, K. R., Mooney, S. M., Zarif, J. C., Coffey, D. S., Taichman, R. S., & Pienta, K. J. (2014). Niche inheritance: A cooperative pathway to enhance cancer cell fitness through ecosystem engineering. *Journal of Cellular Biochemistry, 115*(9), 1478–1485.

Yates, L. R., Knappskog, S., Wedge, D., Farmery, J. H. R., Gonzalez, S., Martincorena, I., Alexandrov, L. B., Van Loo, P., Haugland, H. K., Lilleng, P. K., Gundem, G., Gerstung, M., Pappaemmanuil, E., Gazinska, P., Bhosle, S. G., Jones, D., Raine, K., Mudie, L., Latimer, C., . . . Campbell, P. J. (2017). Genomic evolution of breast cancer metastasis and relapse. *Cancer Cell, 32*(2), 169–184.e7.

Yu, Y., Harris, A. J., & He, X. (2010). S-DIVA (statistical dispersal-vicariance analysis): A tool for inferring biogeographic histories. *Molecular Phylogenetics and Evolution, 56*(2), 848–850.

11 Cancer and the Evolutionary Ecology of Invasions

Joel S. Brown, Sarah R. Amend, and Kenneth J. Pienta

11.1 INTRODUCTION

From cancer initiation to niche filling to tumor growth and then to metastasis, the study of cancer has tight parallels to similar phenomena (speciation, adaptive radiation, range expansion, and invasive species biology) in nature. Each has access to conceptual and modeling frameworks, experimental techniques, technologies, and data that can inform the other. The growing appreciation that cancer evolves by natural selection in response to ecological forces within the tumor and to therapeutic interventions increases the opportunity and need for intellectual cross-fertilization. Cancer exemplifies ecological and evolutionary dynamics that can lead to the progressive evolution and diversification of cancer cell types within the patient. The scariest invasion for most patients is transitioning to metastatic disease, where the primary tumor has given rise to additional tumor burden in other organs of the body. Metastasis becomes the likely culmination of all the invasion processes that happened before, namely cancer initiation, diversification, and tumor growth. In recognition of these parallels, cancer biologists and physicians are taking inspiration and concepts from ecology and evolution, such as invasion ecology. In turn, it is becoming clear that cancer biology has much to offer the fields of ecology and evolution as tests and applications of broader principles and concepts, including those associated with biological invasions.

11.2 INITIATION, GROWTH, AND SPREAD

The initiation, growth, and spread of cancers and their tumor ecosystems have been compared to species invasions in nature. Throughout this chapter, for convenience, we shall refer to "nature" as all of life in the biosphere aside from cancer, even as we note that cancer is a natural process and very much part of nature. From the earliest comparison that we could find, there has been a steady literature seeing aspects of cancer, particularly with respect to metastases, as having striking similarities to biological invasions (Chen & Pienta, 2011; de Groot et al., 2017; Gatenby, 1991; Iwasa et al., 2005; Kareva, 2011; Merlo et al., 2006; Noorbakhsh et al., 2020). Recent reviews cover the ecological and epigenetic/genetic aspects of cancer metastases as biological invasions (Lloyd et al., 2017; Neinavaie et al., 2021). There are even reviews that explore how cancer biology can inform the broader field of invasion ecology (Sepulveda et al., 2012).

Here we aim for a perspective of cancer as a sequence of four different types of biological invasions. The first is **cancer initiation** as the origination of a novel, single-celled protist within the patient. This event, by definition, must happen for all patients. The second is the "invasion" of diverse ecological niches within the cancer's tumor ecosystem. This process of **niche filling through speciation** may be more pronounced in some types of cancers than in others. The third can occur concomitantly and in concert with the second. It involves the **growth of the tumor** as it spreads into adjacent tissue. It can be seen as the invasion of a new habitat through range expansion. In some types of cancers, this growth of the primary tumor ceases. The fourth, **metastasis**, conforms most closely to many aspects of invasive species ecology. Here, cancer cells from a tumor travel through the lymph or blood system to distant sites within the same or different tissues and establish new tumors by colonizing new blood systems and sources or resources. This does not

DOI: 10.1201/9781003307921-11

happen for all cancer cells. A tumor may be benign, grow to some limit, never metastasize, and pose little health risk to the person. In contrast, some primary tumors, such as glioblastoma (a type of brain cancer), do not metastasize but are still lethal to the patient. When metastatic, most cancers become incurable and will eventually lead to patient death.

11.3 CANCER INITIATION: THE INVASION FROM WITHIN

To cancer patients and the public at large, cancer can seem like a loss of control over one's own body. A rebellion of cells grows uncontrollably, and the rebels may even invade other tissues. Most definitions of cancer can be paraphrased as *a disease of unregulated cell proliferation with the capacity to spread to other tissues*. This capacity to spread is even termed invasive cancer. But how does this invasion from within occur, and how should this shape our definition, perspective, and treatment of cancer? We favor defining cancer *as a disease of uncontrolled proliferation by transformed cells subject to evolution by natural selection* (Brown et al., 2023). The process of a cell or cell lineage of a multicellular organism becoming *transformed* in a manner that results in them *evolving by natural selection* is cancer initiation: **the invasion from within.**

Cancer as a disease emanating from within the body has ancient roots. Through the ages, it has variously been seen as a product of the supernatural in ancient Egypt, as an imbalance of the four bodily fluids in 4th century BCE Greece, as something emerging locally within the body in the 17th century, as a cellular multiplication process when first viewed under the microscope, as a result of chromosomal abnormalities in the early 20th century, and as a sequence of driver mutations by the 1980s (Bos, 1989; Faguet, 2015; Haggard & Smith, 1938; Hajdu, 2006, 2011; Hoerni, 2014; Kontomanolis et al., 2020; Manchester, 1995). This brings us to the molecular era of modern genetics and epigenetics, where a cell becomes cancerous through a sequence or combination of mutations to oncogenes that propel carcinogenesis and mutations to tumor suppressor genes that would otherwise eliminate the abnormal cell (Fearon & Vogelstein, 1990).

Cancer initiation represents a jump in levels of selection (Huxley, 1956; Pradeu et al., 2023). A cell goes from being part of a multicellular organism to becoming its own single-celled organism inhabiting what is now its host (Pienta et al., 2020). Natural selection had been operating on the multicellular organism, favoring vast syndromes of traits contributing to its survival and fecundity, including nearly all aspects of the life of that single cell(s) that has now become cancerous. For the emerging cancer cell lineage, it becomes the species, it has survival and proliferation rates, and it has heritable variation. Thus, natural selection now favors traits that maximize the cancer cell's fitness (=survival and proliferation). For cancer cells, the tumor is their ecosystem, and the patient or host is their planet earth (Amend & Pienta, 2015). Therefore, cancer can be seen as a "speciation" event. It seems an unusual speciation event in that it is a multicellular organism spinning off a new single-celled organism, but it is not. In fact, such a clinically detectable event happens to approximately 20 million people worldwide annually and for millions of other multicellular organisms that also get cancer.

As an invasion event, cancer initiation has imperfect parallels with the rare invasion of remote islands by a mainland colonizer. Darwin's finches immediately jump to mind where presumably a single inoculum of a tanager species from what is now Peru or Ecuador led to the initiation, evolution, and subsequent adaptive radiation of about 18 species. Even more spectacular is the invasion of a single fly species from the Order Diptera into the Hawaiian Islands. This invader from the genus *Drosophila* (fruit flies) arrived some 25 million years ago and speciated into over 400 named species of *Drosophila* (and additional ones that formed a new genus) (Coyne & Orr, 1989). Notably, this number represents the lion's share of the approximately 1000 named *Drosophila* species found worldwide. While islands provide amazing examples of invasions and subsequent adaptive radiations, many remote, volcanic island archipelagos come and go. Some, like the Hawaiian Islands, are long-lived due to their location, geology, and near-constant volcanism. Others, like the proverbial lost world of Atlantis, erode into lagoons or underwater mountains as they erode in the absence of continual renewal. Over the history of terrestrial life, how many equivalents of Darwin's finches or Hawaiian picture-winged *Drosophila* have invaded, thrived, and then disappeared forever beneath the waves?

So it is with cancer. Upon initiation, the lifespan of a cancer matches that of the patient (except in the infinitesimally rare cases of transmissible cancers). In some cases, the cancer will not affect the patient's lifespan. Many autopsies reveal the presence of cancer in otherwise perfectly healthy individuals, and patients with potentially lethal cancers may well die from other causes. Thus far, there are no documented cases where cancer has directly increased the patient's lifespan. A striking difference with the island analogy is that the cancer can hasten the patient's demise, whereas geological processes dictate the island's lifespan. While an invasive species can dramatically influence the community of resident species, the sum of erosion, accretion, uplifting, subsidence and volcanism determine the presence and persistence of an island. In fact, the presence of vegetation on the island may hold the soil and slow erosion, possibly extending the "life" of the island.

The clear difference is that the uninhabited island plays no active role in facilitating or preventing colonizations, and these colonizations come from the outside. Here, cancer initiation has no other parallels in nature. Every time two cells of a multicellular organism divide, the organism plays with "evolutionary fire," the result of which can be the initiation of a debilitating cancer. Even as natural selection has produced, favored, and molded the adaptations of the multicellular organism, it can also favor a destroyer of that intricately adapted organism. With cancer, natural selection is playing both sides—one over a multitude of generations of patients and the other within the short lifespan of a patient.

Unsurprisingly, natural selection over generations of the multicellular organism has favored tumor suppressor traits, including what we classify as oncogenes. A highly effective anti-cancer adaptation can be found in the 1 mm long nematode (roundworm) *Caenorhabditis elegans*. Upon maturation, none of its 900–1000 cells retain the ability to divide save for those producing gametes (Corsi et al., 2015). No cell division, no cancer! But it involves a trade-off. *C. elegans* has no capacity for tissue renewal or wound healing. Whereas a frog can regenerate lost toes, among other tissues, a *C. elegans* can regenerate nothing.

Starting in the 1920s and prior to molecular genetics and molecular biology, experiments with mice (and epidemiological studies of people) showed how some chemicals, when applied to the skin, such as various forms and extracts of asphalt and pitch, increased chances of cancer. Furthermore, these effects could be amplified by wounding the skin of the mice (Ames et al., 1975; Bogen & Loomis, 1935; Hieger, 1949; McCann & Ames, 1976a). In some experiments, the combination of wounding and mutagens was necessary to induce cancer; just one alone was not sufficient. By the 1970s, the link was made among gene mutations, carcinogens, and cancer initiation. The Ames test became widely used starting in 1973 to identify mutagenic and potentially carcinogenetic chemicals and substances (McCann & Ames, 1976b).

Animals have evolved diverse safeguards to prevent cancer initiation. The extent of anti-cancer adaptation seems to increase with an animal's size and reproductive lifespan. This co-adaptation between an animal's tumor suppressor traits and its size and life history, likely explains Peto's Paradox (Brown et al., 2015; Peto et al., 1975). Peto's Paradox emerges from a seeming contradiction. Larger animals or those that are long-lived should see many more cell divisions, somatic mutations, and opportunities for cancer initiation. Yet there is no relationship between size and cancer incidence across diverse taxa of animals (Vincze et al., 2022). A striking example are elephants. The *TP53* suppresses cancer initiation when "wildtype," but not so when mutated and defective. Ensuring proper and diverse functioning of the *TP53* gene, African elephants have possibly 19 copies (Abegglen et al., 2015; Tollis et al., 2021).

Thus, animals' multiple layers of defense can include intracellular, tissue architecture, tissue control, and immune control. Cells themselves have traits for repairing DNA as well as "suicide pills" if the cell perceives damage or irregular circumstances (Loftus et al., 2022). The tissue exerts control mechanisms by which neighboring cells signal aberrant cells to initiate apoptosis (programmed cell death). The entire stem cell and progenitor architecture of tissue renewal and healing is an anti-cancer adaptation that discourages long runs of cell divisions via telomere shortening and differentiation pathways that might otherwise allow the accumulation of oncogenic mutations (Hammarlund et al., 2020). Finally, the immune system is crucial in detecting and eliminating damaged, aberrant, mutated, infected, or otherwise out-of-place normal cells (Smyth et al., 2006).

For a cell and its lineage to invade from within and initiate cancer requires the following: 1) genetic and epigenetic mutations that cease differentiation, activate the capacity to divide forever if conditions permit ("replicative immortality"), downregulation of self-destruct apoptotic pathways, deactivation of responses to external tissue control signals, and evasion of destruction by the immune system; and 2) long runs of cell division permitting time to accumulate mutations and epigenetic changes that upregulate and downregulate portions of the genome. Long runs of cell division can result from acute or chronic inflammation and wound healing (Hanahan & Weinberg, 2011). Genetic predispositions that give cells of a tissue a "head-start" toward cancer initiation and lifestyle and environmental exposures that simultaneously inflame, wound, and mutate tissues are mechanisms of how an animal becomes "invaded" by cancer.

We end this section by asking, why is cancer just an invasion from within rather than an invasion beyond the organism? In the rare cases of communicable cancers, the cancer evolves to become an infectious disease (Tasmanian devil facial tumor disease, bivalve transmissible neoplasia, canine transmissible venereal tumor, to name a few) (Rebbeck et al., 2009). While animals with cancer may well shed numerous live cells into the environment—skin sloughing, feces, blood, urine— these cells have numerous strikes against them. First, they have narrow niches regarding temperature, humidity, and diet (e.g., a need for nutrients and growth factors). Second, the outside world is resource poor as it is already inhabited by capable single-celled protists among many other potential competitors. Third, besides desiccating and starving, the cancer cell in the outside world will likely face far more efficient and less restrained predators than immune cells. Despite the failure of cancers to outlive their hosts, it is noteworthy that the successful evolutionary jump from single-celled to a multicellular organism has been rare indeed, with a smattering of examples spread over hundreds of millions of years, whereas the reverse jump, as is observed in cancer initiation (multicellular to single-cellular), happens hundreds of millions of times in a matter of decades.

Currently the most common occurrence of cancers outside of a patient or host involves cell culture lines used for research. Such culture lines, intentionally or inadvertently, see continued evolution through mutations, genetic drift, cell passaging and natural selection. Intentionally, researchers select for drug resistant strains, or other strains of interest. Inadvertently, constant passaging of cancer cells selects for "lab ecotypes" that evolve to become quite different than the original patient-derived cells. The first cancer cell line from Henrietta Lacks, HeLa cells, have evolved diverse "lab morphs," some which mimic and at times contaminate later laboratory derived cancer cells lines (Masters, 2002).

11.4   CANCER CELL DIVERSIFICATION: INVADING NEW TUMOR NICHES

Upon cancer initiation, the cancer cells of the nascent lesion are known to exhibit a rapid increase in genetic and phenotypic diversity. This has been referred to as "cancer's big bang" (Sottoriva et al., 2015). The idea sees genetic instabilities and opportunities for proliferation as producing and allowing for mutations that endure as the cancer cell population grows (Amaro et al., 2016). Alternatively, the heterogeneity that indubitably emerges even in the smallest of tumors may result from the cancer cells diversifying and invading the different ecological niches offered by the tumor's heterogeneities. For instance, even at the earliest stages of breast cancer (ductal carcinoma *in situ*), the cancer cells filling a duct experience a gradient of nutrients, oxygen, and pH whose favorability declines with distance from the ductal wall. Thus, cancer cell heterogeneity may not be so much a "big bang" but rather a "Cambrian Explosion" or, more simply, an adaptive radiation as frequently seen in nature (Pienta et al., 2020).

Darwin's finches, the *Drosophila* of Hawaii, and the Cichlid fishes of the African Rift Valley Lakes are all examples of adaptive radiation and speciation via niche filling (Ngoepe et al., 2023; Yoder et al., 2010). Evolutionarily, the single progenitor species successfully invades different ecological opportunities. All of these represent the diversification of closely related species, many of which coexist within the same environment. Coexisting species that occupy distinct niches do so via mechanisms of coexistence. A mechanism of coexistence involves environmental heterogeneity and a trade-off between the coexisting species such that to become better adapted for some aspect of the heterogeneity necessarily means becoming less apt at some other aspect (Kotler & Brown, 1988; Leibold et al., 2019; Rees et al., 2001). Mechanisms of coexistence of closely related species generally include 1) diet separation, 2) habitat selection, 3) food-safety trade-offs, 4) variance partitioning, and 5) colonization-competition trade-offs (Staples et al., 2016).

The eight squirrel species found around the North American Great Lakes (Family Sciuridae) provide examples of all five of these mechanisms. In terms of **diet,** the woodchuck's (*Marmota monax*) large size facilitates coexistence with the primarily seed-eating thirteen-lined ground squirrel (*Ictidomys tridecemlineatus*) and omnivorous Franklin's ground squirrel (*Spermophilus franklini*). The eastern grey squirrel (*Sciurus carolinensis*) and fox squirrel (*S. niger*) specialize on the hard-shelled nuts of deciduous trees and shrubs (acorns, walnuts, hickory nuts, and hazelnuts), whereas the North American red squirrel (*Tamaisciurus hudsonicus*) favors the softer-shelled seeds of conifers (pine nuts).

In terms of **habitat selection,** several species favor forests (southern flying squirrel—*Glaucomys sabenis*, grey squirrel, red squirrel, and Eastern chipmunk—*Tamias striatus*), while several species favor meadows and grasslands (thirteen-lined and Franklin's ground squirrels). Two species favor ecotones, in this case, the margins between woodlands and meadows (woodchuck and fox squirrel).

In terms of **food-safety trade-offs**, the grey squirrel is a superior resource competitor to the fox squirrel, even as the fox squirrel is better at avoiding predation. The former is favored in the safer, deep woods, while the latter is favored in the more dangerous but less competitive wood margins.

Variance partitioning involves some species exploiting more superficially in times and places of resource plenty while others can profit from places and periods of low resource availability (Lande, 1996; Legendre et al., 2005). The former species are "cream skimmers" and the latter "crumb pickers" (Olsson & Brown, 2010). In this sense, the chipmunk and grey squirrel occupy similar habitats and have similar diets. The chipmunk remains torpid during the winter and emerges to take advantage of the growing season. On the other hand, the grey squirrel remains active throughout the year, enduring the winter by recovering cached nuts and continuing to harvest the growing season's leftovers.

In terms of **colonization-competition trade-offs**, the southern flying squirrel has the greatest dispersal capacity relative to the other forest squirrels with whom it competes, including the fox squirrel, grey squirrel, red squirrel, and chipmunk (Bendel & Gates, 1987; Yu & Wilson, 2001). This species of flying squirrel can agilely glide up to 100 meters with a glide slope as favorable as 10%. The flying squirrel can seek home ranges of high productivity (variance partitioning) and/or unusually low competition from other squirrels or rodents. As such, flying squirrels occur widely but at lower average densities than the other forest squirrels.

As it is with Darwin's finches, Hawaiian *Drosophila*, African cichlids, and North American Great Lake's Sciurids, so it seems with each patient's cancer. Via molecular techniques, it has become clear that the cancer cells within a single tumor manifest considerable genomic, transcriptomic, and phenotypic variability. This variability becomes even greater when comparing cancer cells from different metastatic sites within the same patient. Yet in virtually every case, all cancer cells within a single patient descended from a single cancer-initiating lineage. Hence, they are all closely related with divergence times of much less than a person's lifetime.

More recently, cancer researchers have noted that these diverse cancer cell types seem to diversify in response to heterogeneity within the tumor (Schnipper, 1986). At the largest scale for patients exhibiting metastatic disease, the different tumors may reside in quite different tissues (Spremulli & Dexter, 1983). For instance, in metastatic colorectal cancer, in addition to the cancer cells residing in the original primary tumor in the colon or rectum, there may be descendants of these cancer cells occupying newer tumors in the liver or lung. Each tissue type offers different challenges and opportunities in terms of the quantitative and qualitative properties of the immune system, the delivery of blood, the underlying extracellular matrix, and the structure and presence of host cells. Unsurprisingly, the colorectal cancer cells that invade the liver or lung evolve and diverge as they become better adapted for their new tissue type. Interestingly, breast cancer cells that metastasize to the brain evolve adaptations that make them more similar phenotypically to their new tissue than to their ancestors in the original primary breast tumor (Chen et al., 2015; Cunningham et al., 2015; Park et al., 2011).

At the next spatial scale down, a single tumor offers macrohabitats often described as the tumor's edge, interior, and necrotic zones (Gatenby et al., 2013). Each varies in immune cell infiltration, overall vasculature and blood delivery, pH, oxygen, and composition of normal host cells. In general, the tumor edge may offer the most vasculature, the greatest exposure to immune cells, the highest density of normal cells, and intermediate cancer cell densities. The necrotic zone has the lowest density of all cells due to hypoxia, acidic conditions, no vasculature, and low resource supply. Yet the necrotic zone may select for "extremophiles," cancer cells that are at low abundance, tolerate the unfavorable environmental conditions, and exploit marginal resources. The interior of the tumor may have the highest density of cancer cells and the most intense competition for nutrients and space. Unsurprisingly, the characteristics of cancer cells at the edge of a tumor typically differ strikingly from those in the interior. Like the fox squirrels and grey squirrels, respectively, the edge of primary tumors in breast cancer seems populated by a cancer cell type more adapted to avoid the immune system than competing for resources, while those in the interior favor adaptations for competition over immune evasion.

The smallest spatial scale within a tumor is the distance to the vasculature, which may be approximately six to twelve cell widths (Kim et al., 2012; Zheng et al., 2005), creating strong chemical gradients. Near vasculature sees higher cancer cell densities in response to proximity to the gradient of nutrients and oxygen. There is a striking similarity to riparian habitats in arid and semi-arid landscapes where the density and composition of the vegetation changes with distance from the perennial or seasonal waterways (Alfarouk et al., 2013). Such has been documented in a variety of cancers. It is highly likely, and in some cases documented, that cancer cell types near vasculature are highly competitive, densely packed, and intolerant of hypoxia, low resource supply, and accumulation of toxins, and vice versa for those farther away from a blood vessel.

It remains an exciting, important, and open question whether cancer cell heterogeneity is primarily driven by the invasion of distinct niches as various forms of niche partitioning that lead to definable mechanisms of coexistence. We see this as highly likely. The various scales of habitat heterogeneity, from different tumors within a patient, to macrohabitats within a tumor, to small scale heterogeneity known as the cancer's tumor microenvironment appear to drive the invasion of distinct niches via **habitat selection** and specialization. The coexistence of different cancer cell types within the patient based on habitat heterogeneity seems almost an inevitability.

Cancer cells experience different degrees of exposure to the community of immune cells, and cancer cells exhibit diverse adaptations for immune evasion and suppression of immune threats. Thus, it seems highly likely that many tumors experience the invasion of niches based on **food-safety trade-offs** (Brown & Kotler, 2004). This has been seen at large scales, such as the edge and interior of tumors, but may also manifest at smaller spatial scales where an immune-suppressive but less competitive cancer cell type coexists with the opposite cell type.

Cancer cells forage for a wide range of micro- and macro-molecules primarily based on blood composition. Additionally, via micro- and macro-pinocytosis, cancer cells evolve the capacity to engulf small or large aggregates of debris that may have been released by living cells or lysing dead cells. Thus, the needs and diet of cancer cells, like Darwin's finches or the squirrels of the North American Great Lakes, must be highly overlapping. This would seem to reduce the likelihood of food heterogeneity, allowing for the invasion of distinct niches based on **diet separation** (Klemetsen, 2010). Yet cell culture experiments of different cell lines show differences in their capacity to uptake and utilize nutrients, such as glucose, glutamine, cystine, and their various metabolites. Two to three species of Darwin's finches can coexist on the same island of the Galapagos by partitioning seed size based on different beak sizes. It is not far-fetched to imagine similarly subtle yet profound coexistence mechanisms among cancer cells within a tumor.

Variance partitioning manifests when there are large seasonal or stochastic variabilities in resource supply, spatially or temporally (Ptacnik et al., 2010). Local tumor microenvironments experience large fluctuations that can vary in a matter of minutes to hours as vessels within a tumor experience occlusions, expansion, and changes in pressure within blood vessels and surrounding interstitial fluids. Cancer cell types invading the cream skimmer and crumb picker niches have yet to be clearly documented *in vivo*. However, it is likely that a trade-off between competitive ability at high and low resources exists.

The last mechanism, **competition-colonization trade-offs**, seems likely in cancer (Yu & Wilson, 2001). The epithelial-mesenchymal transition (EMT) describes cancer cells transitioning from a competitive-sessile morph (epithelial) to a motile morph (mesenchymal) in response to stress (Diepenbruck & Christofori, 2016; Ribatti et al., 2020). Whether these morphs are simply phenotypically plastic states of the same cancer type or whether different cancer cell types exhibit different propensities for competitive versus colonizing new space remains an open question. It is noteworthy that the three cancer cell types often used to describe cellular heterogeneity within glioblastomas seem to include coexistence based on colonization-competition trade-offs (Aum et al., 2014). Another way this trade-off has been articulated in cancer research is a trade-off between "grow or go." Cancer cells that invade the ecological niche of colonization aptitudes become the likely candidates for the next two aspects of biological invasions in cancer: tumor growth and metastasis.

11.5 TUMOR GROWTH AS RANGE EXPANSION

The range of a species or population describes its current geographical extent. For cancers, a tumor represents the current range within the patient. When metastatic, there may be more than one tumor throughout the body, just as the populations of a species may be disjunct. All of the rare Amani sunbirds reside either in the Arabuko Sokoke Forest of Kenya or 300 kilometers away in the Usambara highlands of Tanzania (Oyugi et al., 2012). A population can grow either through increased density within its range or by range expansion. With range expansion, a species or, in some cases, a whole community of species invades areas beyond the boundaries of its current range. The same holds true for tumors.

In nature, global climate change and warming are currently drivers of species experiencing range expansions or contractions. In Nepal, the common leopard has been expanding from the lowlands and mid-elevations into higher elevations of the Himalayas, likely at the expense of snow leopards (Baral et al., 2023). Shorter and less severe winters likely explain how these higher elevations have become more favorable for the common leopard. Likewise, tundra habitats in the higher northern latitudes of Europe, Asia, and North America are seeing the northward spread of woody vegetation, conifer forests, and their concomitant communities of invertebrates and vertebrates. The northward expansion of the white-tailed deer in North America and the red fox globally are prime examples.

In cancer, the range expansion (or contraction) of tumors over time is generally of most concern. This perhaps has to do with a time scale separation where tumors expand much more slowly relative to the rate at which cancer cells within the tumor grow (or decline as in the case of necrotic regions) to changes in their local carrying capacity. In this way, the overall dimensions or range size of the cancer provides a proxy for disease burden and the overall population size of cancer cells.

Non-spatially explicit measures of tumor burden can include blood biomarkers, such as PSA for prostate cancer, CEA for colon cancer, and CA-19–9 for pancreatic cancer. When tumor markers go up, it is generally assumed that the population size of cancer cells has grown through the expansion of existing tumors into new space or the occurrence of metastases in which the cancer has colonized a new tissue, such as melanoma establishing in the brain or prostate cancer in bone. These biomarkers work when cancer cells secrete them into the blood at enormously higher rates than normal cells. Current hypotheses for their occurrence include cancer cells voiding themselves of metabolites that otherwise would be degraded or metabolized in normal cells; or these molecules occur at unusual concentrations within the cancer cells and they release when the cancer cell dies (Balk et al., 2003; Scarà et al., 2015; Thompson et al., 1991). Support for this latter hypothesis comes from the observation that some blood biomarkers initially increase with the onset of therapy.

Spatially explicit mappings of tumors are assessed using various radiomic techniques that include CAT scans, MRI imaging, PET CT scans, or even measures of size based on endoscopies or palpation (Parekh & Jacobs, 2017; Rathore et al., 2018). These sorts of scans and metrics can be done with or without ingested or injected contrast agents that alter the images in ways that clarify tumor boundaries, distinguish cancerous regions from non-cancerous, and map macrohabitats within a tumor that may vary in fluid content, cell density, metabolic activity, or regions of necrosis.

An interesting contrast between natural species' ranges and that of tumors is the distribution of individuals across their range. Generally, the average density of individuals is highest near the center of the species' range rather than closer to the boundaries. This is because the population reaches the highest abundance where conditions are best for it, and as one moves away from this core, the conditions gradually (or sometimes abruptly) deteriorate until the habitat is no longer suitable. These boundaries of suitability define the edge of the species' range. While this pattern may hold in some tumors, there are reasons, special to tumors, that likely make it harder to predict the density of cancer cells throughout the tumor (Aktipis et al., 2013, 2015; Kotler & Brown, 2020). As cancer cells spread outward from the core of the tumor, they may recruit additional blood flow and hence nutrients, but these nutrients flow into the tumor from beyond it. The tumor itself provides no additional nutrients beyond the recycling and reformulating of those that flow or diffuse across the boundaries of the tumor. Unlike most ecosystems, it does not have a community of primary producers using photosynthesis or chemosynthesis to provide for a more-or-less continuous renewal of resources. Hence, the interior of the tumor may be so resource poor as to support only a fraction of the density of cancer cells that might occur closer to the tumor's edges. Regions of necrosis are common in tumors, particularly as they become larger, and in large tumors, reference is often made to the necrotic core (Brown, 2002).

Tumors, therefore, are more akin to allochthonous ecological communities where there are no primary producers, and all consumers rely directly or indirectly on nutrients, food sources, and prey that were produced outside of the ecosystem (Huxel et al., 2002). In nature, such systems include the deep ocean ecosystem, communities of cave organisms that rely on the feces and carcasses of bats, and other animals seeking refuge in the caves. For instance, cave crickets occur globally. They use caves for breeding and safety while feeding on the surface. They support other cave species such as cave spiders that prey upon the adults and cave beetles that feed on their eggs.

Unlike most allochthonous systems, the tumor does not necessarily have fixed boundaries such as a cave or beach systems that are reliant on organic flotsam from tides and storms. The tumor itself grows into adjacent normal tissue. In so doing, the cancer cells cohabitate with normal cells, co-opt the functioning of cells such as fibroblasts, evade anti-cancer components of the immune system, and promote the pro-tumor immune cells. How does this range expansion compare to invasive species? The answer is quite slow, even as this "slow" rate can quickly become health- or life-threatening to the patient.

Andow et al. compared actual and potential rates of spread of three invasive pests (Andow et al., 1990). The muskrat (*Ondatra zibethicus*) spread from Bohemia in what is now the Czech Republic at rates of up to 25 km per year and, in many cases, close to 100% of its potential. The habitat as it spread across Europe must have been near ideal. The same applies to the small cabbage white butterfly (*Pieris rapae*) as it spread at rates exceeding 150 km per year upon introduction into eastern North America. Impressively, the cereal leaf beetle *(Oulema melanopus)* as an invasive in Michigan, USA, spread up to 90 km per year and at a rate 10–20 times faster than would be its potential. How could it exceed this potential? This species could be spread through the transportation of grains and crops. Curiously, some invasives do not spread much. An instructive comparison is the house sparrow (*Passer domesticus*) and the Eurasian tree sparrow (*Passer montanus*) (Ramos-Elvira et al., 2023). They are congeners that are widespread in their native ranges, and both were introduced into North America from Eurasia in small numbers during the 19th century. The house sparrow spread rapidly across most of the continent, while the Eurasian tree sparrow still only resides in a few counties adjacent to its release site in St. Louis, Missouri, USA. It is almost as if the former was a highly "metastatic cancer," whereas the latter was a "benign tumor."

Cancers tend to expand the boundaries of their tumors at a much lower rate (<10%) than would be expected from their potential growth rates and dispersal distances. This is true of even the fastest growing tumors such as glioblastomas that can grow 5–6% per day and some pediatric sarcomas that grow up to 8% per day (Eikenberry et al., 2009; Reed et al., 2020). Most solid tissue tumors, even ones that are highly dangerous, grow <2.5% per day (Lorenzo et al., 2016; Norton, 1988). At first glance, the edge of the tumor should be an ideal habitat, much like the landscapes invaded by the muskrat, small cabbage white butterfly, and cereal leaf beetle. Beyond the tumor exists richer vasculature and lower densities of cells. Yet it seems that beyond the tumor is dangerous in terms of the immune system and suppressive fibroblasts. This is because cancers are ecosystem engineers (Myers et al., 2020; Pienta et al., 2008; Yang et al., 2014).

As examples of ecosystem engineering and niche construction, the collective action from neighborhoods of cancer cells results in angiogenesis (recruitment of blood vessels), a favorable extracellular matrix though collogen remodeling, co-option of normal cells such as cancer-associated fibroblasts, and safety in numbers from immune attack. These actions benefit the individual cancer cell while providing public goods to others. The cancer cells' propensity to provide and need public goods likely creates Allee effects (Gerlee et al., 2022; Johnson et al., 2019) where at low population densities, cancer cells see an increase in their fitness with density, at least up to the point where scarcity of resources and space, and buildup of toxic conditions limit a cancer cell's per capita growth rate (Gerlee et al., 2022; Johnson et al., 2019). Furthermore, the resulting microenvironments and structures of the tumor may last for generations of cancer cells. Thus, cancer cells benefit from and inherit the niche construction of their ancestors.

The lack of properly engineered habitats likely makes the normal tissue beyond the tumor's boundary rather inhospitable (Johnson et al., 2019). From a cancer cell's perspective, it represents "wilderness" beyond the boundary of "civilization." For cancer cells, the adjacent tissue has predatory immune cells and normal cells whose functions have yet to be co-opted. If spreading into adjacent tissue is challenging for cancer cells, then spreading into new tissues far from the primary tumor should be even more so.

11.6 METASTASIS

As noted by numerous authors, a tight parallel can be seen between the steps of the metastatic cascade and the introduction and spread of an invasive species (Hanahan & Weinberg, 2011). The metastatic cascade describes a cancer cell or group of cancer cells seeding a new tumor distinct from the primary or original tumor. The metastatic site is usually quite far from the tumor of origin and occupies a distinct tissue or organ from the tissue of origin. Similarly, a successful invasive species involves introduction into a new place, often a continent apart from its native range, and then spread across a landscape that generally has never been occupied by the species. Figure 11.1 of Neinavaie et al. (2021) compares the steps of biological invasions with that of metastasis (Chen &

Pienta, 2011; Neinavaie et al., 2021). The former starts with a species in its native range, the latter with cancer cells in the primary tumor.

For species invasion, transport is the first step whereby individuals of the species find themselves far beyond their native range (Chen & Pienta, 2011). Such transport can be human mediated or not. It can be the active movement of individuals through swimming, flying, or terrestrial locomotion, or it can be passive movement by individuals rafting across the ocean, blown through the air, or carried by another animal. For many species, it may be life stage specific, where seeds, spores, or cysts may disperse and transport more easily than adults. The transport step for biological invasions aligns with what is recognized as the first two steps of the metastatic cascade: intravasation and circulation. Intravasation is the entry of cancer cells into the lymph or circulatory system. This may be passive if the cells are swept up and enter directly through the lymph or ruptures in blood vessels, or it can be active if the cancer cells actively move and invade through the walls of blood vessels. Cells that enter the circulatory system are referred to as circulating tumor cells (CTCs) (Pantel & Speicher, 2016). They may number in the millions over time, and such cells may circulate perhaps just once or perhaps many times, bearing in mind that the blood of the circulatory system makes a complete sojourn every 45–60 seconds in the average adult human (Wilbaux et al., 2015). In this way, the "transport" of the metastatic cascade is a much more prescribed process than the manifold ways seen in nature.

Step 2 of invasion biology, introduction, is analogous to step 3 of the metastatic cascade, extravasation (Ha et al., 2013). Introduction for an invasive species describes how individuals under transport arrive alive at a new locale. The fumigation of bananas arriving from Central America to any number of continents aims to kill non-native insects, fungi, and the like. The cattle egret represents a historical "natural" introduction event from Africa to South America, presumably by birds following air currents. The signal crayfish of North America was introduced into Europe in the 1970s for commercial fisheries. Like many other examples of escapees from aquaculture or animal husbandry, the signal crayfish escaped into natural waterways and have since spread. For cancer, the CTC must actively move out of the circulatory system into a novel site or passively spread into a tissue through bleeding or vessel breakage. Cancer cells may also metastasize to lymph nodes from which they can be introduced into other tissues. Whether such jumps from the lymph to a new organ requires going through the circulatory system remains an open question.

Once introduced into a novel locale, the successful invasive species must establish (step 3 biological invasions) and spread (step 4). For cancers, a successful metastasis sees the establishment of a proliferating population (step 4 of the metastatic cascade) and then tumor formation, angiogenesis, and tumor growth (step 5 of the metastatic cascade) (Geiger & Peeper, 2009). For invasive plants such as the prickly pear cactus, establishment and spread happen swiftly and easily. For both biological invasions and cancer, spread may require evolutionary adaptations to permit the invasive species or cancer to thrive in its new environment that may be more or less similar to its native range. At the very least, invasive species will evolve ecotypic variation, thereby enabling a better fit between the species' phenotype and its new habitats. The same is true for cancer with the additional observation that the new metastatic tumor may see the adaptive radiation of additional cancer cell ecotypes that once again fill the available niches created by the new tumor (Woodhouse et al., 1997). This process is less analogous to an invasive species over ecological time and more analogous to a species invasion of a new adaptive zone that over evolutionary time promotes an adaptive radiation. For example, around 40 million years ago, South America was colonized by primates that rafted over from Africa. It is now thought that there may have been two invasion events. Regardless of whether one or two species arrived in South America, they gave rise to the over 60 extant species of New World primates (Poux et al., 2006).

For biological invasions, there are examples of species that are limited by each step of the biological invasion process. Probably not so for cancer. Once the primary tumor has reached a clinically detectable size, there will be CTCs, and their number becomes large as seen in mouse models and humans (Alix-Panabieres & Pantel, 2014; Williams et al., 2020). Intravasation and circulation seem to come easily. Thus, CTCs and cancer cells through the lymph likely get everywhere and extravasate into just about all tissues of the human body. Yet in line with Paget's seed and soil hypothesis for metastases, the cancer cells from the tissue of origin must in some way align with the recipient tissue (Akhtar et al., 2019; Fidler & Poste, 2008). Prostate cancer invariably metastasizes to the bone, as do many other cancers (Pienta & Loberg, 2005). Colorectal cancer metastasizes frequently to the liver and lung. Whole maps have been created showing the propensity of primary tumors in one organ to metastasize to another. Many cancers will metastasize to the brain, while few brain cancers metastasize elsewhere. Pancreatic cancer relatively quickly metastasizes to the liver, presumably from proximity and similarity of tissue type.

While these are still open questions, certain patterns do seem to apply to the metastatic cascade. First, it is a numbers game (Dujon et al., 2021). The likelihood of any one CTC succeeding to form a metastasis is near zero, but with enough CTCs, the probability becomes near certainty for many cancers when left untreated. Second, while metastases are frequent across many cancer types, there is no selection on cancer cells to metastasize. It is "dumb luck," so to speak. Natural selection can never favor a phenotype for a challenge that has yet to occur. Third, natural selection may favor cancer subtypes that are more likely to succeed at the metastatic cascade. More motile, more immune-evasive, more tumor-edge cancer cell types may be pre-adapted to have a chance at successfully metastasizing. In terms of EMT, generalist cancer cells that retain the plasticity to transition back and forth between epithelial and mesenchymal states may be able to more successfully negotiate the different steps of the metastatic cascade (Jolly et al., 2019, 2017). Fourth, working backward from successful metastases, it may be that success required the extravasation of a group of cancer cells rather than just one or a cancer cell in a particular state.

Recent work has identified cells in the endocycling cancer cell state (also referred to as the poly-aneuploid cancer cell state) as a stress response in several cancers (Pienta et al., 2021). The endocycling state seems to occur at low frequency in cancer cell populations (Trabzonlu et al., 2023). They appear to form through endocycling that creates larger cells with greater and greater ploidy. These cells do not proliferate but are highly resistant to hypoxia, therapy stresses, and conditions generally unfavorable to non-polyploid cancer cells. Upon environmental stress, the frequency of cells in the endocycling state increases in the cancer cell population. In time, these polyaneuploid cells revert to a typical ploidy state through a process of ploidy reduction. It has been proposed that successful metastasis may require the extravasating CTC to enter into a endocycling state while or prior to overcoming the challenges of the novel tissue environment (Mallin et al., 2023).

Another way to view metastases as biological invasions is through the more focused field of host switching in parasites. If the cancer's host acts as its entire planet for purposes of the cancer's ecology and evolution, then the different organ systems of the host represent very different pathogenic landscapes that must be overcome. Such is the case for many pathogens. A pathogen may show host specificity, such as a bat fly species specializing in one or two species of bats. In evolving to be better and better at its present host, it may decrease the likelihood of dispersing to and succeeding on another host species. Models of host switching have much in common with cancer metastases (Araujo et al., 2015; Thines, 2019). Two opposing forces are at work: the first reduces the likelihood of switching because of specialization, and the second increases the likelihood because of accidental and constant exposure to other hosts. Unsurprisingly, like the seed and soil hypothesis for metastases from one organ to another, host switching is most frequent when the new host is closely related ecologically and evolutionarily to the donor host. At an even finer scale, metastasis has some commonality of parasite diversification into different organ systems. Flukes (trematode parasites of vertebrate organ systems) often specialize in specific organs such as liver flukes, blood flukes, intestinal flukes, and lung flukes. These represent different species, and phylogenetically different blood flukes across different species may be more closely related than the tissue-specific flukes of a single species. Like cancer, when invading a new host or a new tissue within the host, flukes must overcome the immune system and tissue-specific defenses as well as acclimate and then adapt to new environmental circumstances.

11.7 CONCLUSIONS AND APPLICATION

Cancer can be viewed as an invasive, diversifying, spreading, and colonizing species that evolves from a cell lineage of its multicellular host. The host is its planet earth, and the current distribution of tumors its range. For purposes of conservation and the protection of economic interests there is considerable interest in preventing and controlling invasive species. Thus, the control measures for invasive species in nature bear parallels to therapeutic strategies for preventing, curing, and controlling cancer.

In the case of cancer initiation and invasive pests, prevention is the best "medicine." Toward that end, the first line of defense against invasive species—particularly those that may be threats to crops, livestock, forestry, and fisheries—involve prevention. Many countries have active programs to fumigate incoming fruits and agricultural produce (Page & Lubatti, 1963; Saccaggi et al., 2016; Xing et al., 2022). Australia has strict border controls on bringing plant and animal products into the country (Black & Bartlett, 2020). Similarly, cancer prevention through lifestyle changes and reductions in environmental carcinogens is seen as the first line of defense against cancer (Lewandowska et al., 2021).

Once the invasion of a noxious species has happened, eliminating it early becomes the next best course of action. This prevents its spread and subsequent evolution (Martinez et al., 2020; Reaser et al., 2020). When an invasive species is still small in numbers and restricted in area, then culling, eradicating, or killing all individuals remains feasible. So it is with cancer. When caught early, the cancer cells generally reside in a small definable area. If so, surgical removal and/or radiation therapy provide cure in the vast majority of cancers that are not yet metastatic (Pienta et al., 2020). Drug therapies may be combined before (neo-adjuvant) or after (adjuvant) surgery to ensure the complete eradication of the cancer (Kaiyin et al., 2023; Paulino & Mansinho, 2023; Tsibulak & Fotopoulou, 2023).

If eradication of the invasive pest is impossible then there are a variety of containment strategies that aim for "functional eradication" defined as controlling the invasive pest at a level where ecological or economic damage is acceptable (Green & Grosholz, 2021). This generally involves using diverse control measures (chemical, habitat modification, biological control agents) to minimize harm and spread (McLaughlin & Dearden, 2019; Weidlich et al., 2020). Most of these containment strategies use principles and approaches from integrated pest management (Deguine et al., 2021). For cancer, when cure is not possible, novel conceptual approaches are emerging that aim to contain the patient's cancer burden at levels that permit a high quality of life for as long as possible. These approaches draw on principles from pest management, and they try to anticipate and steer the ecological and evolutionary dynamics of the disease (Cunningham, 2019; Stankova, 2019; Stankova et al., 2019). Such a containment approach was successful at extending progression free survival for men with a form of metastatic prostate cancer (Zhang et al., 2022). We look forward to similar approaches being utilized in other cancers. It is clear that the cancer community and the patients we treat can benefit from the use of ecological approaches in the treatment of cancer.

REFERENCES

Abegglen, L. M., Caulin, A. F., Chan, A., Lee, K., Robinson, R., Campbell, M. S., Kiso, W. K., Schmitt, D. L., Waddell, P. J., Bhaskara, S., Jensen, S. T., Maley, C. C., & Schiffman, J. D. (2015). Potential mechanisms for cancer resistance in elephants and comparative cellular response to DNA damage in humans. *JAMA*, *314*(17), 1850–1860. https://doi.org/10.1001/jama.2015.13134

Akhtar, M., Haider, A., Rashid, S., & Al-Nabet, A. D. M. (2019). Paget's "seed and soil" theory of cancer metastasis: An idea whose time has come. *Advances in Anatomic Pathology*, *26*(1), 69–74.

Aktipis, C. A., Boddy, A. M., Gatenby, R. A., Brown, J. S., & Maley, C. C. (2013). Life history trade-offs in cancer evolution. *Nature Reviews Cancer*, *13*(12), 883–892.

Aktipis, C. A., Boddy, A. M., Jansen, G., Hibner, U., Hochberg, M. E., Maley, C. C., & Wilkinson, G. S. (2015). Cancer across the tree of life: Cooperation and cheating in multicellularity. *Philosophical Transactions of the Royal Society B: Biological Sciences*, *370*(1673), 20140219.

Alfarouk, K. O., Ibrahim, M. E., Gatenby, R. A., & Brown, J. S. (2013). Riparian ecosystems in human cancers. *Evolutionary Applications*, *6*(1), 46–53. https://doi.org/10.1111/eva.12015

Alix-Panabieres, C., & Pantel, K. (2014). Challenges in circulating tumour cell research. *Nature Reviews Cancer*, *14*(9), 623–631. https://doi.org/10.1038/nrc3820

Amaro, A., Chiara, S., & Pfeffer, U. (2016). Molecular evolution of colorectal cancer: From multistep carcinogenesis to the big bang. *Cancer and Metastasis Reviews*, *35*(1), 63–74. https://doi.org/10.1007/s10555-016-9606-4

Amend, S. R., & Pienta, K. J. (2015). Ecology meets cancer biology: The cancer swamp promotes the lethal cancer phenotype. *Oncotarget*, *6*(12), 9669–9678. https://doi.org/10.18632/oncotarget.3430

Ames, B. N., McCann, J., & Yamasaki, E. (1975). Methods for detecting carcinogens and mutagens with the Salmonella/mammalian-microsome mutagenicity test. *Mutation Research*, *31*(6), 347–364. https://doi.org/10.1016/0165-1161(75)90046-1

Andow, D. A., Kareiva, P. M., Levin, S. A., & Okubo, A. (1990). Spread of invading organisms. *Landscape Ecology*, *4*(2–3), 177–188. https://doi.org/10.1007/Bf00132860

Araujo, S. B., Braga, M. P., Brooks, D. R., Agosta, S. J., Hoberg, E. P., von Hartenthal, F. W., & Boeger, W. A. (2015). Understanding host-switching by ecological fitting. *PLoS One*, *10*(10), e0139225. https://doi.org/10.1371/journal.pone.0139225

Aum, D. J., Kim, D. H., Beaumont, T. L., Leuthardt, E. C., Dunn, G. P., & Kim, A. H. (2014). Molecular and cellular heterogeneity: The hallmark of glioblastoma. *Neurosurgical Focus*, *37*(6), E11.

Balk, S. P., Ko, Y. J., & Bubley, G. J. (2003). Biology of prostate-specific antigen. *Journal of Clinical Oncology*, *21*(2), 383–391. https://doi.org/10.1200/jco.2003.02.083

Baral, K., Adhikari, B., Bhandari, S., Kunwar, R. M., Sharma, H. P., Aryal, A., & Ji, W. (2023). Impact of climate change on distribution of common leopard (Panthera pardus) and its implication on conservation and conflict in Nepal. *Heliyon*, *9*(1).

Bendel, P. R., & Gates, J. E. (1987). Home range and microhabitat partitioning of the southern flying squirrel (Glaucomys volans). *Journal of Mammalogy*, *68*(2), 243–255.

Black, R., & Bartlett, D. M. F. (2020). Biosecurity frameworks for cross-border movement of invasive alien species. *Environmental Science & Policy*, *105*, 113–119. https://doi.org/10.1016/j.envsci.2019.12.011

Bogen, E., & Loomis, R. N. (1935). Comparative carcinogenic potency of common agents. *California and Western Medicine*, *43*(2), 135.

Bos, J. L. (1989). Ras oncogenes in human cancer: a review. *Cancer Research*, *49*(17), 4682–4689. https://www.ncbi.nlm.nih.gov/pubmed/2547513

Brown, J. M. (2002). Tumor microenvironment and the response to anticancer therapy. *Cancer Biology & Therapy*, *1*(5), 453–458.

Brown, J. S., Amend, S. R., Austin, R. H., Gatenby, R. A., Hammarlund, E. U., & Pienta, K. J. (2023). Updating the definition of cancer. *Molecular Cancer Research*, *21*. https://doi.org/10.1158/1541-7786.MCR-23-0411

Brown, J. S., Cunningham, J. J., & Gatenby, R. A. (2015). The multiple facets of Peto's paradox: A life-history model for the evolution of cancer suppression. *Philosophical Transactions of the Royal Society of London. Series B, Biological Sciences*, *370*(1673). https://doi.org/10.1098/rstb.2014.0221

Brown, J. S., & Kotler, B. P. (2004). Hazardous duty pay and the foraging cost of predation. *Ecology Letters*, *7*(10), 999–1014.

Chen, J., Lee, H. J., Wu, X., Huo, L., Kim, S. J., Xu, L., Wang, Y., He, J., Bollu, L. R., Gao, G., Su, F., Briggs, J., Liu, X., Melman, T., Asara, J. M., Fidler, I. J., Cantley, L. C., Locasale, J. W., & Weihua, Z. (2015). Gain of glucose-independent growth upon metastasis of breast cancer cells to the brain. *Cancer Research*, *75*(3), 554–565. https://doi.org/10.1158/0008-5472.CAN-14-2268

Chen, K. W., & Pienta, K. J. (2011). Modeling invasion of metastasizing cancer cells to bone marrow utilizing ecological principles. *Theoretical Biology and Medical Modelling*, *8*, 36. https://doi.org/10.1186/1742-4682-8-36

Corsi, A. K., Wightman, B., & Chalfie, M. (2015). A transparent window into biology: A primer on Caenorhabditis elegans. *Genetics*, *200*(2), 387–407.

Coyne, J. A., & Orr, H. A. (1989). Patterns of speciation in Drosophila. *Evolution*, *43*(2), 362–381.

Cunningham, J. J. (2019). A call for integrated metastatic management. *Nature Ecology and Evolution*, *3*(7), 996–998. https://doi.org/10.1038/s41559-019-0927-x

Cunningham, J. J., Brown, J. S., Vincent, T. L., & Gatenby, R. A. (2015). Divergent and convergent evolution in metastases suggest treatment strategies based on specific metastatic sites. *Evolution, Medicine, and Public Health*, *2015*(1), 76–87. https://doi.org/10.1093/emph/eov006

de Groot, A. E., Roy, S., Brown, J. S., Pienta, K. J., & Amend, S. R. (2017). Revisiting seed and soil: Examining the primary tumor and cancer cell foraging in metastasis. *Molecular Cancer Research*, *15*(4), 361–370. https://doi.org/10.1158/1541-7786.MCR-16-0436

Deguine, J.-P., Aubertot, J.-N., Flor, R. J., Lescourret, F., Wyckhuys, K. A. G., & Ratnadass, A. (2021). Integrated pest management: Good intentions, hard realities. A review. *Agronomy for Sustainable Development*, *41*(3), 38. https://doi.org/10.1007/s13593-021-00689-w

Diepenbruck, M., & Christofori, G. (2016). Epithelial–mesenchymal transition (EMT) and metastasis: Yes, no, maybe? *Current Opinion in Cell Biology*, *43*, 7–13.

Dujon, A. M., Capp, J. P., Brown, J. S., Pujol, P., Gatenby, R. A., Ujvari, B., Alix-Panabieres, C., & Thomas, F. (2021). Is there one key step in the metastatic cascade? *Cancers (Basel)*, *13*(15). https://doi.org/10.3390/cancers13153693

Eikenberry, S. E., Sankar, T., Preul, M. C., Kostelich, E. J., Thalhauser, C., & Kuang, Y. (2009). Virtual glioblastoma: Growth, migration and treatment in a three-dimensional mathematical model. *Cell Proliferation*, *42*(4), 511–528.

Faguet, G. B. (2015). A brief history of cancer: Age-old milestones underlying our current knowledge database. *International Journal of Cancer*, *136*(9), 2022–2036. https://doi.org/10.1002/ijc.29134

Fearon, E. R., & Vogelstein, B. (1990). A genetic model for colorectal tumorigenesis. *Cell*, *61*(5), 759–767. https://doi.org/10.1016/0092-8674(90)90186-i

Fidler, I. J., & Poste, G. (2008). The "seed and soil" hypothesis revisited. *The Lancet Oncology*, *9*(8), 808.

Gatenby, R. A. (1991). Population ecology issues in tumor growth. *Cancer Research*, *51*(10), 2542–2547. https://www.ncbi.nlm.nih.gov/pubmed/2021934

Gatenby, R. A., Grove, O., & Gillies, R. J. (2013). Quantitative imaging in cancer evolution and ecology. *Radiology*, *269*(1), 8–14.

Geiger, T. R., & Peeper, D. S. (2009). Metastasis mechanisms. *Biochimica et Biophysica Acta (BBA)-Reviews on Cancer*, *1796*(2), 293–308.

Gerlee, P., Altrock, P. M., Malik, A., Krona, C., & Nelander, S. (2022). Autocrine signaling can explain the emergence of Allee effects in cancer cell populations. *PLoS Computational Biology*, *18*(3), e1009844. https://doi.org/10.1371/journal.pcbi.1009844

Green, S. J., & Grosholz, E. D. (2021). Functional eradication as a framework for invasive species control. *Frontiers in Ecology and the Environment*, *19*(2), 98–107. https://doi.org/https://doi.org/10.1002/fee.2277

Ha, N.-H., Faraji, F., & Hunter, K. W. (2013). Mechanisms of metastasis. *Cancer Targeted Drug Delivery: An Elusive Dream*, 435–458.

Haggard, H. W., & Smith, G. (1938). Johannes Müller and the modern conception of cancer. *The Yale Journal of Biology and Medicine*, *10*(5), 419, b411.

Hajdu, S. I. (2006). Thoughts about the cause of cancer. *Cancer: Interdisciplinary International Journal of the American Cancer Society*, *106*(8), 1643–1649.

Hajdu, S. I. (2011). A note from history: Landmarks in history of cancer, part 1. *Cancer*, *117*(5), 1097–1102.

Hammarlund, E. U., Amend, S. R., & Pienta, K. J. (2020). The issues with tissues: The wide range of cell fate separation enables the evolution of multicellularity and cancer. *Medical Oncology*, *37*(7), 62. https://doi.org/10.1007/s12032-020-01387-5

Hanahan, D., & Weinberg, R. A. (2011). Hallmarks of cancer: The next generation. *Cell*, *144*(5), 646–674. https://doi.org/10.1016/j.cell.2011.02.013

Hieger, I. (1949). Chemical carcinogenesis: A review. *British Journal of Industrial Medicine*, *6*(1), 1.

Hoerni, B. (2014). Henri-Francois Le Dran and cancer extension. *Oncologie*, *16*(9–10), 462–464.

Huxel, G. R., McCann, K., & Polis, G. A. (2002). Effects of partitioning allochthonous and autochthonous resources on food web stability. *Ecological Research*, *17*, 419–432.

Huxley, J. (1956). Cancer biology: comparative and genetic. *Biological Reviews*, *31*, 474–513.

Iwasa, Y., Michor, F., Komarova, N. L., & Nowak, M. A. (2005). Population genetics of tumor suppressor genes. *Journal of Theoretical Biology*, *233*(1), 15–23. https://doi.org/10.1016/j.jtbi.2004.09.001

Johnson, K. E., Howard, G., Mo, W., Strasser, M. K., Lima, E., Huang, S., & Brock, A. (2019). Cancer cell population growth kinetics at low densities deviate from the exponential growth model and suggest an Allee effect. *PLoS Biology*, *17*(8), e3000399. https://doi.org/10.1371/journal.pbio.3000399

Jolly, M. K., Somarelli, J. A., Sheth, M., Biddle, A., Tripathi, S. C., Armstrong, A. J., Hanash, S. M., Bapat, S. A., Rangarajan, A., & Levine, H. (2019). Hybrid epithelial/mesenchymal phenotypes promote metastasis and therapy resistance across carcinomas. *Pharmacology & Therapeutics*, *194*, 161–184. https://doi.org/10.1016/j.pharmthera.2018.09.007

Jolly, M. K., Ware, K. E., Gilja, S., Somarelli, J. A., & Levine, H. (2017). EMT and MET: Necessary or permissive for metastasis? *Molecular Oncology*, *11*(7), 755–769. https://doi.org/10.1002/1878-0261.12083

Kaiyin, M., Lingling, T., Leilei, T., Wenjia, L., & Bin, J. (2023). Head-to-head comparison of contrast-enhanced mammography and contrast-enhanced MRI for assessing pathological complete response to neoadjuvant therapy in patients with breast cancer: A meta-analysis. *Breast Cancer Research and Treatment*, *202*(1), 1–9. https://doi.org/10.1007/s10549-023-07034-7

Kareva, I. (2011). What can ecology teach us about cancer? *Translational Oncology*, *4*(5), 266–270. https://doi.org/10.1593/tlo.11154

Kim, E., Stamatelos, S., Cebulla, J., Bhujwalla, Z. M., Popel, A. S., & Pathak, A. P. (2012). Multiscale imaging and computational modeling of blood flow in the tumor vasculature. *Annals of Biomedical Engineering*, *40*, 2425–2441.

Klemetsen, A. (2010). The charr problem revisited: Exceptional phenotypic plasticity promotes ecological speciation in postglacial lakes. *Freshwater Reviews*, *3*(1), 49–74.

Kontomanolis, E. N., Koutras, A., Syllaios, A., Schizas, D., Mastoraki, A., Garmpis, N., Diakosavvas, M., Angelou, K., Tsatsaris, G., Pagkalos, A., Ntounis, T., & Fasoulakis, Z. (2020). Role of oncogenes and tumor-suppressor genes in carcinogenesis: A review. *Anticancer Research*, *40*(11), 6009–6015. https://doi.org/10.21873/anticanres.14622

Kotler, B. P., & Brown, J. S. (1988). Environmental heterogeneity and the coexistence of desert rodents. *Annual Review of Ecology and Systematics*, *19*, 281–307. http://www.jstor.org/stable/2097156

Kotler, B. P., & Brown, J. S. (2020). Cancer community ecology. *Cancer Control*, *27*(1). https://doi.org/10.1177/1073274820951776.

Lande, R. (1996). Statistics and partitioning of species diversity, and similarity among multiple communities. *Oikos*, 5–13.

Legendre, P., Borcard, D., & Peres-Neto, P. R. (2005). Analyzing beta diversity: partitioning the spatial variation of community composition data. *Ecological Monographs*, *75*(4), 435–450.

Leibold, M. A., Urban, M. C., De Meester, L., Klausmeier, C. A., & Vanoverbeke, J. (2019). Regional neutrality evolves through local adaptive niche evolution. *Proceedings of the National Academy of Sciences of the United States of America*, *116*(7), 2612–2617. https://doi.org/10.1073/pnas.1808615116

Lewandowska, A. M., Lewandowski, T., Rudzki, M., Rudzki, S., & Laskowska, B. (2021). Cancer prevention—review paper. *Annals of Agricultural and Environmental Medicine*, *28*(1), 11–19. https://doi.org/10.26444/aaem/116906

Lloyd, M. C., Gatenby, R. A., & Brown, J. S. (2017). Ecology of the metastatic process. In *Ecology and evolution of cancer* (pp. 153–165). Elsevier.

Loftus, L. V., Amend, S. R., & Pienta, K. J. (2022). Interplay between cell death and cell proliferation reveals new strategies for cancer therapy. *International Journal of Molecular Sciences*, *23*(9). https://doi.org/10.3390/ijms23094723

Lorenzo, G., Scott, M. A., Tew, K., Hughes, T. J., Zhang, Y. J., Liu, L., Vilanova, G., & Gomez, H. (2016). Tissue-scale, personalized modeling and simulation of prostate cancer growth. *Proceedings of the National Academy of Sciences*, *113*(48), E7663–E7671.

Mallin, M. M., Kim, N., Choudhury, M. I., Lee, S. J., An, S. S., Sun, S. X., Konstantopoulos, K., Pienta, K. J., & Amend, S. R. (2023). Cells in the polyaneuploid cancer cell (PACC) state have increased metastatic potential. *Clinical & Experimental Metastasis*, *40*(4), 321–338. https://doi.org/10.1007/s10585-023-10216-8

Manchester, K. L. (1995). Theodor Boveri and the origin of malignant tumours. *Trends in Cell Biology*, *5*(10), 384–387.

Martinez, B., Reaser, J. K., Dehgan, A., Zamft, B., Baisch, D., McCormick, C., Giordano, A. J., Aicher, R., & Selbe, S. (2020). Technology innovation: Advancing capacities for the early detection of and rapid response to invasive species. *Biological Invasions*, *22*(1), 75–100. https://doi.org/10.1007/s10530-019-02146-y

Masters, J. R. (2002). HeLa cells 50 years on: The good, the bad and the ugly. *Nature Reviews Cancer*, *2*(4), 315–319. https://doi.org/10.1038/nrc775

McCann, J., & Ames, B. N. (1976a). Detection of carcinogens as mutagens in the Salmonella/microsome test: Assay of 300 chemicals: Discussion. *Proceedings of the National Academy of Sciences of the United States of America*, *73*(3), 950–954. https://doi.org/10.1073/pnas.73.3.950

McCann, J., & Ames, B. N. (1976b). A simple method for detecting environmental carcinogens as mutagens. *Annals of the New York Academy of Sciences*, *271*, 5–13. https://doi.org/10.1111/j.1749-6632.1976. tb23086.x

McLaughlin, G. M., & Dearden, P. K. (2019). Invasive insects: Management methods explored. *Journal of Insect Science*, *19*(5). https://doi.org/10.1093/jisesa/iez085

Merlo, L. M., Pepper, J. W., Reid, B. J., & Maley, C. C. (2006). Cancer as an evolutionary and ecological process. *Nature Reviews Cancer*, *6*(12), 924–935. https://doi.org/10.1038/nrc2013

Myers, K. V., Pienta, K. J., & Amend, S. R. (2020). Cancer cells and M2 macrophages: Cooperative invasive ecosystem engineers. *Cancer Control*, *27*(1). https://doi.org/10.1177/1073274820911058

Neinavaie, F., Ibrahim-Hashim, A., Kramer, A. M., Brown, J. S., & Richards, C. L. (2021). The genomic processes of biological invasions: From invasive species to cancer metastases and back again. *Frontiers in Ecology and Evolution*, *9*, 681100.

Ngoepe, N., Muschick, M., Kishe, M. A., Mwaiko, S., Temoltzin-Loranca, Y., King, L., Courtney Mustaphi, C., Heiri, O., Wienhues, G., Vogel, H., Cuenca-Cambronero, M., Tinner, W., Grosjean, M., Matthews, B., & Seehausen, O. (2023). A continuous fish fossil record reveals key insights into adaptive radiation. *Nature*, *622*. https://doi.org/10.1038/s41586-023-06603-6

Noorbakhsh, J., Zhao, Z. M., Russell, J. C., & Chuang, J. H. (2020). Treating cancer as an invasive species. *Molecular Cancer Research*, *18*(1), 20–26. https://doi.org/10.1158/1541-7786.MCR-19-0262

Norton, L. (1988). A Gompertzian model of human breast cancer growth. *Cancer Research*, *48*(24_Part_1), 7067–7071.

Olsson, O., & Brown, J. S. (2010). Smart, smarter, smartest: Foraging information states and coexistence. *Oikos*, *119*(2), 292–303.

Oyugi, J. O., Brown, J. S., & Whelan, C. J. (2012). Foraging behavior and coexistence of two sunbird species in a Kenyan woodland. *Biotropica*, *44*(2), 262–269.

Page, A. B. P., & Lubatti, O. F. (1963). Fumigation of insects. *Annual Review of Entomology*, *8*(1), 239–264. https://doi.org/10.1146/annurev.en.08.010163.001323

Pantel, K., & Speicher, M. (2016). The biology of circulating tumor cells. *Oncogene*, *35*(10), 1216–1224.

Parekh, V. S., & Jacobs, M. A. (2017). Integrated radiomic framework for breast cancer and tumor biology using advanced machine learning and multiparametric MRI. *NPJ Breast Cancer*, *3*(1), 43.

Park, E. S., Kim, S. J., Kim, S. W., Yoon, S. L., Leem, S. H., Kim, S. B., Kim, S. M., Park, Y. Y., Cheong, J. H., Woo, H. G., Mills, G. B., Fidler, I. J., & Lee, J. S. (2011). Cross-species hybridization of microarrays for studying tumor transcriptome of brain metastasis. *Proceedings of the National Academy of Sciences of the United States of America*, *108*(42), 17456–17461. https://doi.org/10.1073/pnas.1114210108

Paulino, J., & Mansinho, H. (2023). Recent developments in the treatment of pancreatic cancer. *Acta Médica Portuguesa*, *36*(10), 670–678. https://doi.org/10.20344/amp.19957

Peto, R., Roe, F. J., Lee, P. N., Levy, L., & Clack, J. (1975). Cancer and ageing in mice and men. *British Journal of Cancer*, *32*(4), 411–426. https://doi.org/10.1038/bjc.1975.242

Pienta, K. J., Hammarlund, E. U., Axelrod, R., Amend, S. R., & Brown, J. S. (2020). Convergent evolution, evolving evolvability, and the origins of lethal cancer. *Molecular Cancer Research*, *18*(6), 801–810.

Pienta, K. J., Hammarlund, E. U., Brown, J. S., Amend, S. R., & Axelrod, R. M. (2021). Cancer recurrence and lethality are enabled by enhanced survival and reversible cell cycle arrest of polyaneuploid cells. *Proceedings of the National Academy of Sciences of the United States of America*, *118*(7). https://doi. org/10.1073/pnas.2020838118

Pienta, K. J., & Loberg, R. (2005). The "emigration, migration, and immigration" of prostate cancer. *Clinical Prostate Cancer*, *4*(1), 24–30. https://doi.org/10.3816/cgc.2005.n.008

Pienta, K. J., McGregor, N., Axelrod, R., & Axelrod, D. E. (2008). Ecological therapy for cancer: Defining tumors using an ecosystem paradigm suggests new opportunities for novel cancer treatments. *Translational Oncology*, *1*(4), 158–164. https://doi.org/10.1593/tlo.08178

Poux, C., Chevret, P., Huchon, D., De Jong, W. W., & Douzery, E. J. (2006). Arrival and diversification of caviomorph rodents and platyrrhine primates in South America. *Systematic Biology*, *55*(2), 228–244.

Pradeu, T., Daignan-Fornier, B., Ewald, A., Germain, P. L., Okasha, S., Plutynski, A., Benzekry, S., Bertolaso, M., Bissell, M., Brown, J. S., Chin-Yee, B., Chin-Yee, I., Clevers, H., Cognet, L., Darrason, M., Farge, E., Feunteun, J., Galon, J., Giroux, E., . . . Laplane, L. (2023). Reuniting philosophy and science to advance cancer research. *Biological Reviews of the Cambridge Philosophical Society*, *98*(5), 1668–1686. https:// doi.org/10.1111/brv.12971

Ptacnik, R., Moorthi, S. D., & Hillebrand, H. (2010). Hutchinson reversed, or why there need to be so many species. In *Advances in ecological research* (Vol. 43, pp. 1–43). Elsevier.

Ramos-Elvira, E., Banda, E., Arizaga, J., Martín, D., & Aguirre, J. I. (2023). Long-term population trends of house sparrow and Eurasian tree sparrow in Spain. *Birds*, *4*(2), 159–170.

Rathore, S., Akbari, H., Doshi, J., Shukla, G., Rozycki, M., Bilello, M., Lustig, R., & Davatzikos, C. (2018). Radiomic signature of infiltration in peritumoral edema predicts subsequent recurrence in glioblastoma: Implications for personalized radiotherapy planning. *Journal of Medical Imaging*, *5*(2), 021219.

Reaser, J. K., Burgiel, S. W., Kirkey, J., Brantley, K. A., Veatch, S. D., & Burgos-Rodriguez, J. (2020). The early detection of and rapid response (EDRR) to invasive species: A conceptual framework and federal capacities assessment. *Biological Invasions*, *22*(1), 1–19. https://doi.org/10.1007/s10530-019-02156-w

Rebbeck, C. A., Thomas, R., Breen, M., Leroi, A. M., & Burt, A. (2009). Origins and evolution of a transmissible cancer. *Evolution*, *63*(9), 2340–2349. https://doi.org/10.1111/j.1558-5646.2009.00724.x

Reed, D. R., Metts, J., Pressley, M., Fridley, B. L., Hayashi, M., Isakoff, M. S., Loeb, D. M., Makanji, R., Roberts, R. D., & Trucco, M. (2020). An evolutionary framework for treating pediatric sarcomas. *Cancer*, *126*(11), 2577–2587.

Rees, M., Condit, R., Crawley, M., Pacala, S., & Tilman, D. (2001). Long-term studies of vegetation dynamics. *Science*, *293*(5530), 650–655. https://doi.org/10.1126/science.1062586

Ribatti, D., Tamma, R., & Annese, T. (2020). Epithelial-mesenchymal transition in cancer: a historical overview. *Translational Oncology*, *13*(6), 100773.

Saccaggi, D. L., Karsten, M., Robertson, M. P., Kumschick, S., Somers, M. J., Wilson, J. R. U., & Terblanche, J. S. (2016). Methods and approaches for the management of arthropod border incursions. *Biological Invasions*, *18*(4), 1057–1075. https://doi.org/10.1007/s10530-016-1085-6

Scarà, S., Bottoni, P., & Scatena, R. (2015). CA 19–9: Biochemical and clinical aspects. *Advances in Experimental Medicine and Biology*, *867*, 247–260. https://doi.org/10.1007/978-94-017-7215-0_15

Schnipper, L. E. (1986). Clinical implications of tumor-cell heterogeneity. *New England Journal of Medicine*, *314*(22), 1423–1431.

Sepulveda, A., Ray, A., Al-Chokhachy, R., Muhlfeld, C., Gresswell, R., Gross, J., & Kershner, J. (2012). Aquatic invasive species: Lessons from cancer research. *American Scientist*, *100*(3), 234–242.

Smyth, M. J., Dunn, G. P., & Schreiber, R. D. (2006). Cancer immunosurveillance and immunoediting: The roles of immunity in suppressing tumor development and shaping tumor immunogenicity. *Advances in Immunology*, *90*, 1–50.

Sottoriva, A., Kang, H., Ma, Z., Graham, T. A., Salomon, M. P., Zhao, J., Marjoram, P., Siegmund, K., Press, M. F., & Shibata, D. (2015). A big bang model of human colorectal tumor growth. *Nature Genetics*, *47*(3), 209–216.

Spremulli, E. N., & Dexter, D. (1983). Human tumor cell heterogeneity and metastasis. *Journal of Clinical Oncology*, *1*(8), 496–509.

Stankova, K. (2019). Resistance games. *Nature Ecology and Evolution*, *3*(3), 336–337. https://doi.org/10.1038/s41559-018-0785-y

Stankova, K., Brown, J. S., Dalton, W. S., & Gatenby, R. A. (2019). Optimizing cancer treatment using game theory: A review. *JAMA Oncology*, *5*(1), 96–103. https://doi.org/10.1001/jamaoncol.2018.3395

Staples, T. L., Dwyer, J. M., Loy, X., & Mayfield, M. M. (2016). Potential mechanisms of coexistence in closely related forbs. *Oikos*, *125*(12), 1812–1823.

Thines, M. (2019). An evolutionary framework for host shifts—jumping ships for survival. *New Phytologist*, *224*(2), 605–617. https://doi.org/10.1111/nph.16092

Thompson, J. A., Grunert, F., & Zimmermann, W. (1991). Carcinoembryonic antigen gene family: Molecular biology and clinical perspectives. *Journal of Clinical Laboratory Analysis*, *5*(5), 344–366. https://doi.org/10.1002/jcla.1860050510

Tollis, M., Ferris, E., Campbell, M. S., Harris, V. K., Rupp, S. M., Harrison, T. M., Kiso, W. K., Schmitt, D. L., Garner, M. M., Aktipis, C. A., Maley, C. C., Boddy, A. M., Yandell, M., Gregg, C., Schiffman, J. D., & Abegglen, L. M. (2021). Elephant genomes reveal accelerated evolution in mechanisms underlying disease defenses. *Molecular Biology and Evolution*, *38*(9), 3606–3620. https://doi.org/10.1093/molbev/msab127

Trabzonlu, L., Pienta, K. J., Trock, B. J., De Marzo, A. M., & Amend, S. R. (2023). Presence of cells in the polyaneuploid cancer cell (PACC) state predicts the risk of recurrence in prostate cancer. *Prostate*, *83*(3), 277–285. https://doi.org/10.1002/pros.24459

Tsibulak, I., & Fotopoulou, C. (2023). Tumor biology and impact on timing of surgery in advanced epithelial ovarian cancer. *The International Journal of Gynecological Cancer*, *33*(10), 1627–1632. https://doi.org/10.1136/ijgc-2023-004676

Vincze, O., Colchero, F., Lemaitre, J. F., Conde, D. A., Pavard, S., Bieuville, M., Urrutia, A. O., Ujvari, B., Boddy, A. M., Maley, C. C., Thomas, F., & Giraudeau, M. (2022). Cancer risk across mammals. *Nature*, *601*(7892), 263–267. https://doi.org/10.1038/s41586-021-04224-5

Weidlich, E. W. A., Flórido, F. G., Sorrini, T. B., & Brancalion, P. H. S. (2020). Controlling invasive plant species in ecological restoration: A global review. *Journal of Applied Ecology*, *57*(9), 1806–1817. https://doi.org/10.1111/1365-2664.13656

Wilbaux, M., Tod, M., De Bono, J., Lorente, D., Mateo, J., Freyer, G., You, B., & Hénin, E. (2015). A joint model for the kinetics of CTC count and PSA concentration during treatment in metastatic castration-resistant prostate cancer. *CPT: Pharmacometrics & Systems Pharmacology*, *4*(5), 277–285.

Williams, A. L., Fitzgerald, J. E., Ivich, F., Sontag, E. D., & Niedre, M. (2020). Short-term circulating tumor cell dynamics in mouse xenograft models and implications for liquid biopsy. *Frontiers in Oncology*, *10*, 601085. https://doi.org/10.3389/fonc.2020.601085

Woodhouse, E. C., Chuaqui, R. F., & Liotta, L. A. (1997). General mechanisms of metastasis. *Cancer: Interdisciplinary International Journal of the American Cancer Society*, *80*(S8), 1529–1537.

Xing, H., Hu, Y., Liupeng, Y., Lin, J., Bai, H., Li, Y., Tanvir, R., Li, L., Bai, M., Zhang, Z., Xu, H., & Wu, H. (2022). Fumigation activity of essential oils of Cinnamomum loureirii toward red imported fire ant workers. *Journal of Pest Science*, *96*, 1–16. https://doi.org/10.1007/s10340-022-01540-1

Yang, K. R., Mooney, S. M., Zarif, J. C., Coffey, D. S., Taichman, R. S., & Pienta, K. J. (2014). Niche inheritance: A cooperative pathway to enhance cancer cell fitness through ecosystem engineering. *Journal of Cellular Biochemistry*, *115*(9), 1478–1485. https://doi.org/10.1002/jcb.24813

Yoder, J. B., Clancey, E., Des Roches, S., Eastman, J. M., Gentry, L., Godsoe, W., Hagey, T. J., Jochimsen, D., Oswald, B. P., Robertson, J., Sarver, B. A. J., Schenk, J. J., Spear, S. F., & Harmon, L. J. (2010). Ecological opportunity and the origin of adaptive radiations. *Journal of Evolutionary Biology*, *23*(8), 1581–1596. https://doi.org/10.1111/j.1420-9101.2010.02029.x

Yu, D. W., & Wilson, H. B. (2001). The competition-colonization trade-off is dead; long live the competition-colonization trade-off. *The American Naturalist*, *158*(1), 49–63.

Zhang, J., Cunningham, J., Brown, J., & Gatenby, R. (2022). Evolution-based mathematical models significantly prolong response to abiraterone in metastatic castrate-resistant prostate cancer and identify strategies to further improve outcomes. *Elife*, *11*. https://doi.org/10.7554/eLife.76284

Zheng, X., Wise, S., & Cristini, V. (2005). Nonlinear simulation of tumor necrosis, neo-vascularization and tissue invasion via an adaptive finite-element/level-set method. *Bulletin of Mathematical Biology*, *67*, 211–259.

12 Unifying Theories in Comparative Oncology

Zachary T. Compton

12.1 INTRODUCTION: A DISEASE IN THE LIGHT OF EVOLUTION

George C. Williams' and Randolph M. Nesse's coining of evolutionary medicine opened a unique inquiry into the ultimate origins of human disease vulnerability (Nesse & Williams, 1998, 1999; Williams & Nesse, 1991). Advancements in modern medicine have largely been driven by leaps forward in biotechnology, drug development, genomics, imaging, and surgical techniques. While these advancements have driven increases in quality of life and lifespan, they have left a remaining gap in our holistic understanding of disease. Evolutionary medicine provides an important complement to these technological advances by providing a unifying scaffolding in which cancer and other diseases can be viewed in the context of evolution and our evolutionary histories. The field has rapidly expanded to create evolutionary frameworks to understand a diversity of human maladies from psychiatric disorders to cardiovascular disease (Nesse, 1994, 2000, 2019, 2022; Swynghedauw, 2016; Swynghedauw et al., 2010). However, perhaps no human disease is more quintessential to the field than cancer. Since 2017 more than 2,000 research articles are published a year discussing cancer in the context of evolution. The deep evolutionary history of cancer and cancer-like phenotypes makes it an ideal candidate disease to be interrogated by evolutionary theory. Not only does evolutionary medicine open a multitude of novel questions about the origin and risk factors associated with cancer, it serves to dispel the stigma of a disease that has now been demonstrated to be a near universal burden across the tree of life (Aktipis et al., 2015).

12.2 CANCER EVOLUTION ON TWO SCALES

An evolutionary framework of cancer has been applied at two scales: 1) the evolutionary principles that describe cancer cell population dynamics and 2) those that govern organismal population level susceptibility to cancer. In back to back years, John Cairns and Peter Nowell provided one of the earliest descriptions of cancer as a process of somatic natural selection, crystallizing the evolutionary dynamics of cancer at the cellular scale (Cairns, 1975; Nowell, 1976). Their nascent framework, now both highly cited, set the crucial precedent for cancer to be viewed through the lens of evolution.

Carlo Maley and Mel Greaves finalized the full integration of evolutionary theory with cancer biology (Greaves & Maley, 2012) to address one of the most significant clinical problems in modern oncology: the evolution of therapeutic resistance.

At the organismal level (namely, *Homo sapiens*), evolutionary medicine has proposed mechanisms for the exceptionally high lifetime cancer incidence in humans.

Antagonistic pleiotropy, first introduced by George C. Williams, provided a compelling explanation for why we seem to have so many disease facilitating genes that manifest late in life. Similarly, Greaves has expanded our understanding on the potential for the evolutionary mismatch hypothesis to have explanatory power in human cancers. While he has largely focused on childhood leukemia (Greaves, 2003, 2007, 2018; Greaves et al., 2021), the mismatch framework has been equally successful in assessing female reproductive scheduling as a determinant of cancer risk (Adami et al., 1994; MacMahon et al., 1970; Trichopoulos et al., 1983). Hormonal changes associated with pregnancy induce dramatic changes to breast ductal cell composition and differentiation. The terminal differentiation of these cells eliminates their potential for cancerous transformation and reduces lifetime risk of breast cancer (Katz, 2016).

DOI: 10.1201/9781003307921-12

Dating its origins back to Rudolf Virchow's descriptions of various pathologies across species, comparative oncology was constrained experimentally to studying cancer in model organisms. The early 2000s saw the expansion of comparative oncology from rodent models of cancer to non-model organisms, typically companion animals (cats and dogs) (Cannon, 2015; Garden et al., 2018; Gordon et al., 2009; Paoloni & Khanna, 2007; Schiffman & Breen, 2015; Somarelli, Gardner, et al., 2020; Somarelli, Rupprecht, et al., 2020; Tuohy et al., 2020). Expanding still, comparative oncology has recently focused on the following key questions: 1) What are the evolutionary origins of cancer; 2) How can fundamental, intrinsic risk factors be stratified across taxa; and 3) What are the mechanisms of enhanced cancer suppression in species, particularly for those with extraordinary life histories? However, before the advent of large cross-species pathology databases, the first strides toward the broad approach of comparative oncology across the tree of life began with influential experiments in rodent models, the most well-known of which is Peto's Paradox.

12.3   THE NATURE OF A PARADOX

It is difficult to imagine that Sir Richard Peto was conscious of what he was beginning when he made his 1977 observation that intrinsic cancer risk does not appear to scale with body mass. Mice, especially laboratory mice, seemed to get cancer at roughly the same rate as humans despite their incredible divergence in lifespan and body size (Peto, 2016; Tollis et al., 2017). Ironically, it should be pointed out that although "Peto's Paradox" almost inevitably invokes images of elephants, the mascot of comparative oncology is completely absent from his landmark text (Peto, 2016). In fact, his remark was built on first principles. It went as follows: If we assign some risk of an individual cell undergoing a cancerous transformation then an individual's chance of getting cancer should scale with the number of somatic cells (body size). Further, the longer an individual must sustain a healthy, functioning soma and endure cell divisions, the greater their cancer risk (lifespan). First coined in 1999 by Leonard Nunney, "Peto's Paradox" refers to the lack of any apparent relationship between a species' level of cancer risk and either its body size or lifespan (Nunney, 1999). Further support for this paradox arrived with the application of mathematical models of cancer initiation via somatic mutation accumulation (Calabrese & Shibata, 2010; Nunney, 1999, 2016, 2020).

Predictions from these mathematical models demonstrated that not only should cancer risk increase with body mass and lifespan, but some species, namely the blue whale (*Balaenoptera musculus*), would have a 100% colon cancer incidence by a certain age (Caulin et al., 2015). We need none of the empirical cross-species cancer data that we have today to know that a 100% cancer mortality in any species is dubious. Which is where we arrive at the question of *where is the paradox*? Which do we expect to be more paradoxical, a large, long-lived species that is riddled with cancerous tumors or one that is relatively free of them? This common difference in perspective can be explained two ways. The first is to see the problem the way Peto described it, cancer risk compounds with additional somatic cells and somatic cell divisions. Through the lens of evolutionary biology, however, one might predict that such a species as the elephant or blue whale could not evolve such extraordinary life history traits without natural selection shaping equally powerful counter measures. As we continue to amass data on cancer prevalence across species, we should find it equally paradoxical when we encounter species with an exceptionally high cancer risk within the confines of their natural lifespan.

12.4   AN ECOLOGICAL PERSPECTIVE ON CANCER SUPPRESSION

How an organism utilizes the energy it extracts from the environment then diversifies its energetic investments between survival and reproduction is largely the domain of life history theory. However, to use life history theory to understand differences in cancer prevalence, we need to understand how environments shape the need for increased body size. Long before Peto's Paradox, evolutionary biologists have attempted to describe the environmental pressures that shape the evolution of gigantism

(Blanckenhorn, 2000; Bonner, 1988; Moran & Woods, 2012; Smith et al., 2010). Predation, resource stability, and population dynamics can all drive body size evolution (Blackburn & Gaston, 1994; Marquet et al., 1995; Maurer et al., 1992). Utilizing these known drivers of body size evolution can allow us to further our predictive framework to identify ecological profiles where we expect the evolution of cancer suppression mechanisms to follow increases in body size. A demonstration of this predictive framework from ecology was initially employed to understand the ecological drivers of the transitions to multicellularity (Aktipis et al., 2015; Herron et al., 2019; Knoll, 2011; Knoll & Lahr, 2016; Ratcliff et al., 2015; Tang et al., 2023; Tong et al., 2022). It has since been expanded to begin making predictions on how ecological conditions stratify species by cancer risk (Kapsetaki et al., 2022).

Much of ecological theory is connected to the principle of energy landscapes, which describes how the sum of energy in a specific ecological landscape moves throughout the landscape space (Bronfenbrenner, 2000; Shepard et al., 2013; Wilson et al., 2012). Energy in this context is typically measured in calories. A classic, yet simplified understanding of species evolution under this model is that natural selection drives phenotypic development to exploit niches within the energy landscape. This dynamic is often bidirectional, with species shaping their environment via niche construction and changes in the environment shifting selective pressures. For a given energy landscape there are countless ways to extract energy from the primary producers (namely, plants and algae), but these myriad methods can be ultimately categorized broadly as direct or indirect consumption. Energy therefore moves vertically through trophic levels as secondary consumers extract (consume) energy from these primary producers. Natural selection can, in this way, be thought of as a process connected to optimizing the extraction of energy from the environment and transferring that energy into individual fitness. The giraffe's elongated neck and the luminescent barbel on the anglerfish both provide extreme examples of this point (Prokofiev, 2020; Wilkinson & Ruxton, 2012). The stupendous diversity and innovation that evolutionary processes have displayed in optimizing these two fronts is what provides modern observers with the mirage of design.

12.5   A LIFE HISTORY THEORY OF CANCER RISK

Life history traits are characterized on a linear spectrum from "slow" to "fast" life history species (traditionally K and r strategists, respectively) (Brommer, 2000; Lika & Kooijman, 2003; Roff, 1993). "Slow" life history species have high energetic investments in body size, longevity, and comparatively slower times to sexual maturity. In contrast, "fast" life history species invest heavily in comparatively rapid rates of reproduction and development. We can conceptualize the scale of life history differences through the mouse and the elephant. The striking difference between their life history traits, namely body size and lifespan, illuminates the ends of this spectrum. This understanding has provoked comparative oncologists to consider how a species' life history strategy may be predictive of their cancer risk (Boddy et al., 2020; Brown et al., 2015; Bulls et al., 2023; Compton, 2023; Compton et al., 2023; Kapsetaki, Basile, et al., 2023; Kapsetaki, Compton, et al., 2023; Vincze et al., 2022). The life history framework of disease risk allows straightforward predictions to be made on where across the tree of life we should expect to find the evolution of enhanced cancer suppression. Chiari et al. laid this out clearly in their proposed natural experiment of identifying large, long-lived species that are within a phylogenetic clade of species with comparatively faster life history traits (Chiari et al., 2018). Genomic differences between the target species and its closest phylogenetic relatives would point to mechanisms that facilitated their divergence in cancer risk. This approach is deeply enhanced when paired with the recent advances in the availability of cross-species cancer prevalence data (Boddy et al., 2020; Bulls et al., 2023; Compton et al., 2023; Madsen et al., 2017; Vincze et al., 2022).

However, while life history theory provides an excellent scaffold for the generation of hypotheses, the simplification of cancer risk to a single life history trait can be complicated by the complex interplay of multiple life history phenotypes. The naked mole rat exemplifies this complexity. The naked mole rat (*Heterocephalus glaber*) has a maximum reported lifespan of more than 30 years (Schneider, 2012) and an average body mass of only 50 grams (Jarvis & Sherman, 2002). The naked mole rat, for its long lifespan, has very little cancer. Uncovering species that do not fit within our linear understanding of the relationship between body size, longevity, and cancer risk could assist our understanding of how these traits drive cancer risk.

A logical extension of this paradigm is to consider how the difference in these energetic investments across life history strategies may be reflected in tissue level expenditures in somatic maintenance. Energetic investments in somatic maintenance may be reflected in a myriad of different phenotypic functions relevant to intrinsic cancer risk, from DNA damage response to tissue level architectures that impede clonal expansions (Cairns, 1975; Garinis et al., 2008; Hasty et al., 2003; Jackson & Bartek, 2009). Regulation of cell cycle control, DNA damage detection and repair, and propensity to apoptosis are all expected to reflect the demands of a species' life history (Brown et al., 2015; Kirkwood, 2008). From an evolutionary perspective, natural selection should balance these mechanisms of somatic maintenance proportionally to their need. The total number of cells and the volume of total cell turnover, like seen in an elephant, demands a much higher investment in these mechanisms than a species with a shorter life history such as a mouse.

Similarly, regulators of differentiation are fundamental to the evolutionary transitions from unicellular to multicellular life, and are crucial to precluding tumor initiation (Brunet & King, 2017; Compton et al., 2022; Michod et al., 2006; Ruiz & Chen, 2008).

Downstream of enhanced somatic maintenance is tumor suppression. We can avoid missing the forest for the trees in our search for mechanisms of tumor suppression by remembering that some life history traits may require enhanced somatic maintenance for reasons beyond purely cancer suppression.

12.6   A GENE'S EYE VIEW OF CANCER SUPPRESSION

The primary intellectual legacy of the late George C. Williams, the one formalized in *Adaptation and Natural Selection* (Williams & Burt, 1997), is that individual genes are the ultimate level that natural selection acts upon. The formidable tome served to refute much of the generalized group selection thinking that was prevalent at the time and while still disputed, remains a mainstay in evolutionary thought. As mentioned in the introduction, Williams' contributions to evolutionary medicine were primarily predicated on this gene-centric view of natural selection, especially within his theory of antagonistic pleiotropy (Nesse & Williams, 1998; Williams & Nesse, 1991). Antagonistic pleiotropy has helped us understand diseases of old age by demonstrating that genes implicated in these diseases confer fitness benefits up until the conclusion of humans' reproductive window (Austad & Hoffman, 2018; Blomquist, 2009; Carter & Nguyen, 2011; Fox, 2018; Kirkwood & Rose, 1991; Thomas et al., 2012). Within oncology it has been especially useful in describing why the human genome is replete with proto-oncogenes. The healthy functioning of proto-oncogenes is essential for a myriad of physiological tasks from development to wound healing (Okadal et al., 1996; Torry & Cooper, 1991).

Across these domains, Williams' concept provides a strong framework to assess the multitude of oncogenic trade-offs vertebrate species have made throughout their evolutionary history. Despite Williams' success, and a resurgence in interest in using the gene-centric view of evolution as a predictive framework (Arvid Ågren, 2021), it has been a neglected framework in comparative oncology.

As comparative genomics has made strides in sequencing and annotating the genomes of an evergrowing number of organisms, the complex interconnectivity of gene networks across species remains less understood, while increasing attention has been paid to how single genes interact within gene networks (Boutelle & Attardi, 2021; He et al., 2007; Serrano & Massagué, 2000). Given this understanding, it has been increasingly controversial to rely on predictions from the gene-centric view of evolutionary biology (Dawkins, 1981; Williams & Burt, 1997). However, the life history framework and this gene-centric view of cancer risk are deeply complementary. David Haig makes this point most saliently in his commentary on an exchange between Amy Boddy and Gunter Wagner. Boddy and Wagner, separately, had spearheaded much of the interest to determine how placentation style could predict cancer risk (Boddy et al., 2020; Wagner et al., 2020). They hypothesized that an organism's immune system would have an increased tolerance for non-self (fetal) cells in the most invasive placentation types, specifically the hemochorial placentation type seen in primates (Furukawa et al., 2014).

This increased tolerance for non-self cells could also hinder the immune system's ability to detect and destroy neoplastic cells. Haig's commentary highlighted that predictions on cancer risk derived from a life history framework can be enhanced by considering genetic conflict. Placentation type and litter size both could have different effects on intrinsic cancer risk depending on the zygosity of the litter, Haig argued. Within this understanding, the effective impact on cancer risk would be similar between species that produce monozygotic quadruplets and monozygotic singletons, but much higher in those that produce dizygotic twins (Haig, 2020). Whether this is true has yet to be confirmed, and intriguing theories on the impact of pregnancy on species' level cancer risk have been put forward (Kapsetaki, Fortunato, et al., 2023; Natri et al., 2019).

Regardless of the outcome of these initial hypotheses, predictions from gene level selection and genetic conflict will be fundamental toward building a complete portrait of evolved cancer risk.

12.7 CONCLUSION

As comparative oncology enters the era of rapid increases in the availability of genomic data, functional cell assays of gene activity, and cross-species cancer prevalence data, we should be ambitious in our pursuit of big picture explanations. Ideally, these big picture explanations would address one of the following central themes in comparative oncology:

1) The current hypothesis is that cancer, or cancer-like phenomena, are ubiquitous across multicellular life. This has been well validated in vertebrate animals but is largely unexplored in more ancient systems of multicellularity. Evidence has emerged for neoplasms in invertebrates but has just begun to be fully explored (Metzger et al., 2015; Robert, 2010; Rosenfield et al., 1994).
2) It remains an open question whether the genes and gene regulatory networks we refer to as tumor suppressive evolved via natural selection against the cancer phenotype *per se*, but rather evolved as a more general need for enhanced maintenance of somatic or germline cells based on species' life history.
3) While the most well-studied mechanisms of tumor suppression are cell cycle control, apoptosis, DNA damage response, and immune surveillance, cross-species models of cancer risk are ripe for investigation into whether there are other mechanisms that may be protective against tumor initiation and progression.

The latter two questions on this list have significant implications for translation to human oncology as they both could serve to identify unique vulnerabilities and therapeutic targets to treat cancer.

The comparative oncology field has seen a flood of exciting theoretical and experimental work to elucidate the mechanisms of evolved cancer resistance. As the field has evolved, researchers from diverse backgrounds and expertise have brought with them unique perspectives and emphasis. These contributions have allowed a deeper understanding of cancer as a disease that is intimately connected to multicellularity. In our continued pursuit to further understand one of the fundamental consequences of multicellular life, let us allow for the distinct possibility, as is the case in so many domains of biology, that a multi-factorial approach will paint the most complete understanding of disease risk.

ACKNOWLEDGMENTS

National Institutes of Health Award Number T32 CA272303

REFERENCES

Adami, H. O., Hsieh, C. C., Lambe, M., Trichopoulos, D., Leon, D., Persson, I., Ekbom, A., & Janson, P. O. (1994). Parity, age at first childbirth, and risk of ovarian cancer. *The Lancet, 344*(8932), 1250–1254.

Aktipis, C. A., Boddy, A. M., Jansen, G., Hibner, U., Hochberg, M. E., Maley, C. C., & Wilkinson, G. S. (2015). Cancer across the tree of life: Cooperation and cheating in multicellularity. *Philosophical Transactions of the Royal Society of London. Series B, Biological Sciences, 370*(1673). https://doi.org/10.1098/rstb.2014.0219

Arvid Ågren, J. (2021). *The gene's-eye view of evolution.* Oxford University Press.

Austad, S. N., & Hoffman, J. M. (2018). Is antagonistic pleiotropy ubiquitous in aging biology? *Evolution, Medicine, and Public Health, 2018*(1), 287–294.

Blackburn, T. M., & Gaston, K. J. (1994). Animal body size distributions: Patterns, mechanisms and implications. *Trends in Ecology & Evolution, 9*(12), 471–474.

Blanckenhorn, W. U. (2000). The evolution of body size: What keeps organisms small? *The Quarterly Review of Biology, 75*(4), 385–407.

Blomquist, G. E. (2009). Trade-off between age of first reproduction and survival in a female primate. *Biology Letters, 5*(3), 339–342.

Boddy, A. M., Abegglen, L. M., Pessier, A. P., Schiffman, J. D., Maley, C. C., & Witte, C. (2020). Lifetime cancer prevalence and life history traits in mammals. *Evolution, Medicine, and Public Health, 2020*(1). https://doi.org/10.1093/emph/eoaa015

Bonner, J. T. (1988). *The evolution of complexity by means of natural selection.* Princeton University Press.

Boutelle, A. M., & Attardi, L. D. (2021). p53 and tumor suppression: It takes a network. *Trends in Cell Biology, 31*(4), 298–310.

Brommer, J. E. (2000). The evolution of fitness in life-history theory. *Biological Reviews of the Cambridge Philosophical Society, 75*(3), 377–404.

Bronfenbrenner, U. (2000). Ecological systems theory. In A. E. Kazdin (Ed.), *Encyclopedia of psychology* (Vol. 3, pp. 129–133). Oxford University Press.

Brown, J. S., Cunningham, J. J., & Gatenby, R. A. (2015). The multiple facets of Peto's paradox: A life-history model for the evolution of cancer suppression. *Philosophical Transactions of the Royal Society of London. Series B, Biological Sciences, 370*(1673). https://doi.org/10.1098/rstb.2014.0221

Brunet, T., & King, N. (2017). The origin of animal multicellularity and cell differentiation. *Developmental Cell, 43*(2), 124–140.

Bulls, S. E., Platner, L., Ayub, W., Moreno, N., Arditi, J.-P., Dreyer, S., McCain, S., Wagner, P., Burgstaller, S., Davis, L. R., Sonsbeek, L. G. R. B., Fischer, D., Lynch, V. J., Claude, J., Glaberman, S., & Chiari, Y. (2023). Cancer prevalence is related to body mass and lifespan in tetrapods and remarkably low in turtles. *bioRxiv, 2022.* https://doi.org/10.1101/2022.07.12.499088

Cairns, J. (1975). Mutation selection and the natural history of cancer. *Nature, 255*(5505), 197–200.

Calabrese, P., & Shibata, D. (2010). A simple algebraic cancer equation: Calculating how cancers may arise with normal mutation rates. *BMC Cancer, 10*, 3.

Cannon, C. M. (2015). Cats, cancer and comparative oncology. *Veterinary Science in China, 2*(3), 111–126.

Carter, A. J. R., & Nguyen, A. Q. (2011). Antagonistic pleiotropy as a widespread mechanism for the maintenance of polymorphic disease alleles. *BMC Medical Genetics, 12*, 160.

Caulin, A. F., Graham, T. A., Wang, L.-S., & Maley, C. C. (2015). Solutions to Peto's paradox revealed by mathematical modelling and cross-species cancer gene analysis. *Philosophical Transactions of the Royal Society of London. Series B, Biological Sciences, 370*(1673). https://doi.org/10.1098/rstb.2014.0222

Chiari, Y., Glaberman, S., & Lynch, V. J. (2018). Insights on cancer resistance in vertebrates: Reptiles as a parallel system to mammals [Review]. *Nature Reviews, Cancer, 18*(8), 525.

Compton, Z. (2023). *The nature of cancer: Unifying evolutionary theory in cancer biology.* https://search.proquest.com/openview/16563110bc0176af661ab6edfca32106/1?pq-origsite=gscholar&cbl=18750&diss=y

Compton, Z., Hanlon, K., Compton, C. C., Aktipis, A., & Maley, C. C. (2022). A missing hallmark of cancer: Dysregulation of differentiation. *arXiv [q-bio.PE], arXiv.* http://arxiv.org/abs/2210.13343

Compton, Z., Harris, V., Mellon, W., Rupp, S., Mallo, D., Kapsetaki, S. E., Wilmot, M., Kennington, R., Noble, K., Baciu, C., Ramirez, L., Peraza, A., Martins, B., Sudhakar, S., Aksoy, S., Furukawa, G., Vincze, O., Giraudeau, M., Duke, E. G., . . . Boddy, A. M. (2023). Cancer prevalence across vertebrates. *bioRxiv: The Preprint Server for Biology,* PMID: 36824942. https://doi.org/10.1101/2023.02.15.527881

Dawkins, R. (1981). In defence of selfish genes. *Philosophy, 56*(218), 556–573.

Fox, M. (2018). "Evolutionary medicine" perspectives on Alzheimer's disease: Review and new directions. *Ageing Research Reviews, 47*, 140–148.

Furukawa, S., Kuroda, Y., & Sugiyama, A. (2014). A comparison of the histological structure of the placenta in experimental animals. *Journal of Toxicologic Pathology, 27*(1), 11–18.

Garden, O. A., Volk, S. W., Mason, N. J., & Perry, J. A. (2018). Companion animals in comparative oncology: One medicine in action. *Veterinary Journal, 240*, 6–13.

Garinis, G. A., van der Horst, G. T. J., Vijg, J., & Hoeijmakers, J. H. J. (2008). DNA damage and ageing: New-age ideas for an age-old problem. *Nature Cell Biology, 10*(11), 1241–1247.

Gordon, I., Paoloni, M., Mazcko, C., & Khanna, C. (2009). The comparative oncology trials consortium: Using spontaneously occurring cancers in dogs to inform the cancer drug development pathway. *PLoS Medicine*, *6*(10), e1000161.

Greaves, M. (2003). Pre-natal origins of childhood leukemia. *Reviews in Clinical and Experimental Hematology*, *7*(3), 233–245.

Greaves, M. (2007). Darwinian medicine: A case for cancer. *Nature Reviews, Cancer*, *7*(3), 213–221.

Greaves, M. (2018). A causal mechanism for childhood acute lymphoblastic leukaemia. *Nature Reviews, Cancer*, *18*(8), 471–484.

Greaves, M., Cazzaniga, V., & Ford, A. (2021). Can we prevent childhood Leukaemia? *Leukemia*, *35*(5), 1258–1264.

Greaves, M., & Maley, C. C. (2012). Clonal evolution in cancer. *Nature*. https://www.nature.com/articles/nature10762?page=29

Haig, D. (2020). Comment on the exchange between Boddy et al. and Wagner et al.: Malignancy, placentation and litter size. *Evolution, Medicine, and Public Health*, *2020*(1), 217–218.

Hasty, P., Campisi, J., Hoeijmakers, J., van Steeg, H., & Vijg, J. (2003). Aging and genome maintenance: Lessons from the mouse? *Science*, *299*(5611), 1355–1359.

Herron, M. D., Borin, J. M., Boswell, J. C., Walker, J., Chen, I.-C. K., Knox, C. A., Boyd, M., Rosenzweig, F., & Ratcliff, W. C. (2019). De novo origins of multicellularity in response to predation. *Scientific Reports*, *9*(1), 2328.

He, X., He, L., & Hannon, G. J. (2007). The guardian's little helper: MicroRNAs in the p53 tumor suppressor network. *Cancer Research*, *67*(23), 11099–11101.

Jackson, S. P., & Bartek, J. (2009). The DNA-damage response in human biology and disease. *Nature*, *461*(7267), 1071–1078.

Jarvis, J. U. M., & Sherman, P. W. (2002). Heterocephalus glaber. *Mammalian Species*, *2002*(706), 1–9.

Kapsetaki, S. E., Basile, A. J., Compton, Z. T., Rupp, S. M., Duke, E. G., Boddy, A. M., Harrison, T. M., Sweazea, K. L., & Maley, C. C. (2023). The relationship between diet, plasma glucose, and cancer prevalence across vertebrates. *bioRxiv: The Preprint Server for Biology*, PMID: 37577544. https://doi.org/10.1101/2023.07.31.551378

Kapsetaki, S. E., Compton, Z., Dolan, J., Harris, V. K., Rupp, S. M., Duke, E. G., Harrison, T. M., Aksoy, S., Giraudeau, M., Vincze, O., McGraw, K. J., Aktipis, A., Tollis, M., Boddy, A. M., & Maley, C. C. (2023). Life history and cancer in birds: Clutch size predicts cancer. *bioRxiv: The Preprint Server for Biology*, PMID: 36824773. https://doi.org/10.1101/2023.02.11.528100

Kapsetaki, S. E., Compton, Z., Rupp, S. M., & Garner, M. M. (2022). The ecology of cancer prevalence across species: Cancer prevalence is highest in desert species and high trophic levels. *bioRxiv*. https://www.biorxiv.org/content/10.1101/2022.08.23.504890.abstract

Kapsetaki, S. E., Fortunato, A., Compton, Z., Rupp, S. M., Nour, Z., Riggs-Davis, S., Stephenson, D., Duke, E. G., Boddy, A. M., Harrison, T. M., Maley, C. C., & Aktipis, A. (2023). Is chimerism associated with cancer across the tree of life? *PloS One*, *18*(6), e0287901.

Katz, T. A. (2016). Potential mechanisms underlying the protective effect of pregnancy against breast cancer: A focus on the IGF pathway. *Frontiers in Oncology*, *6*, 228.

Kirkwood, T. B. L. (2008). Understanding ageing from an evolutionary perspective. *Journal of Internal Medicine*, *263*(2), 117–127.

Kirkwood, T. B. L., & Rose, M. R. (1991). Evolution of senescence: Late survival sacrificed for reproduction. *Philosophical Transactions of the Royal Society of London. Series B, Biological Sciences*, *332*(1262), 15–24.

Knoll, A. H. (2011). The multiple origins of complex multicellularity. *Annual Review of Earth and Planetary Sciences*, *39*(1), 217–239.

Knoll, A. H., & Lahr, D. J. (2016). Fossils, feeding, and the evolution of complex multicellularity. In *Multicellularity, origins and evolution, the Vienna series in theoretical biology* (pp. 1–16). Massachusetts Institute of Technology.

Lika, K., & Kooijman, S. A. L. M. (2003). Life history implications of allocation to growth versus reproduction in dynamic energy budgets. *Bulletin of Mathematical Biology*, *65*(5), 809–834.

MacMahon, B., Cole, P., Lin, T. M., Lowe, C. R., Mirra, A. P., Ravnihar, B., Salber, E. J., Valaoras, V. G., & Yuasa, S. (1970). Age at first birth and breast cancer risk. *Bulletin of the World Health Organization*, *43*(2), 209–221.

Madsen, T., Arnal, A., Vittecoq, M., Bernex, F., Abadie, J., Labrut, S., Garcia, D., Faugère, D., Lemberger, K., Beckmann, C., Roche, B., Thomas, F., & Ujvari, B. (2017). Chapter 2: Cancer prevalence and etiology in wild and captive animals. In B. Ujvari, B. Roche, & F. Thomas (Eds.), *Ecology and evolution of cancer* (pp. 11–46). Academic Press.

Marquet, P. A., Navarrete, S. A., & Castilla, J. C. (1995). Body size, population density, and the energetic equivalence rule. *The Journal of Animal Ecology, 64*(3), 325–332.

Maurer, B. A., Brown, J. H., & Rusler, R. D. (1992). The micro and macro in body size evolution. *Evolution; International Journal of Organic Evolution, 46*(4), 939–953.

Metzger, M. J., Reinisch, C., Sherry, J., & Goff, S. P. (2015). Horizontal transmission of clonal cancer cells causes leukemia in soft-shell clams. *Cell, 161*(2), 255–263.

Michod, R. E., Viossat, Y., Solari, C. A., Hurand, M., & Nedelcu, A. M. (2006). Life-history evolution and the origin of multicellularity. *Journal of Theoretical Biology, 239*(2), 257–272.

Moran, A. L., & Woods, H. A. (2012). Why might they be giants? Towards an understanding of polar gigantism. *The Journal of Experimental Biology, 215*(Pt 12), 1995–2002.

Natri, H., Garcia, A. R., Buetow, K. H., Trumble, B. C., & Wilson, M. A. (2019). The pregnancy pickle: Evolved immune compensation due to pregnancy underlies sex differences in human diseases. *Trends in Genetics: TIG, 35*(7), 478–488.

Nesse, R. M. (1994). Fear and fitness: An evolutionary analysis of anxiety disorders. *Ethology and Sociobiology*. https://www.sciencedirect.com/science/article/pii/0162309594900027

Nesse, R. M. (2000). Is depression an adaptation? *Archives of General Psychiatry*. https://jamanetwork.com/journals/jamapsychiatry/article-abstract/481547

Nesse, R. M. (2019). *Good reasons for bad feelings: Insights from the frontier of evolutionary psychiatry*. https://books.google.ca/books?hl=en&lr=&id=frmHDwAAQBAJ&oi=fnd&pg=PR13& ots=TBHFiHM Fwl&sig=4usb1AswcmAYa-x4Ex8z_v5i9m8

Nesse, R. M. (2022). Why evolutionary do mental disorders persist? *Perspectives on Evolution and Mental Health*. https://books.google.com/books?hl=en&lr=&id=Tph-EAAAQBAJ&oi=fnd&pg=PA84&dq=Why+Evo lutionary+Mental+Disorders+Persist&ots=d6KaREPMi8&sig=bnsD-2 uUjZKVB6BjEd76YEgWgB0

Nesse, R. M., & Williams, G. C. (1998). Evolution and the origins of disease. *Scientific American, 279*(5), 86–93.

Nesse, R. M., & Williams, G. C. (1999). *On Darwinian medicine*. http://www.personal.umich.edu/~nesse/ Articles/OnDarMed-LifeScience-1999.PDF

Nowell, P. C. (1976). The clonal evolution of tumor cell populations. *Science, 194*(4260), 23–28.

Nunney, L. (1999). Lineage selection and the evolution of multistage carcinogenesis. *Proceedings. Biological Sciences / The Royal Society, 266*(1418), 493–498.

Nunney, L. (2016). Commentary: The multistage model of carcinogenesis, Peto's paradox and evolution [Review]. *International Journal of Epidemiology, 45*(3), 649–653. www.academic.oup.com

Nunney, L. (2020). Resolving Peto's paradox: Modeling the potential effects of size-related metabolic changes, and of the evolution of immune policing and cancer suppression. *Evolutionary Applications, 13*(7), 1581–1592.

Okadal, Y., Saika, S., Hashizume, N., Kobata, S., Yamanaka, O., Ohnishi, Y., & Senba, E. (1996). Expression of fos family and jun family proto-oncogenes during corneal epithelial wound healing. *Current Eye Research, 15*(8), 824–832.

Paoloni, M. C., & Khanna, C. (2007). Comparative oncology today. *The Veterinary Clinics of North America. Small Animal Practice, 37*(6), 1023–1032, v.

Peto, R. (2016). Epidemiology, multistage models, and short-term mutagenicity tests. *International Journal of Epidemiology, 45*(3), 621–637.

Prokofiev, A. M. (2020). New data on the morphology and distribution of two rare species of deep-sea ang-lerfish from the families linophrynidae and himantolophidae. *Journal of Ichthyology, 60*(4), 548–554.

Ratcliff, W. C., Fankhauser, J. D., Rogers, D. W., Greig, D., & Travisano, M. (2015). Origins of multicellular evolvability in snowflake yeast. *Nature Communications, 6*, 6102.

Robert, J. (2010). Comparative study of tumorigenesis and tumor immunity in invertebrates and nonmamma-lian vertebrates. *Developmental and Comparative Immunology, 34*(9), 915–925.

Roff, D. (1993). *Evolution of life histories: Theory and analysis* (D. A. Roff (ed.); 1993rd ed.). Springer.

Rosenfield, A., Kern, F. G., & Keller, B. J. (1994). *Invertebrate neoplasia: Initiation and promotion mecha-nisms: Proceedings of international workshop, 23 June 1992* (A. Rosenfield, F. G. Kern, and B. J. Keller (compilers and eds.). https://repository.library.noaa.gov/view/noaa/3196

Ruiz, S. A., & Chen, C. S. (2008). Emergence of patterned stem cell differentiation within multicellular struc-tures. *Stem Cells, 26*(11), 2921–2927.

Schiffman, J. D., & Breen, M. (2015). Comparative oncology: What dogs and other species can teach us about humans with cancer. *Philosophical Transactions of the Royal Society of London. Series B, Biological Sciences, 370*(1673). https://doi.org/10.1098/rstb.2014.0231

Schneider, E. (2012). *Handbook of the biology of aging*. Elsevier.

Serrano, M., & Massagué, J. (2000). Networks of tumor suppressors. *EMBO Reports*, *1*(2), 115–119.

Shepard, E. L. C., Wilson, R. P., Rees, W. G., Grundy, E., Lambertucci, S. A., & Vosper, S. B. (2013). Energy landscapes shape animal movement ecology. *The American Naturalist*, *182*(3), 298–312.

Smith, F. A., Boyer, A. G., Brown, J. H., Costa, D. P., Dayan, T., Ernest, S. K. M., Evans, A. R., Fortelius, M., Gittleman, J. L., Hamilton, M. J., Harding, L. E., Lintulaakso, K., Lyons, S. K., McCain, C., Okie, J. G., Saarinen, J. J., Sibly, R. M., Stephens, P. R., Theodor, J., & Uhen, M. D. (2010). The evolution of maximum body size of terrestrial mammals. *Science*, *330*(6008), 1216–1219.

Somarelli, J. A., Gardner, H., Cannataro, V. L., Gunady, E. F., Boddy, A. M., Johnson, N. A., Fisk, J. N., Gaffney, S. G., Chuang, J. H., Li, S., Ciccarelli, F. D., Panchenko, A. R., Megquier, K., Kumar, S., Dornburg, A., DeGregori, J., & Townsend, J. P. (2020). Molecular biology and evolution of cancer: From discovery to action. *Molecular Biology and Evolution*, *37*(2), 320–326.

Somarelli, J. A., Rupprecht, G., Altunel, E., Flamant, E. M., Rao, S., Sivaraj, D., Lazarides, A. L., Hoskinson, S. M., Sheth, M. U., Cheng, S., Kim, S. Y., Ware, K. E., Agarwal, A., Cullen, M. M., Selmic, L. E., Everitt, J. I., McCall, S. J., Eward, C., Eward, W. C., & Hsu, D. S. (2020). A comparative oncology drug discovery pipeline to identify and validate new treatments for osteosarcoma. *Cancers*, *12*(11). https://doi.org/10.3390/cancers12113335

Swynghedauw, B. (2016). Evolutionary paradigms in cardiology: The case of chronic heart failure. In A. Alvergne, C. Jenkinson, & C. Faurie (Eds.), *Evolutionary thinking in medicine: From research to policy and practice* (pp. 137–153). Springer International Publishing.

Swynghedauw, B., Delcayre, C., Samuel, J.-L., Mebazaa, A., & Cohen-Solal, A. (2010). Molecular mechanisms in evolutionary cardiology failure. *Annals of the New York Academy of Sciences*, *1188*, 58–67.

Tang, S., Pichugin, Y., & Hammerschmidt, K. (2023). An environmentally induced multicellular life cycle of a unicellular cyanobacterium. *Current Biology: CB*, *33*(4), 764–769.e5.

Thomas, F., Elguero, E., Brodeur, J., Roche, B., Missé, D., & Raymond, M. (2012). Malignancies and high birth weight in human: Which cancers could result from antagonistic pleiotropy? *Journal of Evolutionary Medicine*, *1*, 1–5.

Tollis, M., Boddy, A. M., & Maley, C. C. (2017). Peto's paradox: How has evolution solved the problem of cancer prevention? *BMC Biology*, *15*(1), 60.

Tong, K., Bozdag, G. O., & Ratcliff, W. C. (2022). Selective drivers of simple multicellularity. *Current Opinion in Microbiology*, *67*, 102141.

Torry, D. S., & Cooper, G. M. (1991). Proto-oncogenes in development and cancer. *American Journal of Reproductive Immunology*, *25*(3), 129–132.

Trichopoulos, D., Hsieh, C. C., MacMahon, B., Lin, T. M., Lowe, C. R., Mirra, A. P., Ravnihar, B., Salber, E. J., Valaoras, V. G., & Yuasa, S. (1983). Age at any birth and breast cancer risk. *International Journal of Cancer, Journal International Du Cancer*, *31*(6), 701–704.

Tuohy, J. L., Somarelli, J. A., Borst, L. B., Eward, W. C., Lascelles, B. D. X., & Fogle, J. E. (2020). Immune dysregulation and osteosarcoma: Staphylococcus aureus downregulates TGF-β and heightens the inflammatory signature in human and canine macrophages suppressed by osteosarcoma. *Veterinary and Comparative Oncology*, *18*(1), 64–75.

Vincze, O., Colchero, F., Lemaître, J.-F., Conde, D. A., Pavard, S., Bieuville, M., Urrutia, A. O., Ujvari, B., Boddy, A. M., Maley, C. C., Thomas, F., & Giraudeau, M. (2022). Cancer risk across mammals. *Nature*, *601*(7892), 263–267.

Wagner, G. P., Kshitiz, & Levchenko, A. (2020). Comments on Boddy et al. 2020: Available data suggest positive relationship between placental invasion and malignancy. *Evolution, Medicine, and Public Health*, *2020*(1), 211–214.

Wilkinson, D. M., & Ruxton, G. D. (2012). Understanding selection for long necks in different taxa. *Biological Reviews of the Cambridge Philosophical Society*, *87*(3), 616–630.

Williams, G. C., & Burt, A. (1997). *Williams: Adaptation and natural selection*. https://www.sscnet.ucla.edu/comm/steen/cogweb/Abstracts/Williams_66.html

Williams, G. C., & Nesse, R. M. (1991). The dawn of Darwinian medicine. *The Quarterly Review of Biology*, *66*(1), 1–22.

Wilson, R. P., Quintana, F., & Hobson, V. J. (2012). Construction of energy landscapes can clarify the movement and distribution of foraging animals. *Proceedings, Biological Sciences / The Royal Society*, *279*(1730), 975–980.

13 From Evolutionary Biology to Bedside and Beyond

A View of Comparative Oncology throughout the Translational Pipeline

Veronica Colmenares, William C. Eward, and Laurie A. Graves

13.1 INTRODUCTION

Cancer, a state of cellular evolution incited by genetic and epigenetic dysregulation leading to proliferation of a new cell population with adapted behavioral capacities within a host environment, is a ubiquitous phenomenon across nearly all multicellular life (Aktipis et al., 2015). Its pervasiveness throughout the evolution of multicellular life is illustrated both as anciently as the identification of malignant tumor growth in dinosaurs of the Mesozoic era and as universally as the presence of accumulating DNA damage across the life span of humans and animals alike (Ekhtiari et al., 2020; Vincze et al., 2022). With its sweeping influence across time, cancer has the potential to significantly impact mortality, and therefore, to impact species health and conservation as a whole. With acknowledgment of the impact that cancer imposes on all living beings, research on cancer across the tree of life has itself evolved, emerging into the field of comparative oncology. As a broad area of study, comparative oncology is uniquely positioned at the intersection of the fields of evolutionary biology, veterinary oncology, and human oncology (Boddy et al., 2020a, 2020b). By serving as a bridge between each of these fields, comparative oncology propels scientists and clinicians toward a common goal: to collaborate across disciplines to advance knowledge in evolution and cancer biology and to improve health for all species impacted by cancer.

With increased recognition over the last two decades that the field of comparative oncology is a unique and critical area of study, the complexity and breadth of the field has also increased. Today, the field of comparative oncology has expanded to include comprehensive studies of evolutionary cancer biology, differences in cancer risk across species, differential patterns of cancer growth, cancer biomarkers, preclinical drug discovery, drug pharmacokinetics and pharmacodynamics, and translational therapeutic trials across a vast number of cancers and a multitude of species (Boddy et al., 2020a). Across cancer types and species, a comparative oncology approach is particularly advantageous in the study of rare cancer subtypes when a greater incidence may be identified in specific species. This cross-species analysis allows a given type of cancer to be studied as a distinct entity, rather than in the context of a health problem affecting one particular species. In so doing, the field can understand fundamental aspects of disease as those aspects pertain to the cancer type (or subtype) independent of the species in which that cancer type occurs.

As a result of progress in the field, one framework for comparative oncology research that has emerged builds on, and complements, established translational pipelines in which hypotheses generated from biologic concepts drive preclinical study, clinical trials to assess safety and drug activity in patients, and subsequent changes in practice aiming to improve public health. However, while established translational pipelines for human disease are traditionally unidirectional, in that research

 DOI: 10.1201/9781003307921-13

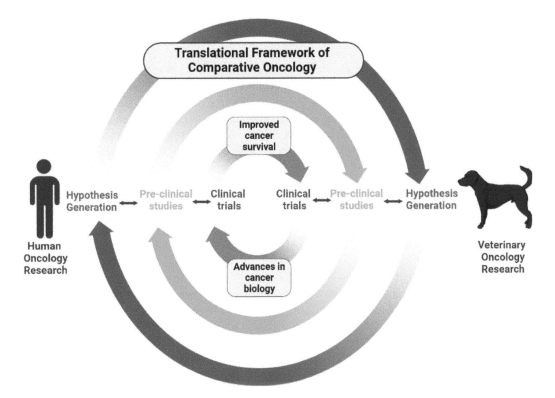

FIGURE 13.1 Multidirectional translational framework of comparative oncology. Combining cancer discoveries across human and veterinary oncology research, the translational framework of comparative oncology is a multidirectional network in which advances at each level of research are easily translated through cross-species platforms to advance knowledge of cancer biology and improve health for humans and animals with cancer.

conducted throughout this framework primarily benefits those individuals at the end of the pipeline, the translational framework of comparative oncology is a multidirectional set of pathways that aim to improve survival for species at multiple levels of research (Figure 13.1). In such a framework, cancer research in animals may continue to benefit humans, but knowledge gained from human studies can also be translated back to improve survival for animals. Among animal species in which comparative oncology studies have the greatest potential to advance knowledge in cancer biology are companion dogs, who have similar immune function and environmental exposures and whose cancers share remarkable genetic similarity to those in humans. Through efforts to also improve survival for animals affected by cancer, the field of comparative oncology can have positive impacts on global animal health and species conservation (Boddy et al., 2020a, 2020b). In this chapter, we aim to further explore the multi-directionality of the translational framework of comparative oncology, highlighting the foundation of research upon which it is established, select recent advances and tools of study, critical components of research, and ongoing challenges within the field.

13.2 COMPARATIVE ONCOLOGY PROVIDES INSIGHT INTO EVOLUTIONARY CANCER BIOLOGY

While once considered a disease of the modern human era, cancer in fact has left its footprints throughout millions of years of evolutionary history across a multitude of species and environments.

Among the earliest known cases of cancer are those of osteosarcoma identified in fossils of a 240-million-year-old stem turtle and a 75-million-year-old *Centrosaurus apertus* (Ekhtiari et al., 2020; Haridy et al., 2019). The study of cancer evolution across species over the millions of years since is the foundation of comparative oncology. This evolution-informed approach to studying cancer provides an opportunity to better understand cancer prevalence, conserved oncogenic drivers, and cancer hallmarks of tumorigenesis (Somarelli et al., 2017). Within the field of comparative oncology, investigators are able utilize these evolutionary insights to generate and explore hypotheses toward discoveries in cross-species cancer biology and novel therapeutic approaches.

One of the most significant insights in the field of evolutionary comparative oncology is the observation that cancer prevalence differs across species (Vincze et al., 2022; Peto et al., 1975). Sir Richard Peto observed that, while individuals of greater size within certain species, including humans and dogs, have a higher lifetime risk of cancer, many species of large body size—like elephants and whales—do not develop cancer with increased frequency (Peto et al., 1975; Caulin & Maley, 2011). This observation is known as Peto's Paradox, in which cancer risk across species does not appear to correlate with either increasing body size, containing more dividing cells at risk of malignant transformation, or a longer life span, traditionally associated with the accumulation of DNA damage (Peto et al., 1975; Caulin & Maley, 2011). The hypothesis of Peto's Paradox has been explored by multiple investigators since, with studies ultimately observing that many species with greater longevity and larger body size have evolved to have either lower rates of somatic mutations per year or enhanced cancer protective mechanisms, such that body size and longevity have a co-evolutionary story with mechanisms for a decreased incidence of cancer (Vincze et al., 2022; Boddy et al., 2020a, 2020b; Cagan et al., 2022; Tollis et al., 2019). Among these studies, one demonstrated that the average number of somatic mutations at the end-of-life is similar across species despite wide variability in lifespan, suggesting that the rate at which animals acquire new somatic mutations is inversely proportional to their lifespan (Cagan et al., 2022). One example of such includes data that the naked mole rat (*Heterocephalus glaber*) and the giraffe, despite significant differences in body size, have similar life spans and thus similar rates of somatic mutations over time (Cagan et al., 2022). Cancer protection has also evolved across multiple species through sequence conservation and copy number expansion of tumor suppressor genes. The tumor suppressor gene *TP53* is amplified in the elephant genome, which confers a lifetime cancer risk in elephants that is less than half that of humans (Abegglen et al., 2015; Caulin et al., 2015; Sulak et al., 2016; Tollis et al., 2021). Within the Proboscidean lineage, evolution over a 25-million-year period has led to copy number expansion of the *TP53* gene, increasing from 3–8 copies in the prehistoric *Mammut americanum*, up to 12–17 copies in the contemporary Asian elephant genome (Sulak et al., 2016). Studies in the African elephant, manatee, and rock hyrax have also observed upregulation of the leukemia inhibitory factor-6 (*LIF6*) gene, which enhances apoptosis in response to DNA damage and likely serves as another cancer protective mechanism that has evolved over time (Vazquez et al., 2018).

In addition to tumor suppressor genes, oncogenes have also been traced back through evolution to some of the earliest eukaryotic genomes, suggesting that these genes likely also played a role in origins of multicellular life (Makashov et al., 2019). Because cancer more commonly occurs after an individual has reproduced, it often manifests without deleterious evolutionary effects, supporting transmission of conserved oncogenes through generations of species and across evolution. A comparative genomic analysis identified multiple evolutionarily ancient genes that are involved in both cancer invasion/metastasis and tissue invasion in placental development, including those in Notch signaling and epidermal growth factor receptor families (Hao et al., 2020). In addition to these genes, it is estimated that nearly 90% of human cancer genes are highly conserved and often duplicated in at least one mammalian genome (Tollis et al., 2020). Among the highly conserved oncogenes are gene fusion events, such as the *BCR-ABL* fusion gene associated with hematologic malignancies in both humans and dogs. The "Raleigh chromosome," a t(9;26) translocation that encodes the *BCR-ABL* fusion protein in dogs shares remarkable homology with the t(9;22) Philadelphia chromosome in humans (Breen & Modiano, 2008). Across multiple other tumor types in humans and dogs, oncogenic drivers and mutations are also conserved. These include *PIK3CA* mutations in human and canine breast cancers and loss of *TP53* and *Rb* regulated pathways or *MYC* amplification in human and canine osteosarcoma (Lee et al., 2019; Varshney et al., 2016). High-resolution oligonucleotide array comparative genomic hybridization (aCGH) in human and canine osteosarcoma identified high levels of genetic instability present in osteosarcoma but also demonstrated a significant number of conserved genomic alterations between species, including genes known to be altered in osteosarcoma and other genes not previously associated with osteosarcoma (Angstadt et al., 2012). Aside from dogs, other species share conserved oncogenic drivers with human tumors. For example, feline oral squamous cell carcinoma (SCC), similar to that in humans, frequently overexpresses the membrane tyrosine kinase receptor epidermal growth factor receptor (EGFR), which contributes to tumorigenesis and is a conserved therapeutic target in this otherwise challenging to treat tumor (Cannon, 2015).

Finally, with further study of cancer prevalence across the tree of life, investigators have identified environmental factors associated with greater cancer risk across species. Studies have observed differences in cancer risk across trophic levels or hierarchical positions within the food chain. Relative to herbivores or those in lower trophic levels, multiple studies have observed an increased cancer prevalence among predators and other species in higher trophic levels (Kapsetaki et al., 2022). Species of higher trophic levels have less diversity in their diets, which may be associated with a less robust microbiome (Kapsetaki et al., 2022). Additionally, the increase in cancer risk in carnivores may be associated with ingestion of carcinogenic viruses and microbes from raw meat, such as *Fusobacteria* and *Peptostreptococcus*, and compounded effects of carcinogenic pollutants (Vincze et al., 2022; Kapsetaki et al., 2022; Munson & Moresco, 2007).

The study of evolutionary cancer biology across species builds the foundation of comparative oncology. Through outlining a fundamental genomic and phylogenetic framework illustrating evolved differences in cancer protective mechanisms, conserved oncogenic drivers, and common carcinogenic environmental exposures across species and time, the principles in this critical first step of the multidirectional translational pipeline serve as the groundwork from which investigators are able to generate hypotheses for ongoing research. Rooted in biologic similarities and differences across species, these hypotheses enable scientists to design preclinical and then ultimately, patient-focused translational research to advance knowledge across cancers for humans and animals.

13.3 PRECLINICAL AND TRANSLATIONAL STUDIES WITHIN COMPARATIVE ONCOLOGY ADVANCE OUR UNDERSTANDING OF CANCER BIOLOGY AND DRUG DISCOVERY

Based on evidence of conserved genomic drivers of cancer across the evolutionary spectrum and the pervasiveness of cancer across species, preclinical and translational research in the field of comparative oncology has expanded, utilizing multi-species models to advance knowledge of biology across a multitude of cancers. While mice have historically served as the standard preclinical animal model to study *in vivo* tumor growth and therapeutic efficacy of novel drugs, many mouse models have limited translational relevance to humans and other species due to altered immune function and incomplete homology of key cancer-related genes (Overgaard et al., 2018; Paoloni & Khanna, 2008). As such, there is increasing interest in the incorporation of animal models with naturally occurring cancers that are more analogous to humans for preclinical and translational comparative oncology studies (Figure 13.2). Most common among these animals are dogs, who spontaneously develop multiple cancers that also affect humans, including hematologic malignancies, melanoma, bladder cancer, intracranial neoplasms, soft tissue sarcoma, and osteosarcoma (Schiffman & Breen, 2015). Human and canine malignancies share remarkable similarities, demonstrating similar gene expression patterns by unsupervised clustering and having comparable immune function (Lee et al., 2019; Varshney et al., 2016; Paoloni et al., 2014; Park et al., 2016; Gingrich et al., 2021).

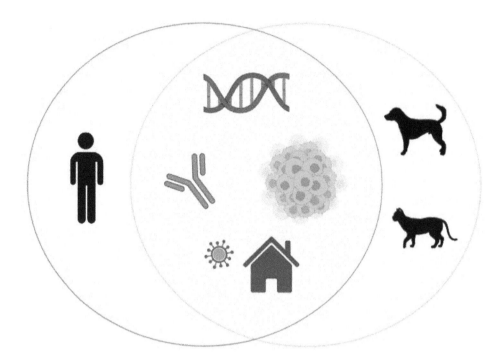

FIGURE 13.2 Parallels between humans, companion dogs, and companion cats in comparative oncology. Humans and companion animals (predominantly dogs and cats) share homology across key cancer-related genes; develop tumors with similar histologies, genetic backgrounds, and clinical outcomes; have comparable immune function; and share similar environments, including exposure to environmental pathogens and pollutants.

Additionally, dogs have a shorter life span and malignancies that progress on a more rapid time scale than humans, enabling study of the natural history or treatment responses of canine tumors in a relatively shorter time period (Schiffman & Breen, 2015). Furthermore, humans and domestic dogs are exposed to similar environmental toxins and pathogens due to shared habitats, which may contribute to the biologic similitude of tumors between the two species. As another comparative oncology model, domestic cats share many of the same advantages as dogs—rapid time scale of malignancy, shared environmental risk factors as humans, and an intact immune system. Several tumor types in cats also share similarity with humans and occur with greater incidence than in dogs, including feline oral squamous cell carcinoma, feline hormone-negative mammary cancer (frequently analogous to triple negative breast cancer in humans), and feline injection site sarcoma (Cannon, 2015; Sommerville et al., 2022). In addition to dogs and cats, cancer naturally occurs in other domestic and companion animals, including rabbits, ferrets, and hens, though cancer in these animals is studied less commonly than in domestic dogs and cats (Johnson & Giles, 2013; van Zeeland, 2017; Schoemaker, 2017).

In human oncology research, the terms "preclinical" and "clinical" research often are meant to apply to research conducted in cancer cells or animal models and in human patients, respectively. "Translational" research is that which bridges the two domains, classically serving to correlate the study of biological features of tumors with patient responses to treatment or patient outcomes. From the perspective of human cancer research, the study of novel cancer therapies in companion animals is often considered "preclinical," but from the veterinary perspective, these studies are appropriately viewed as clinical research. As the study of comparative oncology merges these two perspectives into a multidirectional approach, research across species might be more suitably divided into "preclinical" studies and "translational" trials, with the latter encompassing any clinical trial in humans or animals with results that might translate to advanced knowledge or improve survival in other species with cancer.

13.3.1 Preclinical and Translational Comparative Oncology Models Advance Knowledge in Cancer Biology

Through the study of patient-derived tumor tissue, preclinical and translational studies support advances in our understanding of cancer biology across species and cancer types. A comparative oncology framework more effectively advances knowledge in cancer biology through the comparison of evolved oncogenic drivers across species and by enabling study of a greater number of cross-species patients affected by a particular tumor type, which is particularly advantageous in rare cancers that are challenging to study due to their low incidence.

For example, through the study of genomic aberrations across tumors and species, investigators have identified short in-frame deletions involving the BRAF gene distinct from the more common BRAFV595E mutation in up to half of dogs with urogenital carcinoma (Thomas et al., 2023). Urogenital carcinoma is the most common urogenital cancer affecting dogs, with an incidence of more than 60,000 new cases each year. In contrast, several rare subtypes of human cancer, including certain types of leukemia, pancreatic carcinomas, and Langerhans cell histiocytosis, also harbor equivalent deletions in the BRAF gene, which ultimately affect MAPK pathway signaling and promote tumor growth. While the rarity of these human cancer subtypes has limited further study of these genomic alterations, study of the more common canine urogenital carcinoma can advance knowledge of the biology of these specific BRAF gene deletions and help to guide future therapeutic discovery for these tumors (Thomas et al., 2023).

While next-generation-sequencing (NGS) testing is widely used across human cancers, with personalized treatments available for some patients based on genomic findings, molecularly-based testing and precision medicine have not been broadly incorporated into veterinary oncology. However, broader use of gene expression profiling across canine tumors has provided a rich source of data for investigators to study canine cancer biology, prompting investigators to use this data to begin to identify personalized therapeutic options (Lindblad et al., 2005). To study the feasibility of molecular tumor analysis in the clinical veterinary setting, the Comparative Oncology Trials Consortium (COTC) conducted a proof-of-concept study in which tumors from 31 dogs of varying histology underwent gene expression profiling, yielding a personalized therapy report (Paoloni et al., 2014). The study demonstrated feasibility of generating a personalized medicine report in a clinically relevant time frame. Further analysis of molecular profiling data revealed that gene expression clustered by cancer histologic subtype but was still able to capture heterogeneity in gene expression across patients within each subtype. The resolution with which tumor sequences both reliably clustered by tumor type and identified heterogeneity across tumors provides support for broader use of cross-species molecular-based testing to learn more about the biology of tumor heterogeneity in comparative oncology. Personalized medicine reports identified numerous therapeutic options, ranging from agents used in standard-of-care therapy for each particular tumor type, to those not previously studied in the oncology setting. The range of therapeutic options identified by molecular profiling supports the concept for the future prospective comparative clinical study of such a platform to both select effective and novel treatment options for canine oncology patients (Paoloni et al., 2014).

13.3.2 The Preclinical and Translational Path toward Drug Discovery in Comparative Oncology

For over a century, preclinical studies in cancer biology and drug discovery have utilized cancer cell lines for *in vitro* assays (Fowles et al., 2017). While many cancer cell lines in use are derived from human tumors, a growing number are from non-human species, including canine and feline cancer cell lines (Cannon, 2015; Fowles et al., 2017). Within the scope of comparative oncology, multi-species cell line panels can be used to improve speed in validating drug efficacy across species. One such study exploring drug efficacy in multi-species cell lines demonstrated correlation in growth inhibition to six standard chemotherapy agents between paired multi-cancer panels of human- and canine-derived cell lines, further illustrating similarity between human and canine tumors at the level of their response to chemotherapy (Fowles et al., 2017). Additional drug sensitivity assays conducted with human and canine osteosarcoma cell lines have helped to identify effective drug candidates that inhibit tumor growth (Somarelli et al., 2020c). As another example, feline cell lines have been utilized in *in vitro* studies to test novel therapies for domestic cats. Since feline and human SCC both overexpress the tyrosine kinase receptor EGFR and EGFR-targeting strategies are used in human oral SCC, investigators have begun to translate EGFR targeting therapeutic studies back to domestic cats. Feline SCC cell lines were used to validate the activity of the anti-EGFR monoclonal antibody, cetuximab, and the tyrosine kinase inhibitor (TKI) gefitinib in feline models (Altamura & Borzacchiello, 2022; Bergkvist et al., 2011). Additionally, these preclinical studies explored stem-like properties of SCC cells, further elucidated mechanisms of resistance to gefitinib and strategies to overcome this resistance, and identified strategies to improve radio-sensitivity in feline SCC *in vitro* models (Bergkvist et al., 2011; Pang et al., 2012). These studies in feline SCC cell lines demonstrate the value of the multidirectional translational framework of comparative oncology at the level of preclinical research, as ongoing work in this field may lead to new therapy options for this often-fatal feline tumor.

Comparative oncology models have also been utilized to study novel therapies and repurposed FDA-approved drugs throughout the translational pathway to in-animal and in-human clinical trials. For example, building on multiple preclinical studies of CB-5083, an early generation small molecule inhibitor of valosin-containing protein (VCP/p97), across human and canine cell lines and *in vivo* preclinical models, a COTC sponsored trial studied the efficacy, tolerability, and pharmacokinetic/pharmacodynamic effects of the novel second-generation VCP inhibitor, CB-5339 (Le Moigne et al., 2017; LeBlanc et al., 2022). Across canine cell lines and subsequently in 24 dogs with a range of solid tumors and hematologic malignancies, investigators observed manageable canine toxicity, a pharmacodynamic signal between drug exposure and target modulation, and efficacy in 33% of dogs with multiple myeloma, providing support for ongoing early phase trials of CB5339 in humans with multiple myeloma and acute myeloid leukemia (AML) (LeBlanc et al., 2022).

In canine lymphoma and osteosarcoma cell lines, auranofin, a thioredoxin reductase inhibitor that is FDA approved for use in rheumatoid arthritis, was identified to have single-agent anti-cancer activity (Zhang et al., 2018; Parrales et al., 2018). Furthermore, when repurposed in canine osteosarcoma cell lines, auranofin was observed to have synergistic activity in combination with mammalian target of rapamycin (mTOR) inhibitor, rapamycin, or histone deacetylase (HDAC) inhibitor, vorinostat (Parrales et al., 2018). In human cell line and murine xenograft models of osteosarcoma, treatment with auranofin was associated with decreased incidence of pulmonary metastasis (Topkas et al., 2016). This encouraging preclinical data supported a multicenter Phase I/II clinical trial to study the addition of auranofin to standard amputation and carboplatin in canine osteosarcoma. When compared to a historical cohort similar to that of multiple canine osteosarcoma trials, addition of auranofin to the standard-of-care increased overall survival by 2-fold (240 to 474 days) in male dogs without an increase in grade ≥ 2 toxicity beyond that expected with single-agent carboplatin (Endo-Munoz et al., 2019). Future trials may further study the clinical benefit of auranofin in a larger group of dogs with osteosarcoma.

Losartan, an angiotensin II receptor blocker, approved for the treatment of hypertension in humans, has also more recently been studied as a repurposed oncology agent (Regan et al., 2021). At optimized doses based on pharmacokinetic and pharmacodynamic studies, losartan inhibited CCL2 secretion and subsequently inhibited monocyte migration in *in vitro* models of both human and canine osteosarcoma. The addition of losartan to the multi-tyrosine kinase inhibitor (TKI), toceranib, in 28 canine patients with pulmonary metastases was associated with an objective response rate of 25% and clinical benefit rate of 50% (Regan et al., 2021). This data has supported study of the combination of losartan and sunitinib (in-human TKI with targets equivalent to that of toceranib) in a Phase I clinical trial for the treatment of pediatric and young adult patients with pulmonary metastatic osteosarcoma (NCT03900793).

13.3.3 Comparative Oncology Models for the Preclinical and Translational Study of Tumor-Immune Interactions and Novel Immunotherapy Agents

Immune dysregulation, supporting the ability of cancer cells to evade immune destruction by the body, is likely a universal hallmark of cancer across species. Due to the universal impact of immune dysfunction in cancer development, an understanding of cancer biology and therapeutic vulnerabilities in the field of comparative oncology would be incomplete without an evaluation of immune function across the expanse of evolution and across species. The earliest studies of cross-species immune function first compared innate immunity of humans with that of the fruit fly, *Drosohila melanogaster*, then expanded to those comparing humans and mice, and since have more broadly explored these comparisons across a multitude of species (Bryant & Monie, 2012; Overgaard et al., 2018). While there are similarities in immune function across species, key differences also exist. For example, researchers comparing immune-related gene expression across humans, mice, and pigs identified 188 genes shared across the three species, but the presence of 37, 16, and 174 genes unique to humans, pigs, and mice, respectively. They also identified differences in NK cell receptor number in mice compared to humans or pigs (Overgaard et al., 2018). Conversely, activated NK cells from humans and dogs share remarkable similarity, with comparative transcriptomics of NK related genes from dogs, mice, and humans demonstrating greater similarities between dog and human NK cells than those of mice (Gingrich et al., 2021). In addition to differences in immune function at the gene level, important differences in environmental conditions exist across species, which impact the microbiome and immune function. Studies have demonstrated variability in tumor growth and response to immuno-oncology agents among mice housed at multiple institutions with differences in gut microbiome (Sivan et al., 2015; Hooper et al., 2012; Ivanov & Honda, 2012). However, other studies have demonstrated similarities between the microbiome of humans and their pet dogs that result from shared environmental conditions and frequent contact (Song et al., 2013). Each of these similarities and differences can impact how each species uniquely responds to infection or inflammation and how differences in immunity relate to cancer development, progression, and response to therapy (Bryant & Monie, 2012; Park et al., 2016).

In acknowledging differences in immune function across species, scientists recognize not only the ways in which they impact cancer biology within and across species, but also the impact that these differences can have on the use of cross-species models for preclinical and translational research. To date, *in vivo* preclinical studies have traditionally relied upon mouse models to explore patterns of tumor growth and response to novel therapeutics or therapeutic combinations prior to advancement to in-human clinical trials. However, the majority of these studies utilize mouse models that are immunocompromised in order to support *in vivo* growth of various tumor types of cross-species lineage. As greater recognition of the importance of the immune system in cancer growth has emerged and as immune-modulating therapies are utilized with increasing frequency across cancer types, scientists recognize the need for preclinical and translational models that more faithfully recapitulate tumor-immune and tumor-drug interactions than traditional immunocompromised mouse models. To meet this need, immunocompetent models are also used across preclinical research. Of these, syngeneic mouse models, involving injection of tumor cell lines that are histocompatible to the recipient mouse, are able to model tumor-immune biology and disease progression. The capacity to research tumor-immune interactions and activity of novel immuno-oncology agents in syngeneic models is particularly significant when these syngeneic tumors are grown following orthotopic implantation, which models tumor growth and therapy response in a microenvironment natural to the particular cancer of interest (Overgaard et al., 2018). While immunocompetent mice are better able to model the tumor-immune landscape and response to immunotherapy, differences remain between the immune system and genetic background of the immunocompetent mouse and human patients, potentially limiting the translational impact of findings in these models (Overgaard et al., 2018). Humanized mice, engrafted with human cells and tissue to recapitulate the human immune system, have been developed to bridge this gap in the study of immune interactions with cancer biology and of novel immunotherapies. However, use of humanized mouse models is associated with several limitations, including the risk of graft-versus-host disease, incomplete correlation with the human immune system (lacking immunoglobulin G responses and a fully developed lymphoid system), and a relatively short life span (Allen et al., 2019; Overgaard et al., 2018). As immune function in other mammalian species has evolved to more closely resemble that of humans, application of a comparative oncology framework incorporating animals like companion dogs and cats complements existing studies of tumor-immune interactions and of novel immunotherapy agents in cancer research across species (Park et al., 2016; Gingrich et al., 2021; Overgaard et al., 2018).

For example, the study of tumor infiltrating lymphocytes (TILs) is growing across the field of comparative oncology, with the recognition that across several human breast cancer subtypes, the presence of TILs within the tumor microenvironment is associated with poorer outcomes but has been correlated with a higher incidence of achieving a pathologic complete response to neoadjuvant chemotherapy (Pinard et al., 2022). Across human breast cancer and melanoma, greater TIL infiltration was associated with enhanced response to checkpoint inhibition, including anti-PD1 and anti-CTLA4 therapy (Loi et al., 2017). In canine mammary carcinoma and oral melanoma, higher CD4+ or CD20+ TIL infiltration, respectively, was associated with a poor prognosis (Estrela-Lima et al., 2010; Porcellato et al., 2020). In the aforementioned study of repurposed losartan in canine osteosarcoma, *in vivo* and clinical administration of losartan was associated with decreased CCL2-CCR2 mediated monocyte migration, with impaired monocyte migration observed in 60% of dogs with clinical benefit after treatment with losartan in combination with toceranib (Regan et al., 2021).

As immune checkpoints have been identified as an immune-evading mechanism conserved across cancers of multiple species, there has been greater interest in studying these checkpoints over the last two decades. As a result of these studies, inhibitors of immune checkpoints PD-1, PD-L1, or CLTA4 have demonstrated anti-tumor activity across multiple cancers in humans. Translation of these findings back to animals has led to the identification of upregulated PD-1/PD-L1 across a variety of canine and feline tumors (Maekawa et al., 2014; Stevenson et al., 2021, Coy et al., 2017; Hartley et al., 2018; Nascimento et al., 2020; Nishibori et al., 2023). As a result anti-canine and anti-feline PD-1/PD-L1 antibodies have been developed, with *in vitro* administration associated with enhanced T cell function in canine cell lines and diminished lymphocyte exhaustion in feline peripheral blood lymphocytes (Maekawa et al., 2014; Hartley et al., 2018, Nishibori et al., 2023, Choi et al., 2020). Pantelyushin et al studied whether FDA-approved immune checkpoint inhibitors showed cross-reactivity for canine PD-1/PD-L1 and CTLA4. While four agents demonstrated cross-reactivity, only the PD-L1 inhibitor, atezolizumab showed cross-reactivity with resulting increase in cytokine production of healthy and patient-derived cancer peripheral blood mononuclear cells (PBMCs), particularly those from canine lymphoma patients (Pantelyushin et al., 2021). These findings provide early support for the additional study of species cross-reactivity with anti-human immunotherapies to assess their efficacy in other animals with cancer.

A comparative immune-oncology approach has also been beneficial toward advancing drug discovery across human and canine melanoma and osteosarcoma. Of interest is that of a xenogenic anti-human tyrosinase DNA vaccine that has been studied in several canine melanoma studies, first demonstrating safety and later therapeutic efficacy with documented anti-tyrosinase immune response in a subset of dogs (Barutello et al., 2018). These results led to approval of this anti-human tyrosinase DNA vaccine by the United States Department of Agriculture (USDA) for dogs with melanoma and supported rapid translation of anti-tyrosinase vaccines to human clinical trials. Though the clinical benefit of anti-tyrosinase vaccines has been inconclusive across more recent canine and human clinical trials, this framework serves as an example of the multidirectional translational potential of comparative oncology, particularly in the immune-oncology setting (Barutello et al., 2018). More recently, a recombinant *Listeria monocytogenes* vaccine expressing a chimeric HER2/neu fusion protein has been studied in dogs with osteosarcoma, with a reduction in the incidence of metastatic disease and improved survival as a result of enhanced T cell and antigen-specific IFNγ responses (Mason et al., 2016). These findings led to the development of a Phase II Children's Oncology Group (COG) trial for children and young adults with recurrent osteosarcoma.

13.4 CORNERSTONES OF THE FRAMEWORK: CRITICAL COMPONENTS OF SUCCESSFUL TRANSLATIONAL COMPARATIVE ONCOLOGY RESEARCH

Transdisciplinary cross-species cancer research has provided a framework to advance our current understanding of cancer biology, particularly with regard to cancer-related genes, oncogenic drivers, and signaling pathways conserved across species and throughout evolution (Makashov et al., 2019; Hao et al., 2020; Tollis et al., 2020). Identification of these evolutionarily conserved, potentially therapeutic targets builds a foundation for future cancer drug discovery research. The discovery of novel or repurposed cancer treatments in cross-species models has the potential to serve as an important adjunct to traditional preclinical *in vitro* and *in vivo* models to enhance understanding of 1) drug pharmacokinetics, 2) pharmacodynamic interactions within the tumor and tumor microenvironment, and 3) therapeutic toxicity. When combined, enhanced knowledge in these areas of cancer research may help investigators better nominate novel therapies for clinical trials with the greatest potential translational benefit for patients. As a result of prioritizing agents for clinical trials with the greatest potential translational impact, clinical trial results may be negative or equivocal less often (decreasing costs to trial sponsors and patients), investigators may more quickly identify the most effective therapies for patients, and patient survival may increase more rapidly. As advances are made along this translational pathway, knowledge gained can be continually translated across multidirectional networks to benefit a multitude of species. Therefore, by integrating the translational framework of comparative oncology into ongoing and future research, cancer outcomes may improve for all species. The potential for a comparative oncology framework to directly improve patient survival is greatest perhaps in rare cancers, in which access to patient samples to study tumor biology and clinical trial accrual through traditional research pipelines is notoriously limited.

For future comparative oncology research to advance the field, several cornerstones of the framework must be upheld and expanded upon. First among these includes communication and collaboration between human physicians, evolutionary biologists, zoologists, and veterinarians. Communication between veterinarians and human physicians establishes important partnerships that advances cancer research across species and therefore should be treated as a top priority in the scientific community (Somarelli et al., 2020b). With recognition of this need, multiple comparative oncology consortiums have emerged, including the National Cancer Institute's Comparative Oncology Program, the National Cancer Institute's Comparative Oncology Trials Consortium, and the Clinical and Translational Science Award One Health Alliance (COHA) (Somarelli et al., 2020a). Expanding large-scale cancer data registries that merge human and animal cancer history, treatment information, and tumor genomics is also integral to continue advancing the field of evolutionary cancer biology and comparative oncology. Among established databases include the Exotic Species Cancer Research Alliance (ESCRA), which collects data on cancer cases and treatments from multiple facilities caring for non-domestic and aquatic institutions, and the Zoologic Information Management System (ZIMS), a large database of zoologic and aquatic animals with cancer (Vincze et al., 2022; Duke et al., 2022; Hopewell et al., 2020; Abegglen et al., 2022; Conde et al., 2019).

Second among these cornerstones is the need to expand the breadth of—and reduce the cost of—important tools and reagents used in cross-species studies. These include cross-species antibodies, genomic platforms, and CRISPR/Cas9 reagents that would enable comparative oncology investigators to more efficiently study tumor biology across species and tumors (Somarelli et al., 2020a). More recent recognition of the role of comparative oncology in cancer and drug discovery has led to an increase in the study, production, and availability of cross-species reagents and tools thus far, but ongoing advances across these platforms will be required for the comparative oncology framework to have greatest potential impact on species health.

Third, efforts should be made to increase animal access to specialized oncology care and research. Although the treatment options available to pets largely parallel those options available to humans, limited access to specialized care and research results from unique factors such as geographical constraints (fewer veterinary schools and centers compared to medical schools and hospitals), owner financial limitations affecting ability to pay for diagnostic or therapeutic cancer care, and a lack of owner awareness of available comparative oncology studies and trials (Thamm & Vail, 2015). Furthermore, conducting comparative oncology research in animals in the wild is challenged by limited data on cancer prevalence and inherent difficulties in tissue sampling and clinical outcome follow up.

Finally, and of utmost importance, careful attention should be paid to ethical and regulatory guidelines when conducting clinical trials in humans and animals. Just as human clinical trials are rigorously monitored with built-in safety guidelines, so must a set of standard guidelines for monitoring and safety in veterinary oncology trials be upheld. Analogous to the human Institutional Review Board (IRB), the Guide for the Care and Use of Laboratory Animals and the Animal Welfare Act (AWA) requires institutions to maintain an Institutional Animal Care and Use Committee (IACUC) to oversee animal research (Regan et al., 2018). Both biology-focused and therapeutic clinical trials enrolling domestic animals should require informed consent by pet owners to maintain ethical standards for conducting animal research. As part of informed consent, animal subject owners should be made aware of the potential risks, benefits, and alternative treatment options. Additional ethical considerations in the study of animals includes ensuring adequate access to monitoring health and well-being and minimizing any potential discomfort or suffering during the study. Ultimately, broad participation of companion animals in veterinary clinical trials has the potential to significantly advance the field of comparative oncology to improve outcomes for all species with cancer. It is vital that safety monitoring and ethical review in veterinary clinical trials are continually evolving and that trial investigators remain up-to-date with regulations to ensure that the field of comparative oncology can continue to grow in a scientifically rigorous and ethical manner (Thamm & Vail, 2015).

13.5   CONCLUSION

The field of comparative oncology advances research in cancer biology and drug discovery for humans and animals alike. To achieve this mission, multidirectional translational pathways promote transdisciplinary collaboration across the fields of evolutionary cancer biology and veterinary and human oncology. Ranging from large-scale, cross-species phylogenetic and genomic studies to preclinical exploration of novel therapeutic compounds, and ultimately translating discoveries to early and late-phase clinical trials across species, the framework of comparative oncology has evolved—and will continue to evolve—as access to new cross-species tools and reagents expands and as transdisciplinary collaboration grows. As research and collaboration increases, there will remain an ongoing need to generate and maintain large-scale databases to compile or share data across the field and to continue efforts to build upon existing cross-species research platforms. These sustained efforts from scientists and clinicians across disciplines to expand upon and further integrate the existing frameworks of comparative and evolutionary oncology have the potential to make a profoundly positive impact on the health of animals and humans alike.

REFERENCES

Abegglen LM, Caulin AF, Chan A, et al. Potential mechanisms for cancer resistance in elephants and comparative cellular response to DNA damage in humans. *JAMA*. 2015;314(17):1850–1860. doi:10.1001/jama.2015.13134

Abegglen LM, Harrison TM, Moresco A, et al. Of elephants and other mammals: A comparative review of reproductive tumors and potential impact on conservation. *Animals (Basel)*. 2022;12(15):2005. Published 2022 Aug 8. doi:10.3390/ani12152005

Aktipis CA, Boddy AM, Jansen G, et al. Cancer across the tree of life: Cooperation and cheating in multicellularity. *Philos Trans R Soc Lond B Biol Sci*. 2015;370(1673):20140219. doi:10.1098/rstb.2014.0219

Allen TM, Brehm MA, Bridges S, et al. Humanized immune system mouse models: Progress, challenges and opportunities. *Nat Immunol*. 2019;20(7):770–774. doi:10.1038/s41590-019-0416-z

Altamura G, Borzacchiello G. Anti-EGFR monoclonal antibody Cetuximab displays potential anti-cancer activities in feline oral squamous cell carcinoma cell lines. *Front Vet Sci*. 2022;9:1040552. Published 2022 Nov 17. doi:10.3389/fvets.2022.1040552

Angstadt AY, Thayanithy V, Subramanian S, Modiano JF, Breen M. A genome-wide approach to comparative oncology: High-resolution oligonucleotide aCGH of canine and human osteosarcoma pinpoints shared microaberrations. *Cancer Genet*. 2012;205(11):572–587. doi:10.1016/j.cancergen.2012.09.005

Barutello G, Rolih V, Arigoni M, et al. Strengths and weaknesses of pre-clinical models for human melanoma treatment: Dawn of dogs' revolution for immunotherapy. *Int J Mol Sci*. 2018;19(3):799. Published 2018 Mar 10. doi:10.3390/ijms19030799

Bergkvist G.T., Argyle D.J., Pang L.Y., Muirhead R., Yool D.A. Studies on the inhibition of feline EGFR in squamous cell carcinoma: Enhancement of radiosensitivity and rescue of resistance to small molecule inhibitors. *Cancer Biol Therapy*. 2011;11:927–937. doi:10.4161/cbt.11.11.15525

Boddy AM, Abegglen LM, Pessier AP, et al. Lifetime cancer prevalence and life history traits in mammals. *Evol Med Public Health*. 2020a;2020(1):187–195. Published 2020 May 25. doi:10.1093/emph/eoaa015

Boddy AM, Harrison TM, Abegglen LM. Comparative oncology: New insights into an ancient disease. *iScience*. 2020b;23(8):101373. doi:10.1016/j.isci.2020.101373

Breen M, Modiano JF. Evolutionarily conserved cytogenetic changes in hematological malignancies of dogs and humans—man and his best friend share more than companionship. *Chromosome Res*. 2008;16(1):145–154. doi:10.1007/s10577-007-1212-4

Bryant CE, Monie TP. Mice, men and the relatives: Cross-species studies underpin innate immunity. *Open Biol*. 2012;2(4):120015. doi:10.1098/rsob.120015

Cagan A, Baez-Ortega A, Brzozowska N, et al. Somatic mutation rates scale with lifespan across mammals. *Nature*. 2022;604(7906):517–524. doi:10.1038/s41586-022-04618-z

Cannon CM. Cats, cancer and comparative oncology. *Vet Sci*. 2015;2(3):111–126. Published 2015 Jun 30. doi:10.3390/vetsci2030111

Caulin AF, Graham TA, Wang LS, Maley CC. Solutions to Peto's paradox revealed by mathematical modelling and cross-species cancer gene analysis. *Philos Trans R Soc Lond B Biol Sci*. 2015;370:1850–1860.

Caulin AF, Maley CC. Peto's paradox: Evolution's prescription for cancer prevention. *Trends Ecol Evol*. 2011;26(4):175–182. doi:10.1016/j.tree.2011.01.002

Choi JW, Withers SS, Chang H, Spanier JA, De La Trinidad VL, Panesar H, Fife BT, Sciammas R, Sparger EE, Moore PF, et al. Development of canine PD-1/PD-L1 specific monoclonal antibodies and amplification of canine T cell function. *PLoS One*. 2020;15:e0235518. doi:10.1371/journal.pone.0235518

Conde DA, et al. Data gaps and opportunities for comparative and conservation biology. *Proc. Natl Acad Sci USA*. 2019;116:9658–9664. doi:10.1073/pnas.1816367116

Coy J, Caldwell A, Chow L, Guth A, Dow S. PD-1 expression by canine T cells and functional effects of PD-1 blockade. *Vet Comp Oncol*. 2017;15:1487–1502. doi:10.1111/vco.12294

Duke EG, Harrison SH, Moresco A, Trout T, Troan BV, Garner MM, Smith M, Smith S, Harrison TM. A multi-institutional collaboration to understand neoplasia, treatment and survival of snakes. *Animals*. 2022;12:258. doi:10.3390/ani12030258

Ekhtiari S, Chiba K, Popovic S, et al. First case of osteosarcoma in a dinosaur: A multimodal diagnosis. *Lancet Oncol*. 2020;21:1021–1022. doi:10.1016/S1470-2045(20)30171-6

Endo-Munoz L, Bennett TC, Topkas E, Wu SY, Thamm DH, Brockley L, Cooper M, Sommerville S, Thomson M, O'Connell K, et al. Auranofin improves overall survival when combined with standard of care in a pilot study involving dogs with osteosarcoma. *Vet Comp Oncol*. 2019;18:206–213.

Estrela-Lima A, Araújo MS, Costa-Neto JM, Teixeira-Carvalho A, Barrouin-Melo SM, Cardoso SV, Martins-Filho OA, Serakides R, Cassali GD. Immunophenotypic features of tumor infiltrating lymphocytes from mammary carcinomas in female dogs associated with prognostic factors and survival rates. *BMC Cancer*. 2010;10:256. doi:10.1186/1471-2407-10-256

Fowles JS, Dailey DD, Gustafson DL, Thamm DH, Duval DL. The flint animal cancer center (FACC) canine tumour cell line panel: A resource for veterinary drug discovery, comparative oncology and translational medicine. *Vet Comp Oncol*. 2017;15(2):481–492. doi:10.1111/vco.12192

Gingrich AA, Reiter TE, Judge SJ, et al. Comparative immunogenomics of canine natural killer cells as immunotherapy target. *Front Immunol*. 2021;12:670309. Published 2021 Sep 14. doi:10.3389/fimmu.2021.670309

Hao Y, Lee HJ, Baraboo M, Burch K, Maurer T, Somarelli JA, Conant GC. Baby genomics: Tracing the evolutionary changes that gave rise to placentation. *Genome Biol Evol*. 2020 Mar 1;12(3):35–47. doi:10.1093/gbe/evaa026. PMID: 32053193; PMCID: PMC7144826.

Haridy Y, Witzmann F, Asbach P, Schoch RR, Fröbisch N, Rothschild BM. Triassic cancer-osteosarcoma in a 240-million-year-old stem-turtle. *JAMA Oncol*. 2019;5(3):425–426. doi:10.1001/jamaoncol.2018.6766

Hartley G, Elmslie R, Dow S, Guth A. Checkpoint molecule expression by B and T cell lymphomas in dogs. *Vet Comp Oncol*. 2018;16:352–360. doi:10.1111/vco.12386

Hooper LV, Littman DR, Macpherson AJ. Interactions between the microbiota and the immune system. *Science*. 2012;336(6086):1268–1273. doi:10.1126/science.1223490

Hopewell E, Harrison SH, Posey R, Duke EG, Troan B, Harrison T. Analysis of published amphibian neoplasia case reports. *J Herpetol Med Surg*. 2020;30:148–155. doi:10.5818/19-09-212.1

Ivanov II, Honda K. Intestinal commensal microbes as immune modulators. *Cell Host Microbe*. 2012;12(4):496–508. doi:10.1016/j.chom.2012.09.009

Johnson PA, Giles JR. The hen as a model of ovarian cancer. *Nat Rev Cancer*. 2013;13(6):432–436. doi:10.1038/nrc3535

Kapsetaki SE, Marquez Alcaraz G, Maley CC, Whisner CM, Aktipis A. Diet, microbes, and cancer across the tree of life: A systematic review. *Curr Nutr Rep*. 2022;11(3):508–525. doi:10.1007/s13668-022-00420-5

LeBlanc AK, Mazcko CN, Fan TM, et al. Comparative oncology assessment of a novel inhibitor of valosin-containing protein in tumor-bearing dogs. *Mol Cancer Ther*. 2022;21(10):1510–1523. doi:10.1158/1535-7163.MCT-22-0167

Lee KH, Hwang HJ, Noh HJ, Shin TJ, Cho JY. Somatic mutation of *PIK3CA* (H1047R) is a common driver mutation hotspot in canine mammary tumors as well as human breast cancers. *Cancers (Basel)*. 2019;11(12):2006. Published 2019 Dec 12. doi:10.3390/cancers11122006

Le Moigne R, Aftab BT, Djakovic S, et al. The p97 inhibitor CB-5083 is a unique disrupter of protein homeostasis in models of multiple myeloma. *Mol Cancer Ther*. 2017;16(11):2375–2386. doi:10.1158/1535-7163.MCT-17-0233

Lindblad-Toh K, Wade C, Mikkelsen T, et al. Genome sequence, comparative analysis and haplotype structure of the domestic dog. *Nature*. 2005;438:803–819. https://doi.org/10.1038/nature04338

Loi S, Adams S, Schmid P, Cortés J, Cescon DW, Winer EP, Toppmeyer DL, Rugo HS, Laurentiis MD, Nanda R, et al. Relationship between tumor infiltrating lymphocyte (TIL) levels and response to pembrolizumab (Pembro) in metastatic triple-negative breast cancer (MTNBC): Results from KEYNOTE-086. *Ann Oncol* 2017;28:v608. doi:10.1093/annonc/mdx440.005

Maekawa N, Konnai S, Ikebuchi R, Okagawa T, Adachi M, Takagi S, Kagawa Y, Nakajima C, Suzuki Y, Murata S, et al. Expression of PD-L1 on canine tumor cells and enhancement of IFN-γ production from tumor-infiltrating cells by PD-L1 blockade. *PLoS One*. 2014;9:e98415. doi:10.1371/journal.pone.0098415

Makashov AA, Malov SV, Kozlov AP. Oncogenes, tumor suppressor and differentiation genes represent the oldest human gene classes and evolve concurrently. *Sci Rep*. 2019;9(1):16410. Published 2019 Nov 11. doi:10.1038/s41598-019-52835-w

Mason NJ, Gnanandarajah JS, Engiles JB, et al. Immunotherapy with a HER2-targeting listeria induces HER2-specific immunity and demonstrates potential therapeutic effects in a phase I trial in canine osteosarcoma. *Clin Cancer Res*. 2016;22(17):4380–4390. doi:10.1158/1078-0432.CCR-16-0088

Munson L, Moresco A. Comparative pathology of mammary gland cancers in domestic and wild animals. *Breast Dis*. 2007;28:7–21. doi:10.3233/bd-2007-28102

Nascimento C, Urbano AC, Gameiro A, Ferreira J, Correia J Ferreira F. Serum PD-1/PD-L1 Levels, tumor expression and PD-L1 somatic mutations in HER2-positive and triple negative normal-like feline mammary carcinoma subtypes. *Cancers*. 2020;12:1386. doi:10.3390/cancers12061386

Nishibori S, Kaneko MK, Nakagawa T, et al. Development of anti-feline PD-1 antibody and its functional analysis. *Sci Rep*. 2023;13(1):6420. Published 2023 Apr 24. doi:10.1038/s41598-023-31543-6

Overgaard NH, Fan TM, Schachtschneider KM, Principe DR, Schook LB, Jungersen G. Of mice, dogs, pigs, and men: Choosing the appropriate model for immuno-oncology research. *ILAR J*. 2018;59(3):247–262. doi:10.1093/ilar/ily014

Pang LY, Bergkvist GT, Cervantes-Arias A, Yool DA, Muirhead R, Argyle DJ. Identification of tumour initiating cells in feline head and neck squamous cell carcinoma and evidence for gefitinib induced epithelial to mesenchymal transition. *Vet J*. 2012;193:46–52.

Pantelyushin S, Ranninger E, Guerrera D, et al. Cross-reactivity and functionality of approved human immune checkpoint blockers in dogs. *Cancers (Basel)*. 2021;13(4):785. Published 2021 Feb 13. doi:10.3390/cancers13040785

Paoloni M, Khanna C. Translation of new cancer treatments from pet dogs to humans. *Nat Rev Cancer*. 2008;8(2):147–156. doi:10.1038/nrc2273

Paoloni M, Webb C, Mazcko C, et al. Prospective molecular profiling of canine cancers provides a clinically relevant comparative model for evaluating personalized medicine (PMed) trials. *PLoS One*. 2014;9(3):e90028. Published 2014 Mar 17. doi:10.1371/journal.pone.0090028

Park JS, Withers SS, Modiano JF, et al. Canine cancer immunotherapy studies: Linking mouse and human. *J Immunother Cancer*. 2016;4:97. Published 2016 Dec 20. doi:10.1186/s40425-016-0200-7

Parrales A, McDonald P, Ottomeyer M, et al. Comparative oncology approach to drug repurposing in osteosarcoma. *PLoS One*. 2018;13(3):e0194224. Published 2018 Mar 26. doi:10.1371/journal.pone.0194224

Peto R, Roe FJ, Lee PN, Levy L, Clack J. Cancer and ageing in mice and men. *Br J Cancer*. 1975;32(4):411–426. doi:10.1038/bjc.1975.242

Pinard CJ, International immuno-oncology biomarker working group, Lagree A, et al. Comparative evaluation of tumor-infiltrating lymphocytes in companion animals: Immuno-oncology as a relevant translational model for cancer therapy. *Cancers (Basel)*. 2022;14(20):5008. Published 2022 Oct 13. doi:10.3390/cancers14205008

Porcellato I, Silvestri S, Menchetti L, Recupero F, Mechelli L, Sforna M, Iussich S, Bongiovanni L, Lepri E, Brachelente C. Tumour-infiltrating lymphocytes in canine melanocytic tumours: An investigation on the prognostic role of CD3+ and CD20+ lymphocytic populations. *Vet Comp Oncol*. 2020;18:370–380. doi:10.1111/vco.12556

Regan DP, Chow L, Das S, Haines L, Palmer E, Kurihara JN, Coy JW, Mathias A, Thamm DH, Gustafson DL, et al. Losartan blocks osteosarcoma-elicited monocyte recruitment, and combined with the kinase inhibitor toceranib, exerts significant clinical benefit in canine metastatic osteosarcoma. *Clin. Cancer Res*. 2021;28:662–676. doi:10.1158/1078-0432.CCR-21-2105

Regan DP, Garcia K, Thamm D. Clinical, pathological, and ethical considerations for the conduct of clinical trials in dogs with naturally occurring cancer: A comparative approach to accelerate translational drug development. *ILAR J*. 2018;59(1):99–110. doi:10.1093/ilar/ily019

Schiffman JD, Breen M. Comparative oncology: What dogs and other species can teach us about humans with cancer. *Philos Trans R Soc Lond B Biol Sci*. 2015;370(1673):20140231. doi:10.1098/rstb.2014.0231

Schoemaker NJ. Ferret oncology: Diseases, diagnostics, and therapeutics. *Vet Clin North Am Exot Anim Pract*. 2017;20(1):183–208. doi:10.1016/j.cvex.2016.07.004

Sivan A, Corrales L, Hubert N, Williams JB, Aquino-Michaels K, Earley ZM, Benyamin FW, Lei YM, Jabri B, Alegre ML, et al. Commensal bifidobacterium promotes antitumor immunity and facilitates anti-PD-L1 efficacy. *Science*. 2015;350(6264):1084–1089. doi:10.1126/science.aac4255

Somarelli JA, Boddy AM, Gardner HL, et al. Improving cancer drug discovery by studying cancer across the tree of life. *Mol Biol Evol*. 2020a;37(1):11–17. doi:10.1093/molbev/msz254

Somarelli JA, Gardner H, Cannataro VL, Gunady EF, Boddy AM, Johnson NA, Fisk JN, Gaffney SG, Chuang JH, Li S, Ciccarelli FD, Panchenko AR, Megquier K, Kumar S, Dornburg A, DeGregori J, Townsend JP. Molecular biology and evolution of cancer: From discovery to action. Mol Biol Evol. 2020b Feb 1;37(2):320–326. doi:10.1093/molbev/msz242. PMID: 31642480; PMCID: PMC6993850.

Somarelli JA, Rupprecht G, Altunel E, et al. A comparative oncology drug discovery pipeline to identify and validate new treatments for osteosarcoma. *Cancers (Basel)*. 2020c;12(11):3335. Published 2020 Nov 11. doi:10.3390/cancers12113335

Somarelli JA, Ware KE, Kostadinov R, et al. PhyloOncology: Understanding cancer through phylogenetic analysis. *Biochim Biophys Acta Rev Cancer*. 2017;1867(2):101–108. doi:10.1016/j.bbcan.2016.10.006

Sommerville L, Howard J, Evans S, Kelly P, McCann A. Comparative gene expression study highlights molecular similarities between triple negative breast cancer tumours and feline mammary carcinomas. *Vet Comp Oncol*. 2022;20(2):535–538. doi:10.1111/vco.12800

Song SJ, Lauber C, Costello EK, Lozupone CA, Humphrey G, Berg-Lyons D, Caporaso JG, Knights D, Clemente JC, Nakielny S, et al. Cohabiting family members share microbiota with one another and with their dogs. *eLife*. 2013;2:e00458.

Stevenson VB, Perry SN, Todd M, Huckle WR, LeRoith T. PD-1, PD-L1, and PD-L2 gene expression and tumor infiltrating lymphocytes in canine melanoma. *Vet Pathol.* 2021;58:692–698. doi:10.1177/03009858211011939

Sulak M, Fong L, Mika K, Chigurupati S, Yon L, Mongan NP, Emes RD, Lynch VJ. 2016. TP53 copy number expansion is associated with the evolution of increased body size and an enhanced DNA damage response in elephants. *Elife.* 5.

Thamm DH, Vail DM. Veterinary oncology clinical trials: design and implementation. *Vet J.* 2015;205(2):226–232. doi:10.1016/j.tvjl.2014.12.013

Thomas R, Wiley CA, Droste EL, Robertson J, Inman BA, Breen M. Whole exome sequencing analysis of canine urothelial carcinomas without BRAF V595E mutation: Short in-frame deletions in BRAF and MAP2K1 suggest alternative mechanisms for MAPK pathway disruption. *PLoS Genet.* 2023;19(4):e1010575. Published 2023 Apr 20. doi:10.1371/journal.pgen.1010575

Tollis M, Ferris E, Campbell MS, et al. Elephant genomes reveal accelerated evolution in mechanisms underlying disease defenses. *Mol Biol Evol.* 2021;38(9):3606–3620. doi:10.1093/molbev/msab127)

Tollis M, Robbins J, Webb AE, Kuderna LFK, Caulin AF, Garcia JD, Berube M, Pourmand N, Marques-Bonet T, O'Connell MJ, et al. 2019. Return to the sea, get huge, beat cancer: an analysis of cetacean genomes including an assembly for the humpback whale (Megaptera novaeangliae). *Mol Biol Evol.* 36(8):1746–1763.

Tollis M, Schneider-Utaka AK, Maley CC. The evolution of human cancer gene duplications across mammals. *Mol Biol Evol.* 2020;37(10):2875–2886. doi:10.1093/molbev/msaa125

Topkas E, Cai N, Cumming A, et al. Auranofin is a potent suppressor of osteosarcoma metastasis. *Oncotarget.* 2016;7(1):831–844. doi:10.18632/oncotarget.5704

van Zeeland Y. Rabbit oncology: Diseases, diagnostics, and therapeutics. *Vet Clin North Am Exot Anim Pract.* 2017;20(1):135–182. doi:10.1016/j.cvex.2016.07.005

Varshney J, Scott MC, Largaespada DA, Subramanian S. Understanding the osteosarcoma pathobiology: A comparative oncology approach. *Vet Sci.* 2016;3(1):3. Published 2016 Jan 18. doi:10.3390/vetsci3010003

Vazquez JM, Sulak M, Chigurupati S, Lynch VJ. A zombie LIF gene in elephants is upregulated by TP53 to induce apoptosis in response to DNA damage. *Cell Rep.* 2018;24(7):1765–1776. doi:10.1016/j.celrep.2018.07.042

Vincze O, Colchero F, Lemaître JF, et al. Cancer risk across mammals. *Nature.* 2022;601(7892):263–267. doi:10.1038/s41586-021-04224-5

Zhang H, Rose BJ, Pyuen AA, Thamm DH. In vitro antineoplastic effects of auranofin in canine lymphoma cells. *BMC Cancer.* 2018;18:522. doi:10.1186/s12885-018-4450-2

14 What Do We Gain from Viewing Cancer through an Eco-Evo Lens?

Jason A. Somarelli and Norman A. Johnson

14.1 UNDERSTANDING CANCER ACROSS SCALES OF BIOLOGICAL ORGANIZATION

Examination of the tree of life reveals an extraordinarily long history of cancer ([1]). This history of the cancer species is interwoven within the multicellular form and can be considered akin to a return to unicellular behavior from within a many-celled organism (Chapter 4, Helenek et al.; [1]). Indeed, genes that are commonly activated within solid tumors are more conserved within unicellular organisms while genes of metazoan origin are more often silenced in cancer [2]. These cancer speciation events also occur repeatedly, at least once per tumor. The repeated returns to a unicellular stage in cancer are mirrored by the multiple independent origins of multicellularity [3]and experiments demonstrating formation of multicellular behavior on rapid time scales in response to oxygen deprivation and other factors [4–6]. As we have begun to understand the origins of cancer species as unicellular actors within a multicellular context, we contend that cancer research must also consider the ecological niches within which the cancer species reside. Proper consideration should include:

- a deep understanding of the complexity of "players" that interact with the cancer cell population, including the immune system, cancer-associated fibroblasts, endothelial cells, adipocytes, and other cell types (Chapter 11, Brown et al.)
- geographic barriers, such as extracellular matrices of varying stiffness and composition, different cell types, and vasculature, in and around cancer cell populations (Chapter 9, Landguth and Johnson)
- architecture of these geographic boundaries (e.g., islands vs. archipelagos) (Chapter 9, Landguth and Johnson; Chapter 10, Chroni)
- more complete characterization of the interplay between mutation order, mutation effect size [7], the underlying genome (ancestry) of the non-cancer cell ancestor, and the role of karyotype dynamics (e.g., extrachromosomal DNA [8]) in cancer progression (Chapter 3, Janivara and Lachance; Chapter 6, Hazra and Lachance; and Chapter 7, Kasperski and Heng)
- the competition between cancer cells that is driven by the underlying genotypic and phenotypic diversity of the cancer cell population and the selective forces mediated by the interaction between these populations and their ecological niche(s) (Chapter 2, Ataya et al.).

Advancements in spatial omics [9], longitudinal sampling methods, rare cell sampling [10, 11], liquid biopsy technologies, single cell profiling [12], and patient-derived models of cancer [13, 14] are indeed paving the way for researchers and oncologists to gain a clearer picture of the cancer-organism ecology and the dynamics of these interactions during treatment and the emergence of resistance and metastasis. Novel treatment paradigms, such as immune checkpoint blockade, discussed by Kareva and Brown (Chapter 8) that capitalize on these interactions between the cancer and its environment exemplify the importance of understanding these interactions. Likewise, innovative strategies that leverage evolutionary and ecological approaches to steer cancer cell populations to double binds or that take advantage of competition between cancer cells (e.g., adaptive therapy), as described by Ataya et al., are illustrative of the utility of understanding cancer evolution and ecology for new and untapped treatments.

14.2 MISSING INGREDIENTS

Our shared efforts to define, characterize, and treat cancer have led to many successes as well as numerous continued failures, underscoring the "two truths" of cancer research and treatment [15]. Put simply, while we are witnessing some of the most impactful new treatments in decades with the emergence of immunotherapy and novel targeted therapies along with a corresponding increase in survivorship [16], we are also continuing to fall short when it comes to treatment of therapy-resistant, and metastatic disease. Overcoming the challenges in treating these stages of cancer progression may benefit from continued and expanded efforts at illuminating remaining key challenges in applying eco-evo principles to improve cancer outcomes. There is much that remains to learn about the parallels between cancer cell populations and populations of organisms in other natural systems. Some of these key gaps include the following:

- How do predator/prey interactions in non-cancer natural systems mirror interactions between immune cells and cancer cells (Chapter 8, Kareva and Brown)? How do we maximize these relationships for therapeutic benefit?
- Does cancer cell dormancy and quiescence in the face of therapy-induced resource depletion mirror hibernation/turpor at the organismal level, which often occurs in the face of resource depletion? Can we leverage these similarities for therapeutic benefit, by either selectively targeting the metabolic features of the dormant/quiescent cells and/or "trapping" the dormant phenotype?
- How can the "two-phased" cancer evolution model be used against the cancer cell population (Chapter 7, Kasperski and Heng)? A better understanding of mutation at micro and macro scales and how or whether this parallels other systems may provide insights.
- How can insights from molecular evolution, landscape genetics, and other branches of evolutionary biology be applied to better predict which cancer precursors will evolve into invasive cancer? For instance, most incidences of ductal carcinoma *in situ* (DCIS) do not progress to invasive breast cancer [17]. Better tools to distinguish between progressive and non-progressive forms of DCIS will likely lead to faster treatment of progressive DCIS and reduce over-treatment of non-progressive cases.
- Can we apply concepts from invasion ecology to studies of cancer invasion? Biotic resistance, one such concept, occurs when the presence of a healthy, diverse native community of species thwarts invasions. For instance, in their analysis of a large dataset of plant communities across the United States, Beaury et al. (2020) found that the presence of invasive plants strongly and negatively correlated with the species richness of native plants [18]. The implication is that the more diverse communities are better able to resist invasive non-native plants. A parallel to biotic resistance may occur in some forms of cancer wherein individuals that have more diverse (or certain types) of microbiota may resist invasive cancer better than those with depauperate or unhealthy microbiota.
- How do we turn this knowledge into action? Paraphrasing Dr. Mark Dewhirst, a pioneer in the field of radiation oncology, cancer hypoxia, and comparative oncology, who also happened to be a thoughtful mentor and eloquent writer: When writing grant proposals, his advice was always to "finish the thought", meaning that one should not assume the reader understood the point but to make it absolutely clear that the main points were explicitly stated. Likewise, our collective pursuit to eradicate cancer as a cause of mortality should

look beyond the academic enterprise of understanding and seek to "finish the thought" by translating this knowledge into action. While it is interesting to think of cancer through an eco-evo lens, we as a field should strive to connect the knowledge gained from this ecological and evolutionary framework with an improvement in clinical outcomes through new targets, translational platforms (Chapter 12, Compton; Chapter 13, Colmenares et al.), and treatment strategies that capitalize on this new knowledge (Chapter 2, Ataya et al.).

14.3   THE PATH AHEAD

While cancer continues to be a terrible burden on society, there are also reasons for optimism. First, the cancer research community has witnessed rapid advances in technologies that have provided an unprecedented viewpoint into the dynamics of cancer evolution at multiple biological scales. Second, along with these advancements, there is a growing recognition by researchers and physicians that cancer is rooted in the same fundamental processes that govern all life: evolution by natural selection. Third, the incorporation of evolutionary and ecological principles into novel treatment paradigms demonstrates the potential value of leveraging these approaches. The convergence of these factors in recent decades seems to have elevated the cancer and evolution field as a key component in our collective efforts to reduce cancer incidence and mortality. Building upon this momentum we should, as a community, seek to address the following key challenges and questions:

1) How can communication and collaboration further both cancer biology and evolutionary biology?
2) How do we create a new science and medical curriculum in evolutionary medicine to train the next generation of physicians and researchers in cancer eco-evo dynamics?
3) How can we best leverage and integrate spatial omics with longitudinal sampling and rare cell characterization (liquid biopsy, cfDNA, single-cell approaches) to understand the dynamics of cancer evolution during treatment, the evolution of resistance, and metastasis?
4) How can we effectively and reproducibly harness this new knowledge to develop evolutionarily-informed treatment plans that move the field from retroactive (responsive) to proactive (predictive)?

As we enter the path ahead, we are reminded of Dr. Robert Weinberg's essay in 2014 on the "full circle" of cancer research over the years, from the overwhelming complexity of early cancer research to reductionist approaches ushered in by molecular biologists, and back to an overwhelming complexity in the volumes of data we are able to generate [19]. While it is true that we are still grappling with the complexity of cancer in many ways, the understanding of cancer through an ecological and evolutionary lens provides a unifying framework to understand the convergent evolution of the cancer hallmark phenotypes despite seemingly limitless genetic and genomic complexity within diverse host environments. Beyond this conceptual understanding, this eco-evo perspective on cancer also provides a series of well-characterized models to generate testable hypotheses, a host of analytic platforms in which to integrate biological scales (sequence, morphology, etc.), and the capacity to steer population dynamics using the guiding principles of eco-evo population dynamics. Fully capitalizing on these advantages will require a truly transdisciplinary "team science" approach through deep interconnections between cancer researchers, physicians, ecologists, evolutionary biologists, mathematicians, and physicists. Beyond the value to cancer research, the study of cancer within organisms as ecological systems may illuminate new features of evolutionary biology. We hope this volume serves as a motivating force for students and practitioners in their quest to end cancer to think big, proceed with optimism, maintain an openness for growth, learning, and collaboration across disciplines, and carry on with courage to think beyond the confines for the betterment of patients and their families.

REFERENCES

1. Aktipis CA, Boddy AM, Jansen G, Hibner U, Hochberg ME, Maley CC, Wilkinson GS. Cancer across the tree of life: cooperation and cheating in multicellularity. Philos Trans R Soc Lond B Biol Sci. 2015;370(1673). Epub 2015/06/10. doi: 10.1098/rstb.2014.0219. PubMed PMID: 26056363; PMCID: PMC4581024.

2. Trigos AS, Pearson RB, Papenfuss AT, Goode DL. Altered interactions between unicellular and multicellular genes drive hallmarks of transformation in a diverse range of solid tumors. Proc Natl Acad Sci USA. 2017;114(24):6406–11. Epub 2017/05/10. doi: 10.1073/pnas.1617743114. PubMed PMID: 28484005; PMCID: PMC5474804.

3. Niklas KJ, Newman SA. The many roads to and from multicellularity. J Exp Bot. 2020;71(11):3247–53. Epub 2019/12/11. doi: 10.1093/jxb/erz547. PubMed PMID: 31819969; PMCID: PMC7289717.

4. Ratcliff WC, Fankhauser JD, Rogers DW, Greig D, Travisano M. Origins of multicellular evolvability in snowflake yeast. Nat Commun. 2015;6:6102. Epub 2015/01/21. doi: 10.1038/ncomms7102. PubMed PMID: 25600558; PMCID: PMC4309424.

5. Bozdag GO, Zamani-Dahaj SA, Day TC, Kahn PC, Burnetti AJ, Lac DT, Tong K, Conlin PL, Balwani AH, Dyer EL, Yunker PJ, Ratcliff WC. De novo evolution of macroscopic multicellularity. Nature. 2023;617(7962):747–54. Epub 2023/05/11. doi: 10.1038/s41586-023-06052-1. PubMed PMID: 37165189; PMCID: PMC10425966.

6. Bozdag GO, Libby E, Pineau R, Reinhard CT, Ratcliff WC. Oxygen suppression of macroscopic multicellularity. Nat Commun. 2021;12(1):2838. Epub 2021/05/16. doi: 10.1038/s41467-021-23104-0. PubMed PMID: 33990594; PMCID: PMC8121917.

7. Cannataro VL, Gaffney SG, Townsend JP. Effect sizes of somatic mutations in cancer. J Natl Cancer Inst. 2018;110(11):1171–7. Epub 2018/10/27. doi: 10.1093/jnci/djy168. PubMed PMID: 30365005; PMCID: PMC6235682.

8. Wu S, Bafna V, Chang HY, Mischel PS. Extrachromosomal DNA: An emerging hallmark in human cancer. Annu Rev Pathol. 2022;17:367–86. Epub 2021/11/10. doi: 10.1146/annurev-pathmechdis-051821-114223. PubMed PMID: 34752712; PMCID: PMC9125980.

9. Bressan D, Battistoni G, Hannon GJ. The dawn of spatial omics. Science. 2023;381(6657):eabq4964. Epub 2023/08/03. doi: 10.1126/science.abq4964. PubMed PMID: 37535749; PMCID: PMC7614974.

10. Nagrath S, Sequist LV, Maheswaran S, Bell DW, Irimia D, Ulkus L, Smith MR, Kwak EL, Digumarthy S, Muzikansky A, Ryan P, Balis UJ, Tompkins RG, Haber DA, Toner M. Isolation of rare circulating tumour cells in cancer patients by microchip technology. Nature. 2007;450(7173):1235–9. Epub 2007/12/22. doi: 10.1038/nature06385. PubMed PMID: 18097410; PMCID: PMC3090667.

11. Edd JF, Mishra A, Smith KC, Kapur R, Maheswaran S, Haber DA, Toner M. Isolation of circulating tumor cells. iScience. 2022;25(8):104696. Epub 2022/07/27. doi: 10.1016/j.isci.2022.104696. PubMed PMID: 35880043; PMCID: PMC9307519.

12. Heumos L, Schaar AC, Lance C, Litinetskaya A, Drost F, Zappia L, Lucken MD, Strobl DC, Henao J, Curion F, Single-cell Best Practices C, Schiller HB, Theis FJ. Best practices for single-cell analysis across modalities. Nat Rev Genet. 2023;24(8):550–72. Epub 2023/04/01. doi: 10.1038/s41576-023-00586-w. PubMed PMID: 37002403; PMCID: PMC10066026 Janssen, and received consulting fees from Chan-Zuckerberg Initiative. F.J.T. consults for Immunai Inc., Singularity Bio B.V., CytoReason Ltd and Omniscope Ltd, and has ownership interest in Dermagnostix GmbH and Cellarity. M.G.J. consults for and has ownership interests in Vevo Therapeutics. L. Heumos has received speaker's honorarium from Vesalius Therapeutics. Single-Cell Best Practices Consortium: M.G.J. consults for and has ownership interests in Vevo Therapeutics. R.P. is co-founder of Ocean Genomics, Inc. The other authors declare no competing interests.

13. Byrne AT, Alferez DG, Amant F, Annibali D, Arribas J, Biankin AV, Bruna A, Budinska E, Caldas C, Chang DK, Clarke RB, Clevers H, Coukos G, Dangles-Marie V, Eckhardt SG, Gonzalez-Suarez E, Hermans E, Hidalgo M, Jarzabek MA, de Jong S, Jonkers J, Kemper K, Lanfrancone L, Maelandsmo GM, Marangoni E, Marine JC, Medico E, Norum JH, Palmer HG, Peeper DS, Pelicci PG, Piris-Gimenez A, Roman-Roman S, Rueda OM, Seoane J, Serra V, Soucek L, Vanhecke D, Villanueva A, Vinolo E, Bertotti A, Trusolino L. Interrogating open issues in cancer precision medicine with patient-derived xenografts. Nat Rev Cancer. 2017;17(4):254–68. Epub 2017/01/21. doi: 10.1038/nrc.2016.140. PubMed PMID: 28104906.

14. Wensink GE, Elias SG, Mullenders J, Koopman M, Boj SF, Kranenburg OW, Roodhart JML. Patient-derived organoids as a predictive biomarker for treatment response in cancer patients. NPJ Precis Oncol. 2021;5(1):30. Epub 2021/04/14. doi: 10.1038/s41698-021-00168-1. PubMed PMID: 33846504; PMCID:

PMC8042051 are employed by the foundation Hubrecht organoid technology (HUB). The remaining authors declare no conflict of interest. M.K. Institutional financial instructs (IFI): Amgen, Bayer, BMS, Merck-Serono, Nordic Farma, Roche, Servier, Sirtex, Sanofi-Aventis; O.W.K. IFI: Genmab, Johnson & Johnson; J.M.L.R. IFI: Servier, Merck, Bayer. All grants were unrelated to the study and paid to the individual's institution. All authors are involved in an ongoing prospective clinical trial evaluating the predictive role of organoids in mCRC patients.

15. Somarelli JA, DeGregori J, Gerlinger M, Heng HH, Marusyk A, Welch DR, Laukien FH. Questions to guide cancer evolution as a framework for furthering progress in cancer research and sustainable patient outcomes. Med Oncol. 2022;39(9):137. Epub 2022/07/06. doi: 10.1007/s12032-022-01721-z. PubMed PMID: 35781581; PMCID: PMC9252949.

16. Miller KD, Nogueira L, Devasia T, Mariotto AB, Yabroff KR, Jemal A, Kramer J, Siegel RL. Cancer treatment and survivorship statistics, 2022. CA Cancer J Clin. 2022;72(5):409–36. Epub 2022/06/24. doi: 10.3322/caac.21731. PubMed PMID: 35736631.

17. Casasent AK, Almekinders MM, Mulder C, Bhattacharjee P, Collyar D, Thompson AM, Jonkers J, Lips EH, van Rheenen J, Hwang ES, Nik-Zainal S, Navin NE, Wesseling J, Grand Challenge PC. Learning to distinguish progressive and non-progressive ductal carcinoma in situ. Nat Rev Cancer. 2022;22(12):663–78. Epub 2022/10/20. doi: 10.1038/s41568-022-00512-y. PubMed PMID: 36261705.

18. Beaury EM, Finn JT, Corbin JD, Barr V, Bradley BA. Biotic resistance to invasion is ubiquitous across ecosystems of the United States. Ecol Lett. 2020;23(3):476–82. Epub 2019/12/26. doi: 10.1111/ele.13446. PubMed PMID: 31875651.

19. Weinberg RA. Coming full circle-from endless complexity to simplicity and back again. Cell. 2014;157(1):267–71. Epub 2014/04/01. doi: 10.1016/j.cell.2014.03.004. PubMed PMID: 24679541.

Index

Milton Keynes UK
Ingram Content Group UK Ltd.
UKHW052027141024
449569UK00016B/730